ADVANCES IN DOSIMETRY FOR FAST NEUTRONS AND HEAVY CHARGED PARTICLES FOR THERAPY APPLICATIONS

The following States are Members of the International Atomic Energy Agency:

AFGHANISTAN
ALBANIA
ALGERIA
ARGENTINA
AUSTRALIA
AUSTRIA
BANGLADESH
BELGIUM
BOLIVIA
BRAZIL
BULGARIA
BURMA
BYELORUSSIAN SOVIET SOCIALIST REPUBLIC
CANADA
CHILE
CHINA
COLOMBIA
COSTA RICA
CUBA
CYPRUS
CZECHOSLOVAKIA
DEMOCRATIC KAMPUCHEA
DEMOCRATIC PEOPLE'S REPUBLIC OF KOREA
DENMARK
DOMINICAN REPUBLIC
ECUADOR
EGYPT
EL SALVADOR
ETHIOPIA
FINLAND
FRANCE
GABON
GERMAN DEMOCRATIC REPUBLIC
GERMANY, FEDERAL REPUBLIC OF
GHANA
GREECE
GUATEMALA
HAITI
HOLY SEE
HUNGARY
ICELAND
INDIA
INDONESIA
IRAN, ISLAMIC REPUBLIC OF
IRAQ
IRELAND
ISRAEL
ITALY
IVORY COAST
JAMAICA
JAPAN
JORDAN
KENYA
KOREA, REPUBLIC OF
KUWAIT
LEBANON
LIBERIA
LIBYAN ARAB JAMAHIRIYA
LIECHTENSTEIN
LUXEMBOURG
MADAGASCAR
MALAYSIA
MALI
MAURITIUS
MEXICO
MONACO
MONGOLIA
MOROCCO
NAMIBIA
NETHERLANDS
NEW ZEALAND
NICARAGUA
NIGER
NIGERIA
NORWAY
PAKISTAN
PANAMA
PARAGUAY
PERU
PHILIPPINES
POLAND
PORTUGAL
QATAR
ROMANIA
SAUDI ARABIA
SENEGAL
SIERRA LEONE
SINGAPORE
SOUTH AFRICA
SPAIN
SRI LANKA
SUDAN
SWEDEN
SWITZERLAND
SYRIAN ARAB REPUBLIC
THAILAND
TUNISIA
TURKEY
UGANDA
UKRAINIAN SOVIET SOCIALIST REPUBLIC
UNION OF SOVIET SOCIALIST REPUBLICS
UNITED ARAB EMIRATES
UNITED KINGDOM OF GREAT BRITAIN AND NORTHERN IRELAND
UNITED REPUBLIC OF CAMEROON
UNITED REPUBLIC OF TANZANIA
UNITED STATES OF AMERICA
URUGUAY
VENEZUELA
VIET NAM
YUGOSLAVIA
ZAIRE
ZAMBIA

The Agency's Statute was approved on 23 October 1956 by the Conference on the Statute of the IAEA held at United Nations Headquarters, New York; it entered into force on 29 July 1957. The Headquarters of the Agency are situated in Vienna. Its principal objective is "to accelerate and enlarge the contribution of atomic energy to peace, health and prosperity throughout the world".

Printed by the IAEA in Austria
February 1984

PANEL PROCEEDINGS SERIES

ADVANCES IN DOSIMETRY FOR FAST NEUTRONS AND HEAVY CHARGED PARTICLES FOR THERAPY APPLICATIONS

PROCEEDINGS OF AN ADVISORY GROUP MEETING
ON ADVANCES IN DOSIMETRY
FOR FAST NEUTRONS AND HEAVY CHARGED PARTICLES
FOR THERAPY APPLICATIONS
ORGANIZED BY THE
INTERNATIONAL ATOMIC ENERGY AGENCY
AND HELD IN VIENNA, 14–18 JUNE 1982

INTERNATIONAL ATOMIC ENERGY AGENCY
VIENNA, 1984

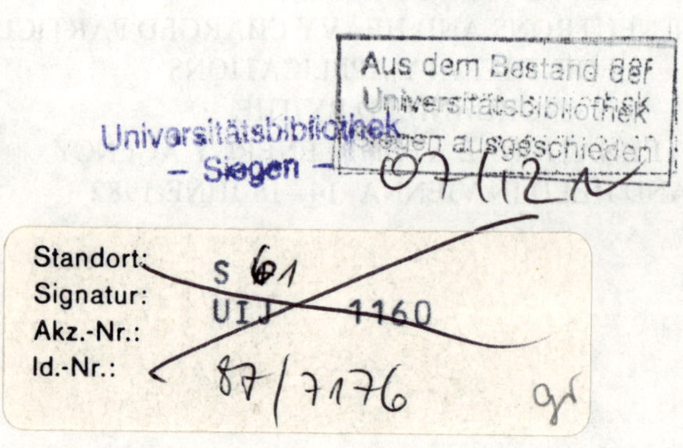

ADVANCES IN DOSIMETRY
FOR FAST NEUTRONS AND HEAVY CHARGED PARTICLES
FOR THERAPY APPLICATIONS
IAEA, VIENNA, 1984
STI/PUB/643
ISBN 92–0–111084–7

FOREWORD

There has been great progress in the treatment of cancer by ionizing radiation since large-scale radioisotope sources were made available by the development of nuclear energy. High-energy accelerators producing photons and electrons have supplemented, and to some extent replaced, ^{60}Co gamma rays, and more recently fast neutrons, protons and other heavy charged particles have also been used in radiotherapy. A significant part of the progress in all these methods has been due to improvements in the accuracy of the dosimetry. This is important in a process where errors as small as ±5% can affect patient survival.

The attainment and the maintenance of the necessary level of accuracy are difficult. The Dosimetry Section of the IAEA has for many years had programmes aimed at helping to improve dosimetry in radiotherapy. These programmes have been developed particularly for hospitals without access to calibration laboratories or other means of verifying their dosimetry. They include a postal dose service using thermoluminescent dosimeters and the establishment of a network of secondary standard dosimetry laboratories.

The determination of absorbed dose in irradiations with fast neutrons and heavy charged particles is more difficult than with photons or electrons. The stopping powers in general are much more dependent on energy and atomic number and thus the dose distributions are complex. In addition, as there are usually photons and electrons present and as the biological response to the heavy particles is different, the dose contribution of both types of radiation must be determined.

The material presented in this report results from an Advisory Group Meeting on Advances in Dosimetry for Fast Neutrons and Heavy Charged Particles for Therapy Applications that was held at the IAEA headquarters in Vienna, 14 to 18 June 1982. The objectives in convening this meeting were to bring together experts in the dosimetry of these types of particle, to review and discuss progress and problems and to propose recommendations for future work. In addition to those responsible for measurements in clinical applications, representatives of primary standard laboratories were invited because of the importance of traceability and international standardization.

It is hoped that not only will this publication – which contains 18 papers presented at the meeting and a summary report of the discussions – serve as a reference to those directly concerned with fast neutron or heavy charged particle dosimetry for radiotherapy, but also that the ideas and techniques presented here will be of use to all radiological physicists in primary and secondary standard laboratories as well as in hospitals.

EDITORIAL NOTE

The papers and discussions have been edited by the editorial staff of the International Atomic Energy Agency to the extent considered necessary for the reader's assistance. The views expressed and the general style adopted remain, however, the responsibility of the named authors or participants. In addition, the views are not necessarily those of the governments of the nominating Member States or of the nominating organizations.

Where papers have been incorporated into these Proceedings without resetting by the Agency, this has been done with the knowledge of the authors and their government authorities, and their cooperation is gratefully acknowledged. The Proceedings have been printed by composition typing and photo-offset lithography. Within the limitations imposed by this method, every effort has been made to maintain a high editorial standard, in particular to achieve, wherever practicable, consistency of units and symbols and conformity to the standards recommended by competent international bodies.

The use in these Proceedings of particular designations of countries or territories does not imply any judgement by the publisher, the IAEA, as to the legal status of such countries or territories, of their authorities and institutions or of the delimitation of their boundaries.

The mention of specific companies or of their products or brand names does not imply any endorsement or recommendation on the part of the IAEA.

Authors are themselves responsible for obtaining the necessary permission to reproduce copyright material from other sources.

CONTENTS

PRIMARY STANDARD LABORATORY ACTIVITIES IN NEUTRON DOSIMETRY

DOSIMETRY OF HEAVY CHARGED PARTICLES IN RADIATION THERAPY AND RADIATION BIOLOGY

REPORT AND RECOMMENDATIONS OF THE ADVISORY GROUP MEETING ON ADVANCES IN DOSIMETRY FOR FAST NEUTRONS AND HEAVY CHARGED PARTICLES FOR THERAPY APPLICATIONS

1. INTRODUCTION

During the last three decades there has been a rapid development of radiation therapy techniques, partly due to the developments in the different branches of modern technology, but also due to collaboration between therapists, physicists and engineers.

At present one can identify several levels of complexity and cost of megavoltage radiotherapy equipment:

(1) *Cobalt units* are the simplest and the basic equipment of a radiotherapy department.

(2) *Electron accelerators* (linear accelerators, betatrons and microtrons) can be used in the photon or electron therapy mode. These accelerators are certainly more complex and, to be used properly, require additional controls, but the improved dose distribution brings a definite benefit for the patients, mainly from the aspect of normal tissue tolerance.

(3) Based on radiobiological findings (mainly the radioresistance of anoxic cancer cells) neutron therapy has been initiated in several countries throughout the world. Most of the first cyclotrons used for therapeutic applications were "low-energy" cyclotrons ($E_d \leqslant 16$ MeV) with poor beam penetration; consequently, the largest series of patients treated with neutrons on which evaluations are now being based were treated with non-optimal dose distribution (and/or technical conditions). However, a *new generation of cyclotrons* more suitable for medical applications are now becoming available (proton or deuteron energies in excess of 40 MeV).

For *neutron therapy* these cyclotrons provide dose distributions and technical treatment conditions (isocentric mounting, availability) which are similar to those obtained with 6–8 MeV electron accelerators.

The same type of accelerators could produce *proton beams* which are of low linear energy transfer (LET) and have the advantage of an excellent dose distribution.

Protons with energies of 60 MeV are sufficient to treat eye tumours, and energies in excess of 160 MeV are usually required for deep-seated tumours. Such equipment is at present as reliable and as stable as modern electron accelerators.

Such a cyclotron costs considerably more than an electron accelerator, depending on the specifications. However, although either accelerator with suitable multiple beam facilities could in principle treat a large number of patients, the cyclotron offers the additional advantage of efficient radionuclide production.

(4) The last group of highly sophisticated and expensive equipment for therapy applications consists of three pion facilities and one heavy ion accelerator. The radiation available at these facilities combines the property of high LET with the advantage of an adjustable dose distribution (similar to proton beams).

The main clinical experience with high LET radiation has been obtained with fast neutrons. In fact, clinical neutron therapy has been performed routinely in more than 20 centres with several thousand patients treated and some series having follow-up as long as 10 years.

The clinical data from the different centres do not always agree and the interpretaton sometimes leads to different conclusions. However, there is sufficient evidence and a rather general consensus in favour of continuing the high LET radiotherapy programmes to:

(i) Identify the types of patient and/or tumour for which high LET radiotherapy offers the maximum probability of improving the results;
(ii) Define the best procedures for applying high LET radiotherapy (e.g. in some instances, mixed schedules or high LET boost).

To take maximum advantage of high LET beams it is essential to optimize the technical treatment conditions (including energy, dose rate, isocentric mounting, variable collimator), and the dosimetric conditions.

The complexity of the equipment, and of the treatment with pi-mesons and particles heavier than protons, has limited the number of facilities; therefore, the clinical experience with these modalities is less than that for neutrons. The value of pi-mesons and heavy ions in cancer therapy will become established only after continued research on the application of these particles to patients.

2. MAIN LINES OF RESEARCH AND DEVELOPMENT

2.1. Instrumentation

The use of calibrated A-150 plastic tissue-equivalent (TE) ionization chambers with TE gas filling is recommended as the practical method of obtaining the tissue kerma in air and the absorbed tissue dose in a TE phantom. This recommendation is based on the fact that TE chambers have been used as the principal dose measuring instrument by the neutron and heavy charged particle therapy groups in Europe, the United States of America and Japan where currently patients are being treated regularly.

At present TE ionization chambers should have applied to them a calibration factor for ^{60}Co, ^{137}Cs or 2 MV X-rays. This calibration should be directly traceable to a primary or a secondary standards laboratory. A direct calibration in a standard neutron or heavy charged particle field as appropriate would be preferable.

In establishing the conversion of the reading of a TE ionization chamber to the total absorbed dose, an absolute dosimetry system, e.g. a TE calorimeter, could be used.

The basic physical parameters which are used to convert a specific dose meter reading to tissue kerma in free air, or to absorbed tissue dose in a phantom, should be determined with a higher accuracy.

The highest priority should be given to the determination of kerma factors for carbon and oxygen for fission neutrons and for neutrons with energies greater than 8 MeV. More accurate values of W_n for methane-based tissue equivalent gas are needed for these neutron energies. Improvements in gas-to-wall absorbed dose conversion factors are also necessary for the chamber types clinically employed.

The corrections for finite size of the chamber for the measurements or calibrations carried out in air or in a phantom, should be determined experimentally and theoretically for the chamber types and neutron energies employed in practice.

The TE ionization chamber should be checked regularly and, if significant changes in the measurement procedure are introduced, e.g. a change of the TE gas, it is recommended that the whole system be recalibrated.

In addition to the generally employed A-150 plastic walled ionization chambers, flushed with the conventional TE gases, it is recommended that other tissue-like materials for wall and gas should be investigated for measuring total absorbed dose.

To achieve a precision of better than ± 2% in total absorbed dose measurement with TE ionization chambers, it is necessary to correct for differences in measurement conditions between calibration and the absorbed dose determination.

In establishing the photon component of the total absorbed dose, non-hydrogenous ionization chambers (e.g. Mg-Ar chamber), or energy-compensated GM counters, are usually employed. The determination of k_U values, specific for a particular chamber type and irradiation conditions, is recommended.[1] Changes in k_U in a phantom with depth or with field size should be investigated. If possible, different methods in establishing the photon dose fraction should be intercompared. When possible, at least two different methods for measuring photon dose should be used.

For measurements in a phantom irradiated by a neutron beam, only a rough estimate of the photon spectrum is necessary to derive the effective k_U values. During measurements in air, however, more detailed information on the photon spectrum is required for accurate photon dose determination.

Both TE and carbon proportional counters can be helpful in solving specific problems related to the dosimetry of the separate photon and neutron absorbed dose components in a mixed beam.

Future research on the use of chemical dose meters and small integrating detectors such as silicon diodes, activation detectors or TLD (the latter especially in establishing separate dose components), should be encouraged. Such detectors may also be used for postal dose intercomparisons between institutes performing high LET irradiations.

2.2. Dose distribution and radiation quality

The main goal in treatment planning is to deliver the absorbed dose as homogeneously as possible to the target volume while keeping it as low as reasonably achievable in surrounding critical tissues. The objective, following from this, is to assess the neutron and gamma-ray spatial dose distributions separately in the patient as accurately as possible. Calculational procedures, doing this, are normally based upon experimentally determined depth and lateral dose data in homogeneous phantoms, irradiated under standard conditions.

Mathematical models are based on extended physical measurements, or on the analytical expression of a limited number of central axis depth dose and lateral profiles. In this respect the isodose generating techniques are completely comparable to those applied in conventional therapy. However, as neutron dose distributions may show considerable variations, even for similar equipment and target reactions, the basic depth dose and lateral data must be measured at each facility.

Non-standardized irradiation conditions, e.g. oblique incidence, must either be avoided by arranging suitable bolus material or verified experimentally. The same holds for irradiation conditions with incomplete scatter, e.g. tangential irradiation of the breast. Corrections have to be applied finally to the homogeneous-case dose distributions for the inhomogeneities encountered in the clinical situation, such as cavities, lung, fat, bone and their associated interfaces. As a basis for this,

---

[1] k_U: the relative neutron sensitivity of dosimeters used for photon-fraction determinations.

more measurements and calculations for non-homogeneous phantom arrangements are required.

Calculational concepts, accounting not only for different densities and kerma factors, but also for different scattering by such inhomogeneities in a three-dimensional mode, still have to be developed. In this approach separate algorithms would be required for the neutron and photon components of the beam. The information contained in X-ray and NMR CT[2] scans may be of great assistance in accomplishing this.

The application of radiation transport codes as SN or Monte Carlo codes can help, not only to verify dosimetric measurements in homogeneous phantoms but also to elucidate the importance of corrections being applied in realistic representative treatment situations. For that purpose the basic neutron interaction data, such as cross-sections and kerma factors, must be evaluated, especially above 8 MeV. They should extend up to about 100 MeV.

The referring libraries should be written in appropriate formats so as to suit standard radiation transport codes. Laboratories interested in the application of such codes should agree upon clinically oriented benchmark problems to intercompare with each other and with experimental data.

More complex treatment schemes (multiple and/or moving fields), as well as the use of non-homogeneous phantoms, necessitate dosimetric verifications of spatial dose distribution calculations by means of small integrating dosimetric systems, such as thermoluminescence dosimetry (TLD), silicon diodes or activation detectors.

The same technique may be applied in vivo, i.e. on the patient's surface, or inside body cavities. In-vivo dosimetry should be carried out, especially during the first period of clinical use of a new irradiation modality.

The development of multi-purpose anthropoid phantoms for neutron irradiations, similar to those existing for X-ray and gamma-ray studies, is needed.

Radiation quality plays a more important role in neutron therapy than in therapy with X- and gamma-rays because there are greater variations in biological effectiveness with the former. At present, radiation quality is described in terms of the neutron energy spectrum and the proportion of gamma-ray radiation, or in terms of microdosimetric parameters. Neither is fully adequate for a prediction of biological effect; improved methods of specifying radiation quality would be useful. Radiobiological studies are needed for a comparison of different neutron therapy installations and for assessing the validity of physical measurements.

Many of the statements and specific recommendations regarding neutron therapy are also applicable, either in total or in part, to heavy charged particle therapy.

---

[2] NMR = nuclear magnetic resonance; CT = computerized tomography.

Many of the advances made (both clinical and dosimetric) with either neutron or heavy charged particle therapy have the potential for improving the application of other forms of radiation therapy.

2.3. The role of standards laboratories

(1) The national primary standards laboratories generally provide the following services:

(a) Provision and maintenance of the national radiation standards, including research programmes for improved and new standards;
(b) Provision of dosimeter calibrations;
(c) Arrangement of national intercomparisons and quality assurance programmes;
(d) Establishment and promotion of secondary standards laboratories;
(e) Participation in international dosimetry intercomparisons with other standards laboratories and the BIPM.[3]

(2) With respect to neutron dosimetry, national standards laboratories should provide ionization chamber calibrations in at least one neutron field of known quality, free in air or at one depth in a phantom, and should provide the best available information for transferring the calibration to other qualities. Both the calibration and the transfer information should take into consideration the neutron spectra involved. Calorimetric techniques have been suggested for establishing these standards.

(3) National standards laboratories should improve the physical data base for neutron dosimetry, as well as compile, evaluate and disseminate data produced by other research laboratories.

3. RECOMMENDATIONS

With respect to the developing countries, priority should be given to providing basic health care of acceptable standards. In particular, in the field of radiotherapy, conventional radiation sources should be available to the majority of the population in a country before more sophisticated equipment is introduced. A reasonable proportion should always be kept between the number of conventional radiation facilities and the more complex installations, much in the same way as in the industrialized countries. It is important, however, in that phase of development that the new techniques be adapted to the existing clinical and scientific conditions,

[3] BIPM = Bureau International des Poids et Mesures.

and that trained personnel and a proper technical substructure are available. The rate of development should be guided by a proper cost-benefit analysis.

In the whole process, it should be remembered that the developing countries should have the opportunities for close contacts with the frontier of radiotherapeutic research and with the very best clinical activities. Such contacts, in the field of non-conventional radiotherapy, could be made for example in the form of prolonged visits or research work at host institutions by students, clinicians, hospital physicists and other scientists from developing countries. The developing countries can benefit, however, only if research, training and information are provided in areas of their specific needs, such as improved therapy planning and dosimetry with a view towards reliability and standardization of methods.

It is equally important that representatives of developing countries be given the opportunity to take part in international scientific meetings and planning to promote a mutual exchange of information.

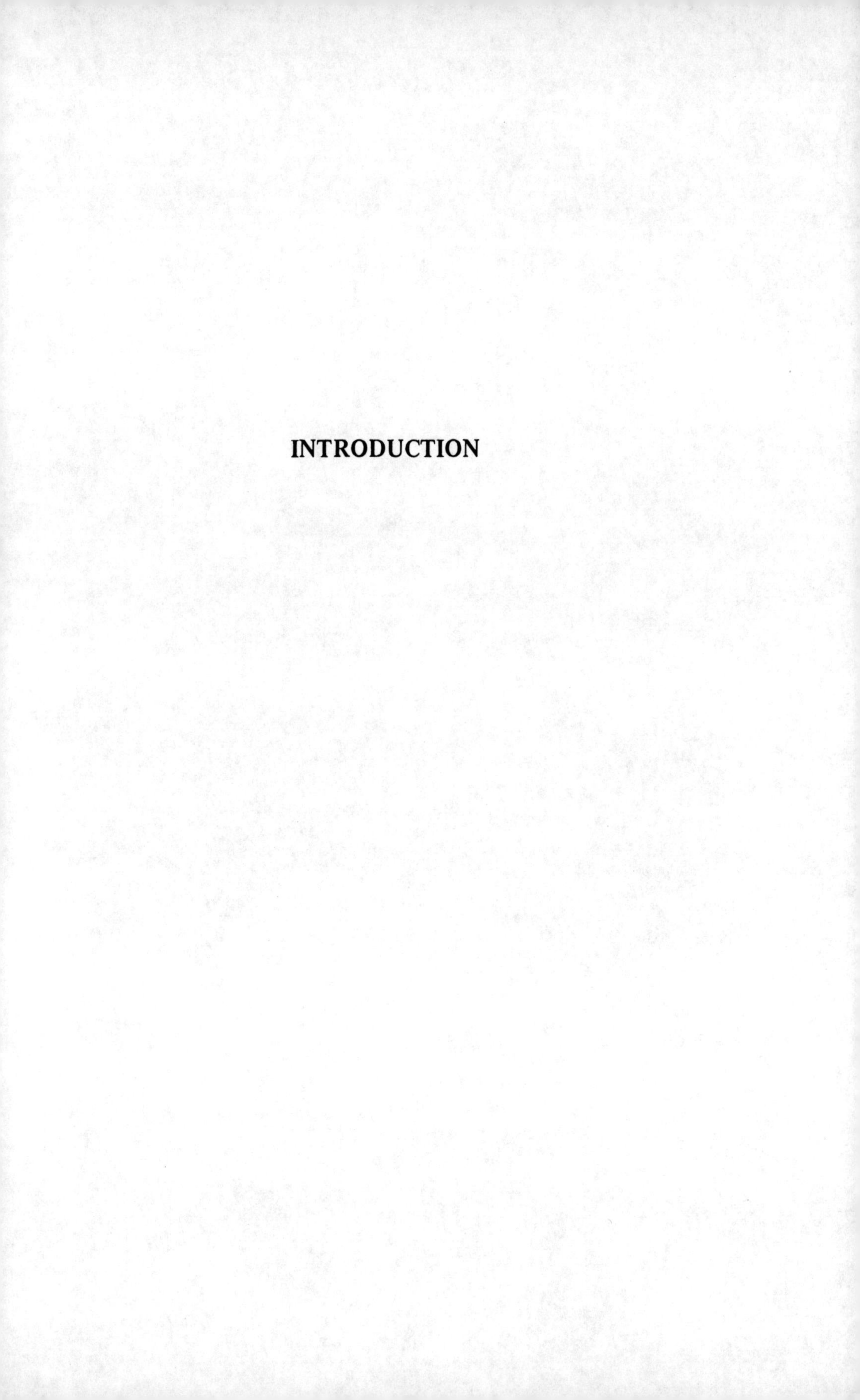

INTRODUCTION

ACCURACY REQUIRED IN RADIOTHERAPY AND IN NEUTRON THERAPY

A. WAMBERSIE, J. GUEULETTE
Unité de Radiothérapie et de Neutronthérapie,
UCL – Cliniques Universitaires Saint-Luc,
Brussels, Belgium

Abstract

ACCURACY REQUIRED IN RADIOTHERAPY AND IN NEUTRON THERAPY.

A review of the available data concerning local tumour control and normal tissue complications, after high-energy photon or electron irradiation, tends to indicate that an accuracy of 5% is needed in the dose delivered in the target volume when tumour eradication is intended and when the dose range approaches the normal tissue tolerance. Although little clinical information is at present available as far as neutron therapy is concerned, it seems reasonable to assume that a similar level of accuracy is needed. However, in neutron therapy, an additional problem is raised by the fact that relative biological effectiveness (RBE) critically depends on neutron energy. It is therefore essential that the beam quality be defined as completely as possible. In this respect, besides physical intercomparisons, biological experiments are useful, and micro-dosimetric and radiobiological intercomparisons are recommended between the different neutron therapy centres.

1. INTRODUCTION

Dose-effect relationships for tumour control and complications are typically sigmoid in shape (Fig.1); and the accuracy required in the radiotherapeutic applications then depends:

- on the *steepness* of the dose-effect relationships (the maximum steepness corresponds to tumour control rates or complication rates of about 30–70%); and
- on the *separation* between the dose-response relationships for tumour control and normal tissue tolerance.

The upper part of Fig.1 illustrates a typical favourable situation where a high tumour control rate can be achieved without inducing a high complication rate. In such a situation the choice of the dose level, and of the actual delivered dose, may not be very critical, provided it is maintained within certain limits. Examples of such situations can be skin carcinomas with limited extension, bone metastases for breast cancer, etc.

Figure 1(B) illustrates a (hypothetical) much less favourable situation: it is impossible to select a dose level which would achieve a high tumour control rate without inducing a high complication rate. The choice of the prescribed target dose will then depend on the clinical status and on several factors:

(1) If the complications are to be avoided altogether (e.g., when spinal cord tolerance is concerned), the prescribed target dose cannot exceed D1 and the tumour control probability (TCP) is low. As pointed out by Withers and Peters [1], all complications may come into this category if the present growth of malpractice suits continues.

(2) By accepting a relatively small probability of normal tissue injury, the radiation dose can be increased to D2 and the TCP significantly improved (e.g., a low percentage of medium severity complications such as mandibular necrosis or soft tissue necrosis may be accepted if a high rate of non-complicated tumour control is obtained).

(3) A further increase of radiation dose – up to D3 – will result in an increased complication rate which in some situations may be accepted (e.g., a localized lung fibrosis may be acceptable, if it is necessary, to achieve high control rates).

In the situation illustrated in Fig.1(B), the selection of the prescribed target dose (*and also of course the dosimetric requirements*) is critical since any variation of the prescribed (or delivered) dose will modify the tumour control rate *and/or* the complication rate.

In this paper, some data are reported which indicate that, at least in some situations or patient series, a 5% variation of absorbed dose significantly modifies the tumour control rate *and/or* the complication rate: the accuracy required in radiotherapy then logically results from these observations.

The accuracy required in neutron therapy, as compared with radiotherapy, will depend on:

- the relative steepness of the dose-effect curves for neutrons compared with γ-rays; and
- the relative separation between the dose-effect curves for neutrons compared with γ-rays.

2. SURVEY OF SOME AVAILABLE DATA IN RADIOTHERAPY WITH PHOTONS AND ELECTRONS

2.1. Radiobiological data

The shape of the *dose-effect curves* for local tumour control can be predicted from radiobiological data, and in particular from the cellular models and from the

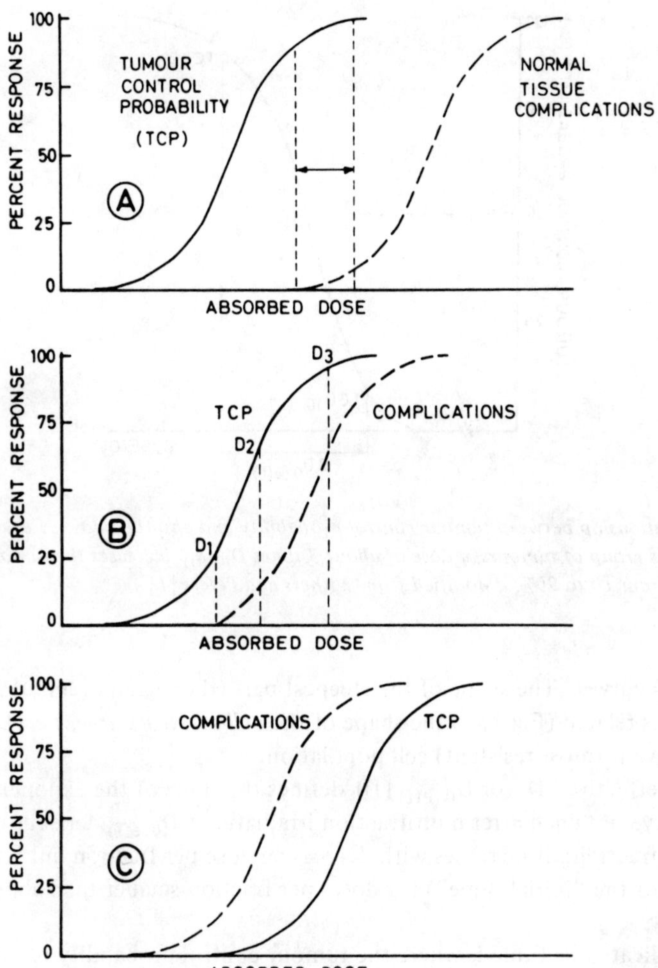

FIG.1. Dose-effect relationships for local tumour control and normal tissue complications. Both curves are sigmoid in shape and, on the figures, are assumed to have the same shape and steepness, although this point can be questionable (see text).

A. Favourable situation: A high local tumour control rate can be achieved without inducing a high complication rate. Some degree of freedom lies in the choice of the target absorbed dose due to the separation between the two curves (↔).

B. Less favourable situation: Any dose level which could achieve a rather high tumour control rate would also induce a high complication rate. The selection of the dose level (D1, D2, D3) is critical as far as both tumour control rate and complication rate are concerned.

C. The situation is still worse: The curve corresponding to the complications is at the left-hand side of that for tumour control. Complications are already seen for dose levels which are not yet able to locally control the tumours. Depending on the situation and on the type of complications, radiotherapy does not seem to bring any benefit at all.

(Modified from Holthusen [18]; Withers and Peters [1].)

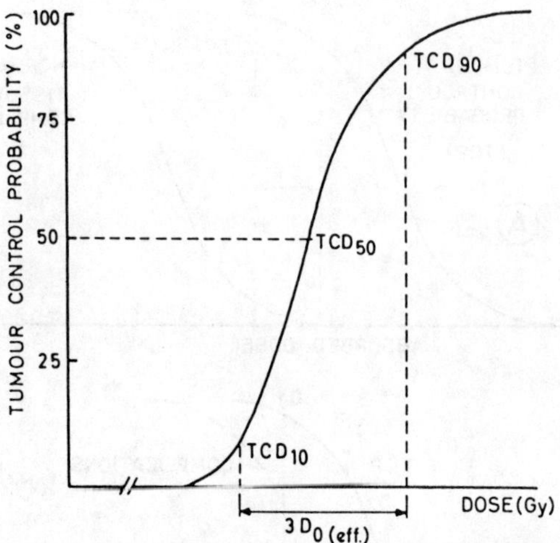

FIG.2. Relationship between tumour control probability and absorbed dose. For a perfectly homogeneous group of tumours, a dose of about 3 times $D_{0(eff)}$ increases the tumour control probability from 10 to 90%. (Modified from Withers and Peters [1].)

cell survival curves. The shape of the steepest part (the region of about 30–70% responses) is related (Fig.2) to the shape of the *cell survival curve,* i.e. (D_0 or $D_{0(eff)}$) for the relevant (most resistant) cell population.

The "effective" D_0 (or $D_{0\,(eff)}$ [1]) defines the slope of the exponential cell survival curve obtained after multifraction irradiation. $D_{0\,(eff)}$ depends on the size of the fraction; it increases with decreasing dose per fraction but reaches its limit $1D_0$ (or the "initial slope") for doses per fraction smaller than about 1.5–2.5 Gy.

As indicated in Table I, where the tumour control probability is expressed as a function of dose, and where the 50% tumour control probability (TCP) is assumed to be 65 Gy, it decreases down to 5% for 60 Gy (dose reduction less by 0.92) and reaches 85% at 70 Gy (dose increase by 1.08).

In practical situations, however, the observed steepness of the dose-control curve will be as steep as the predicted one, only if (1) the group of patients, or tumours, are homogeneous, and to the extent, of course, that the (2) dosimetric and (3) technical treatment conditions are reproducible.

2.2. Clinical observations on dose-effect curves for local tumour control

As far as clinical observations are concerned, it is probably the merit of Shukovsky [2] to have drawn the attention to the steepness of the dose-effect curve for local control rate. For supraglottic squamous cell carcinomas T2 and

TABLE I. VARIATION OF TUMOUR CONTROL PROBABILITY COMPUTED AS A FUNCTION OF DOSE[a]

Absorbed dose (Gy)	Tumour control probability (%)
50	0.0000
55	0.0006
60	5.54
65	50.00
70	84.70
75	96.10
80	99.05
100	99.997

[a] Based on an initial clonogenic cell number of 8×10^7 and $D_{0(eff)}$ for fractionated irradiation of 3.5 Gy, giving a 50% tumour control probability at 65 Gy. (From Withers and Peters [1].)

T3, a dose reduction of 10% (from 1900 to 1710 "rets") reduces the control probability from 70 to around 10%.

In some data, which are cited in this paper, the "nominal single doses" (rets) are used instead of absorbed doses in order to normalize for differences in the fractionation schemes.

The data reported by Stewart and Jackson [3] from Manchester, also on larynx carcinoma, lead to similar conclusions. For T3 tumours, the local control rates decrease from about 70% to 30% for a 10% decrease in dose; the steepness of the dose-effect relationship is similar to that reported by Shukovsky. However, in the same publication [3], for T1 and T2 tumours the control rates are higher and the dose level is less critical since one reaches the upper part (or upper plateau) of the dose-effect "s-shaped" curves (Fig.1).

More recently, Johansson [4] reviewed the dose-effect curves for local control rate of different types of tumour. The steepness of the dose-response curves (Table II) is expressed by the difference in the absorbed dose (in per cent) producing a change in the relative frequency of recurrences from 25 to 75%. Large differences exist between the different types of tumour, but the steepest dose-effect curve is observed for laryngeal carcinoma and its slope is similar to that reported in the two former sets of data [2, 3] which were also obtained for laryngeal carcinomas.

TABLE II. STEEPNESS OF THE DOSE-RESPONSE CURVES FOR LOCAL TUMOUR CONTROL
The steepness is expressed by the difference in absorbed dose (in %) producing a change in recurrence rate from 25 to 75%.
(From Johansson [4].)

Types of tumour	Difference in dose (%)
Laryngeal carcinoma (Hjelm-Hansen, 1979)[a]	± (7–11)
Nasopharynx (Moench & Phillips, 1972)[b]	± 20
Tonsil (Shukovsky, 1974)[c]	± 16
Bladder (Morrison, 1975)[d]	± 33
Lymphoma (Fuks & Kaplan, 1973)[e]	± 27
Skin and lip (Strandqvist, 1944)[f]	± 17

[a] Acta Radiol. Oncol. **18** (1979) 385.
[b] Am. J. Surg. **124** (1972) 515.
[c] In Time-Dose Relationship in Clinical Therapy (CALDWELL, W.L., TOLBERT, D.D., Eds), University of Wisconsin, Madison (1974) 118.
[d] Clin. Radiol. **26** (1975) 67.
[e] Radiology **108** (1973) 675.
[f] Acta Radiol. Suppl. No.55 (1944).

Another aspect of the problem was raised by a randomized clinical trial on tonsil carcinoma performed in 1964 at the Institut Gustave-Roussy in Villejuif [5, 6]. The aim of this study was to compare the effectiveness of high-energy electron and photon beams, and the selected criterion was the tumour regression during the course of the treatment. The planned fractionation scheme was 3 fractions of 2.5 Gy each per week for each radiation quality. Because of dosimetric difficulties, which were not completely solved at the time when the trial was performed, the photon group received a dose higher by 10%, and this difference in the "nominal target dose" [7] resulted in a faster tumour regression in the photon group; the difference reached a significant level for a rather small number of patients (Fig.3). As a difference in RBE could no longer be invoked to explain this observation [6], the conclusion had to be drawn that a difference in dose of only 10% could be detected clinically. It can be assumed that the differences in individual tumour regression were compensated for because the two groups of patients were randomized and the average tumour regression could then be con-

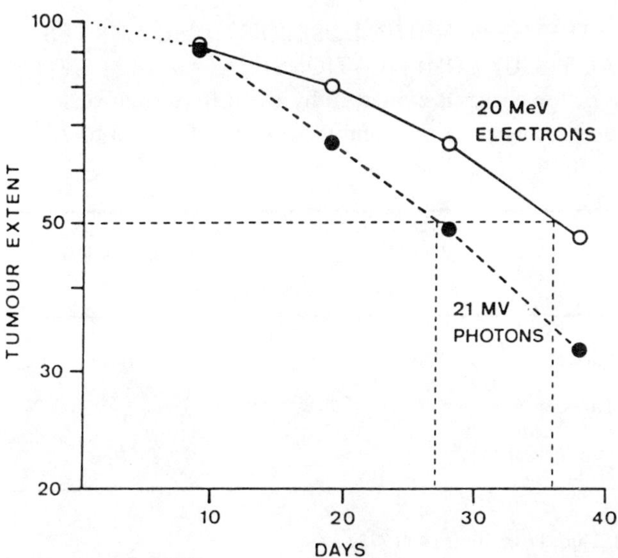

FIG.3. Average tumour regression observed in two randomized groups of patients with tonsillar epithelioma treated with 20 MeV electrons (19 patients) or 21 MV photon beams (23 patients). The tumour regression is faster with photons, the time delays to reach 50% of the initial tumour extent are 36 and 27 days, respectively. This difference in the tumour regression rate, related to a difference of 10% in the target dose, is statistically significant in spite of the rather small size of the patient groups [6].

sidered as a relevant evaluation of the efficiency of the treatment, as discussed elsewhere [8]. However, as different beam qualities were compared, it will never be possible to exclude the influence of a small difference in dose distribution throughout (and outside) the target volume for the same "nominal target absorbed dose", although particular care was taken to achieve, with both techniques, an homogeneous dose distribution within the target volume [6]. We shall come back, when discussing the neutron problems, to the difficulty of separating completely the influence of physical dose distribution and of relative biological effectiveness (RBE).

2.3. Clinical observations on dose-effect curves for normal tissue complications

As far as normal tissue tolerance is concerned, rather sharp dose-effect relationships were derived by Herring and Compton [9] from the data of Fletcher for laryngeal oedema and sigmoiditis and from the data of Phillips and Buschke for myelitis. More recently [4], Johansson reported for normal tissue tolerance dose-effect relationships which were even somewhat steeper than for local tumour control (Table III).

TABLE III. STEEPNESS OF THE DOSE-RESPONSE CURVES FOR NORMAL TISSUE COMPLICATIONS
The steepness of the curves is expressed by the difference in absorbed dose (in %) producing a change of the complication rate from 25 to 75%. (From Johansson [4].)

Normal tissue	Difference in dose (%)
Brachial plexus (Svensson et al., 1975)[a]	± 5
Skin reaction (Turesson & Notter, 1980)[b]	± 6
Skin and lip (Strandqvist, 1944)[c]	± 10

[a] Acta Radiol. Ther. Phys. Biol. **14** (1975) 228.
[b] In Proc. 2nd Rome Int. Symp. on Biological Bases and Clinical Implications of Tumour Resistance, Rome, 1980.
[c] Acta Radiol. Suppl. No.55 (1944).

Another way of evaluating the steepness of the dose-effect curve is to consider the percentage of cases in which a given difference in absorbed dose (e.g., 5–10–20%) could be detected clinically. In a systematic study of skin reactions after high-energy electron irradiation (in the dose range corresponding to erythema and dry desquamation), Wambersie et al. [10] found that, on adjacent areas in the same patient, a difference in dose of 10% could be detected in 80% of the cases, and a difference in dose of 20% could be detected in 90% of the cases. In symmetrical areas, however, much smaller differences in dose (about 5%) could be detected clinically (e.g., comparison of skin reactions in supraclavicular areas).

3. SPECIFIC PROBLEMS IN NEUTRON THERAPY

As pointed out earlier, the accuracy required in neutron therapy compared with that required in therapy with low-LET radiation depends on the relative steepness of the dose-effect relationships and on the relative separation between the dose-effect curves for tumour and normal tissue responses.

3.1. Steepness of the dose-effect relationships

As the dose-effect relationships, both for local tumour control and normal tissue complications, are related to the shape of the cell survival curves, no dif-

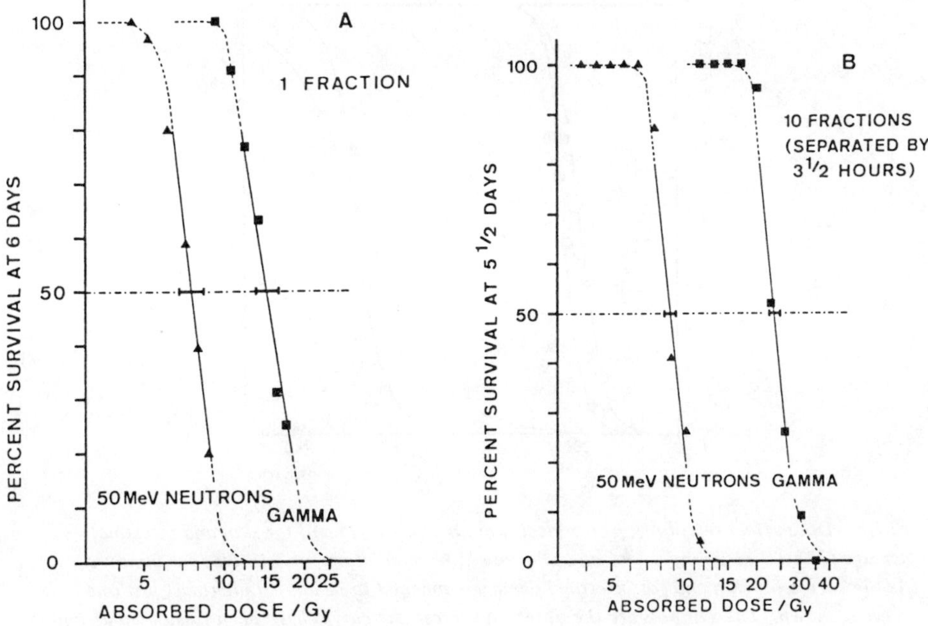

FIG.4. Dose-effect relationships for intestinal tolerance in mice after selective abdominal irradiation with ^{60}Co-gamma and d(50)+Be neutron irradiation. Intestinal tolerance is assessed from the survival of the animals scored at 6 days after a single fraction irradiation (Fig.4A) or after 10 fraction irradiations (Fig.4B). In this latter case, the successive fractions are separated by 3.5 h and the survival is scored at $5\frac{1}{2}$ d after the "middle" of the irradiation (i.e. after the fifth fraction). The computed slopes of the dose-effect relationships ($LD_{90} - LD_{10})/LD_{50}$ are not significantly different. For a single fraction irradiation, $LD_{50_{Co}} = (14.35 \pm 1.45)$ Gy and $LD_{50_n} = (7.67 \pm 0.60)$ Gy which leads to an RBE value of 1.9 ± 0.2. For an irradiation performed in 10 fractions (interval i = 3.5 h), $LD_{50_{Co}} = (23.7 \pm 1.1)$ Gy and $LD_{50_n} = (8.8 \pm 0.4)$ Gy, which leads to an RBE value of 2.7 ± 0.2. (From Gueulette [11].)

ference has to be expected in principle and in a first approximation, between neutrons and photons (after normalization for RBE).

As an example Fig.4 shows dose-effect relationships for intestinal tolerance in mice after single and fractionated gamma and neutron irradiation [11]. No significant difference in shape can be detected. Van der Kogel [12] reached similar conclusions for late effects on the spinal cord.

From a theoretical point of view, it could be argued that when dose-effect relationships are obtained by using the same fraction sizes but by increasing the fraction number, the slopes of the dose-effect curves should in principle be equally steep for neutrons and γ-rays. In fact, for any level of effect, the dose ratio is the RBE of the neutrons for the selected fraction size.

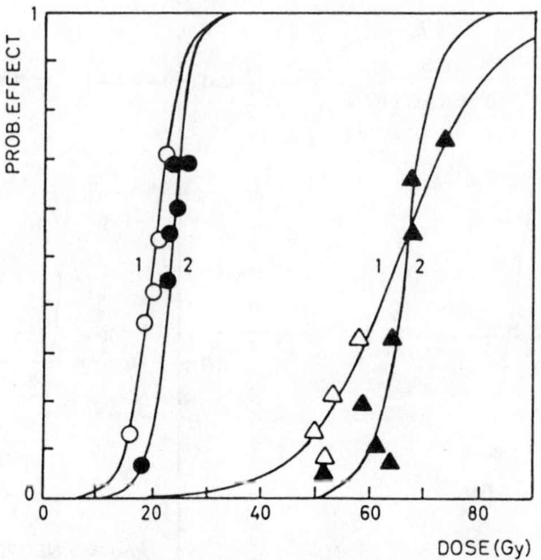

FIG.5. Dose-effect relationships for local tumour control (1) and for skin and intestinal damage (2) after neutron (○, ●) and photon (△, ▲) irradiation for T4B bladder tumours. As the curves for skin and for intestinal damage happened to be almost identical, just one curve is shown. The symbols are the observed values; the curves have been calculated assuming the following equation: $P = \exp(a + bD)/\{1 + \exp(a + bD)\}$. *From the available data, no significant conclusion can be derived concerning the relative slopes of the dose-effect relationships for neutrons and γ-rays or for effects on tumours and normal tissues. (From Battermann [19].)*

On the contrary, when the dose-effect curves are obtained by increasing the fraction sizes, the γ-curve should in principle become somewhat steeper to the extent that the effective D_0 ($D_{0\,(eff)}$) decreases with fraction size. The steepness of the neutron dose-effect relationship is nearly independent of the neutron fraction size, and it is well known that neutron RBE decreases when fraction size increases.

In practical situations, however, the explored dose ranges are generally too narrow, and the precision of the experimental or clinical data too small, to allow one to identify these differences.

As far as the clinical data are concerned, little information is at present available to compare the shapes and the slopes of the γ and neutron dose-effect curves. No significant difference is expected and none has been detected so far.

3.2. Compared differential effect after neutron and photon irradiations

It is well known that neutron RBE relative to photon varies with the biological system, and any therapeutic gain due to neutron irradiation can only be related to a neutron RBE being larger for effects on tumours than for effects on normal tissues (Fig.5).

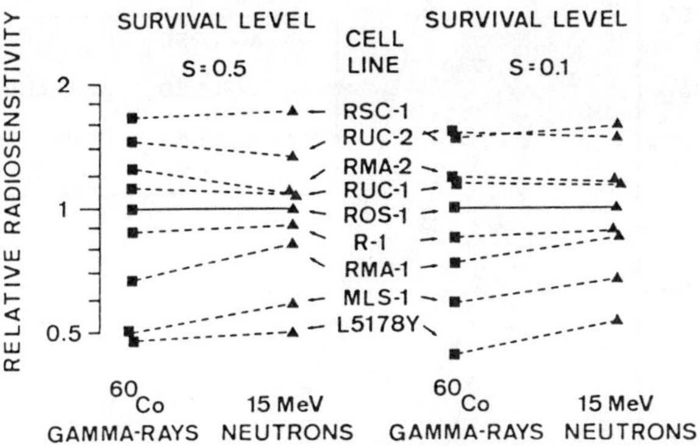

FIG.6. Relative radiosensitivity of nine cell lines assessed in vitro after 15 MeV neutrons and ^{60}Co gamma irradiation. The radiosensitivity of cell line ROS-1 (median value) is taken as reference (≡ 1). The differences in the relative radiosensitivities (assessed from the dose ratios corresponding to 0.5 and 0.1 survival levels) are in general narrower after neutron irradiation, mainly for S = 0.5. (From Barendsen [20].)

In this respect, there is now evidence indicating that variations in neutron RBE are due more to variations in radiosensitivity to γ-rays than to variations in radiosensitivity to neutron irradiation. In other words, the range of radiosensitivity is narrower for neutrons than for photons. Two examples are presented in Figs 6 and 7 to illustrate the point.

As far as the particular case of the OER (oxygen enhancement ratio) gain factor is concerned, one can consider that it results from the fact that neutrons are reducing a difference in radiosensitivity between oxygenated and anoxic cells more than from the fact that neutrons should be specifically efficient against anoxic cells.

In these conditions, the separation between the different dose-effect relationships is expected, in general, to be smaller for neutrons than for photons [13] and, consequently, the required accuracy should be in principle better or at least equal in neutron therapy than in photon therapy. It seems then reasonable to say that in neutron therapy, as in photon therapy [14], an accuracy of 5% should be required.

4. CONCLUSION AND PRACTICAL CONSIDERATIONS

The conclusion which emerges from the present review is that in high-energy photon or electron beam therapy, there is evidence indicating that for certain types of tumour, an accuracy of ± 5% in the delivery of an absorbed dose to a

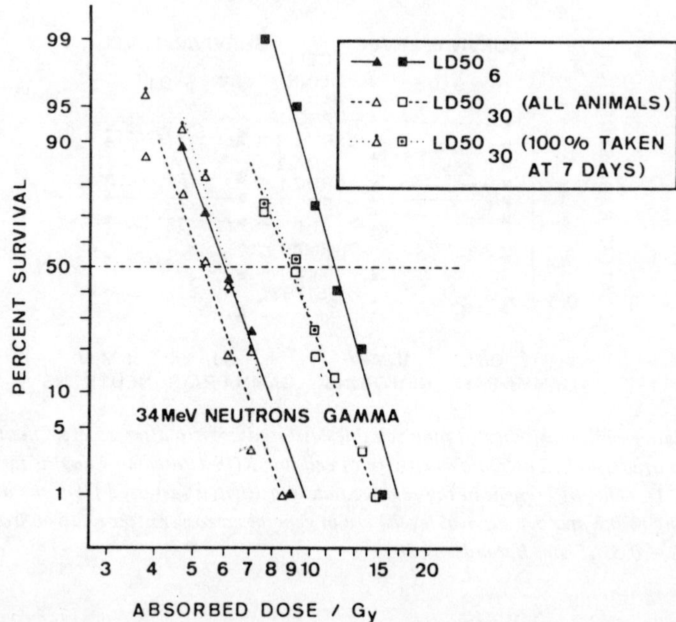

FIG. 7. LD_{50} for bone marrow syndrome ($LD_{50/30}$) and for intestinal syndrome ($LD_{50/6}$) after total body irradiation with ^{60}Co gamma rays and p(34) + Be neutrons. After ^{60}Co irradiation, $LD_{50/30}$ and $LD_{50/6}$ are clearly separated (9.3 and 11.4 Gy, respectively; ratio = 0.82). For ^{60}Co, the computed $LD_{50/30}$ values are similar whether the percentages of survival are scored, taking into account all irradiated mice or only those surviving at 7 days. After neutron irradiation, $LD_{50/6}$ = 5.9 Gy. On the other hand, $LD_{50/30}$ = 5.3 Gy when all the irradiated animals are taken into account (100% survival taken at day zero), but $LD_{50/30}$ = 6.3 Gy when only the animals surviving at 7 days are taken into account. These data indicate that the difference in radiosensitivity between bone marrow and intestinal syndrome is reduced after neutron irradiation as compared to photon irradiation. (From Gueulette [11].)

target volume is necessary if the eradication of the primary tumour is sought. This recommendation was already made in ICRU Report 24 [14].

It can be assumed that the RBE of the beams used in high-energy photon or electron beam therapy is not significantly different from unity (relative to ^{60}Co) in the current energy range [7]. Consequently, when the target absorbed dose and dose distribution are known, the clinical information and the therapeutic protocols can be transferred from one Centre to another and the clinical (or biological) effects in one Centre can normally be predicted from the experience accumulated in other Centres.

To maintain a given treatment policy within a department, or to compare treatments performed in different institutions, an *uncertainty in the absolute value* of the absorbed dose is of relatively little importance as long as the dosimetry is reproducible, and as long as the institutions participating in a comparison use a consistent dosimetry.

This can be achieved in a very efficient way by calibration locally or regionally against a reference beam. The situation is more complex when different beam qualities are available and are used interchangeably, such as high-energy photons and electrons. When these beam qualities are used interchangeably on the same patient or similar patient groups, it is also essential that the *relative dosimetry* at these different beam qualities be consistent. However, on an international level, there is no doubt that a considerable advantage could be gained by using a self-consistent absolute dosimetry for all beam qualities.

As far as neutron therapy is concerned, it seems reasonable to assume that a similar level of accuracy is needed. However, in neutron therapy, an additional problem is raised by the fact that *RBE critically depends on neutron energy*. In these conditions, the clinical experience and the therapeutic protocols cannot be transferred from one Centre to another without taking the RBE problem into account, and the determination of the (physical) dose distribution is not sufficient. Three practical approaches can *a priori* be considered.

4.1. A pure physical approach

As the number of neutron therapy Centres increases as well as the amount of available information concerning their beams, a first simple method could be an interpolation between the beams for which enough information is available. This interpolation could be made, taking into account, for example, the energy of the incident particles. However, this method is strictly valid only when the same type of reaction is used (e.g., d → Be in cyclotrons). With other types of machine (e.g., (d,T)), or when other types of reaction are used (e.g., p → Be in cyclotrons), the neutron energy spectra are different and furthermore may be modified by the addition of filters.

Besides the energy of the incident particles, a rather rough estimate of a "mean" neutron energy could be derived from the beam penetration (e.g.: depth of the 50% isodose measured in defined conditions).

Microdosimetry certainly opens new possibilities, and indeed the complete microdosimetric spectra unambiguously define the "quality" of the beam at the point of interest. However, if one can assume that similar microdosimetric spectra will produce similar biological effects, it is generally difficult, when the spectra are different in shape, to predict what will be (and how large will be) the differences in the biological effects. One way to deal with this problem is to divide the microdosimetric spectra into several "intervals", and to predict the biological effectiveness of the beam by weighting the contributions from the different spectra intervals [15].

Another possible approach, which also takes advantage of microdosimetry, is to select one (or a few) microdosimetric parameter(s) (such as $\bar{y}_D$, $\bar{y}_D^*$, $\bar{y}_F$) which could be considered as relevant for characterizing the quality of the beam, at least for therapeutic applications. Microdosimetry seems to play an increasing role in the comparisons performed between different neutron therapy Centres. Systematic microdosimetric intercomparisons are at present carried out (or planned) between the European neutron therapy Centres within the EORTC High-LET Therapy Group.

4.2. Radiobiological approach

To avoid confusion between "RBE of a neutron beam versus γ-rays" and "RBE of a neutron beam versus a neutron beam of another energy", the expression "RBE ratio" will be used when comparing different neutron beams. Direct determination of RBE ratio *provides an "overall" check* of the dosimetric system, and in particular includes a correction for an eventual *systematic error* which could be introduced from the selection of some physical parameters, such as $\overline{W}_\gamma/\overline{W}_n$, kerma ratios, etc.

It is well known that the RBE between two different radiation qualities depends on several factors: biological system and endpoint, dose, etc. This has clearly been illustrated when comparing, for example, neutrons and photons. However, when comparing different radiation qualities closer to each other (for example, the different fast neutron beams used in therapy) it could be sufficient, for practical purposes, to determine one (or a few) RBE ratios for relevant systems. This implies the assumption that a single conversion factor is sufficient to take into account the RBE differences between the fast neutron beams to be compared. This attitude seems reasonable since RBE ratios are not very different from unity for the relevant effects and for the neutron beams at present used in neutron therapy.

This conversion factor could be called "neutron RBE conversion factor" (NCF) or "neutron intercomparison factor" (NIF). It has to be borne in mind that in the strict sense it is not an experimental RBE ratio, but a kind of "average value" derived from several determinations, which can be used *only* when the information obtained in one Centre with one neutron beam quality has to be transferred to another Centre using another neutron beam quality.

RBE ratios between different neutron beams used for therapeutic applications have been determined for several well-codified biological systems (e.g.,intestinal crypt cell colony method, mammalian cell survival in vitro, etc.). Such intercomparisons were performed between United States Centres, between European Centres, and between United States and European Centres.

4.3. Clinical intercomparisons

Finally, information on the RBE ratio can be derived from clinical intercomparison, mainly on normal tissue tolerance using well-codified scoring systems.

Unfortunately, this method is not very precise, and always implies subjective appreciation. It requires a large series of patients who are not always perfectly comparable.

A final point has to be stressed. Normal tissue reactions depend on beam energy, not only because energy influences the RBE, but also because energy determines the (physical) dose distribution (volume of irradiated tissue or organ). When comparing the clinical effects produced by different beam qualities, it is sometimes very difficult to separate what is related to RBE (radiobiology) and what is due to dose distribution. A similar type of debate happened in the 'sixties about the RBE of high-energy electrons [16, 17].

REFERENCES

[1] WITHERS, H.R., PETERS, L.J., "Basic principles of radiotherapy – Biologic aspects of radiation therapy", Textbook of Radiotherapy – 3rd Edn (FLETCHER, G.H., Ed.), Lea and Febiger (1980) 103–80.

[2] SHUKOVSKY, L.J., Dose, time, volume relationships in squamous cell carcinoma of the supraglottic larynx, Am. J. Roentgenol. **108** (1970) 27–29.

[3] STEWART, J.G., JACKSON, A.W., The steepness of the dose response curve both for tumor cure and normal tissue injury, Laryngoscope **85** (1975) 1107–11.

[4] JOHANSSON, K.A., Studies of different methods of absorbed dose determination and a dosimetric intercomparison at the Nordic radiotherapy centres, Thesis, University of Gothenburg, Sweden (1982).

[5] FLAMANT, R., MALAISE, E.P., DUTREIX, A., DUTREIX, J., HAYEM, M., LAZAR, P., PIERQUIN, B., TUBIANA, M., Un essai thérapeutique clinique sur l'irradiation des cancers amygdaliens par faisceaux de photons ou d'électrons de 20 MeV, Eur. J. Cancer **3** (1967) 169–81.

[6] WAMBERSIE, A., Contribution à l'Etude de l'Efficacité Biologique Relative des Faisceaux de Photons et d'Electrons de 20 MeV du Bétatron, J. Belge Radiol., Monographie No.1, (1967) 1–135.

[7] ICRU (International Commission on Radiation Units and Measurements), Dose Specification for Reporting External Beam Therapy with Photons and Electrons, ICRU Rep. 29 (1978).

[8] WAMBERSIE, A., DUTREIX, J., Analyse de la régression tumorale après irradiation. Interprétation des résultats cliniques d'un essai thérapeutique, J. Eur. Radiother. **1** (1980) 159–70.

[9] HERRING, D.F., COMPTON, D.M.J., "The degree of precision required in the radiation dose delivered in cancer therapy", Computers in Radiotherapy (GLICKSMAN, A.S., COHEN, M., CUNNINGHAM, J.R., Eds), Brit. J. Radiol., Special Rep. No.5, British Institute of Radiology (1971) 51–58.

[10] WAMBERSIE, A., ZREIK, H., PRIGNOT, M., VAN DORPE, J.C., Variation of RBE as a function of depth in a high energy electron beam in the first millimeters of the irradiated tissues determined by the observation of skin reactions on patients (a clinical trial), Strahlentherapie **148** (1974) 279–87.

[11] GUEULETTE, J., Efficacité Biologique Relative (EBR) des neutrons rapides pour la tolérance de la muqueuse intestinale chez la souris, Thèse de Doctorat de l'Université Paul Sabatier de Toulouse (1982).

[12] VAN DER KOGEL, A.J., Late effects of radiation on the spinal cord, Thesis, University of Amsterdam (1979).

[13] WAMBERSIE, A., The European experience in neutron therapy at the end of 1981, Int. J. Radiat. Oncol. Biol. Phys. 8 (1982).

[14] ICRU (International Commission on Radiation Units and Measurements), Determination of Absorbed Dose in a Patient Irradiated by Beams of X- or Gamma-rays in Radiotherapy Procedures, ICRU Rep. 24 (1976).

[15] ICRU (International Commission on Radiation Units and Measurements), Microdosimetry, ICRU Rep. (in press).

[16] DUTREIX, J., WAMBERSIE, A., "Clinical radiobiology of high-energy electrons", Proc. Symp. Electron Beam Therapy, Sloan-Kettering Memorial Cancer Centre, 1979 (CHU, F.C.H., LAUGHLIN, J.S., Eds), Aubrion Press, Inc. (1981) 121–28.

[17] ZUPPINGER, A., PORETTI, G. (Eds), Proc. Symp. High-Energy Electrons, Montreux, Sept. 1964, Springer-Verlag (1965).

[18] HOLTHUSEN, H., Erfahrungen über die Verträglichkeitsgrenze für Roentgenstrahlen und deren Nutzanwendung zur Verhütung von Schäden, Strahlentherapie **57** (1936) 254–69.

[19] BATTERMANN, J.J., Clinical application of fast neutrons: the Amsterdam Experience, Thesis, University of Amsterdam (1981).

[20] BARENDSEN, G.W., "RBE values of fast neutrons for responses of experimental tumours", High-LET Radiations in Clinical Radiotherapy (BARENDSEN, G.W., BROERSE, J.J., BREUR, K., Eds), Pergamon Press, Oxford (1979) 175–79.

DOSIMETRY OF FAST NEUTRON BEAMS

DETERMINATION OF ABSORBED DOSE AND RADIATION QUALITY WITH TISSUE-EQUIVALENT (TE) IONIZATION CHAMBERS

J.J. BROERSE, J. ZOETELIEF
Radiobiological Institute TNO,
Rijswijk, Netherlands

Abstract

DETERMINATION OF ABSORBED DOSE AND RADIATION QUALITY WITH TISSUE-EQUIVALENT (TE) IONIZATION CHAMBERS.

The characteristics of four tissue-equivalent (TE) ionization chambers including those employed as common dosimeters for neutron therapy applications are summarized. These include ion collection, radiation sensitivity of cables, connectors and stem, density and composition of tissue-equivalent (TE) gas, wall thickness, stem scatter and angular dependence of response. The displacement of the effective point of measurement of spherical TE ionization chambers has been investigated. Displacement correction factors are reported for in-phantom measurements of neutron beams with energies from 1 to 2 MeV. Operation of a TE ionization chamber at increased gas pressures (up to 80 bar) and with different collecting voltages provides information about the quality of neutron beams.

1. INTRODUCTION

A prerequisite for biomedical studies is that the energy dissipation in the irradiated material be determined with a sufficient degree of accuracy and precision. Arguments can be advanced on the basis of dose-effect relations that, for biomedical purposes, the absorbed dose in the irradiated object should be determined with an accuracy (overall uncertainty) not larger than ±5 per cent [1]. This is a difficult task, considering the complexity of dose determination due to inhomogeneities in tissue composition and density in the patient, e.g. at interfaces of matter of different composition and/or density. Because of the differences in relative biological effectiveness of the two radiation components, it is necessary to determine separately the neutron absorbed dose, D_N, as well as the photon absorbed dose, D_G, of the mixed field.

Specification of absorbed dose does not account for the microscopic distribution of the energy deposition, i.e. the radiation quality. This microscopic distribution, however, has to be considered, since the degree of damage depends not only on the total amount of energy but also on the spatial distribution of energy deposition.

The use of calibrated A-150 plastic tissue-equivalent (TE) ionization chambers with TE gas filling is generally considered to be the most practical method for determining the total absorbed dose in mixed neutron-photon fields for biomedical applications [2–4].

During the past two years a considerable amount of information has been collected on the basic physical parameters for neutron dosimetry and the characteristics of tissue-equivalent ionization chambers [5–7]. These activities have further resulted in the drafting of protocols for neutron dosimetry for external beam radiotherapy [8, 9]. In this paper some operational characteristics of four ionization chambers employed for biomedical applications are summarized. A specific aspect of dosimetry in phantoms is the location of the effective point of measurement. It is shown that the introduction of a gas-filled cavity influences the total process of attenuation and scattering. Finally, some recent results are given on the response of a TE ionization chamber in dependence on gas pressure and collecting voltage as a means to assess radiation quality.

2. TOTAL ABSORBED DOSE DETERMINATION WITH TE IONIZATION CHAMBERS IN MIXED FIELDS

The ionization chambers applied in neutron dosimetry for radiobiology and radiotherapy show a great variety in design and construction. Some of the chambers have been developed at research institutes; others are commercially available. The design sketches and X-ray radiographs of four ionization chambers are given in Fig.1. The 1 cm^3 TNO chamber [6] with cavity diameter of 12 mm and wall thickness of 1 mm is shown in the figure. Two types of TE ionization chamber, which are widely used for dosimetry of neutrons in biology and medicine, namely the Exradin 0.53 cm^3 thimble type and the Far West Technology (FWT; previously EG&G) spherical chambers, are also shown. The first chamber is used as a common instrument by the European clinical neutron dosimetry groups; the second is used as a reference chamber by United States clinical neutron dosimetrists. More extensive information on the Exradin and FWT chambers is given by Meeker [10] and Kantz [11], respectively. The fourth chamber in Fig.1 is that of the Centre d'Etudes Nucléaires de Fontenay-aux-Roses (CENF) [12].

The charge produced within the cavity is derived from the reading R of the chamber multiplied by several correction factors k_R including those for incomplete charge collection, leakage current and density and composition of the gas in the cavity [9]. If measurements are carried out in a phantom, the quantity to be assessed is the absorbed dose at the centre of the chamber when the chamber is replaced by phantom material. For this purpose a displacement factor k_d, which is defined as the ratio of the absorbed dose for an infinitesimally small cavity to the absorbed dose measured, has to be introduced.

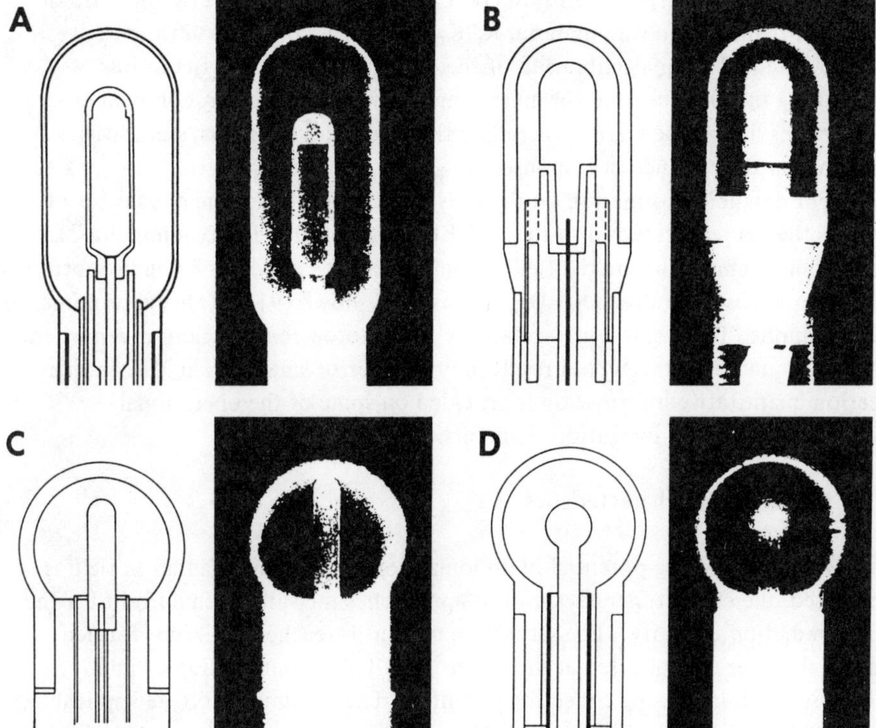

FIG.1. Design sketches and X-ray photographs of four tissue-equivalent ionization chambers manufactured by:
A. Centre d'Etudes Nucléaires, Fontenay-aux-Roses, France (CENF);
B. Exradin, Warrenville, Illinois, USA;
C. Far West Technology, Goleta, California, USA (FWT, EG&G);
D. Radiobiological Institute TNO, Rijswijk, the Netherlands (TNO).

The absorbed dose in tissue adjacent to the cavity of the chamber D_t can be derived from the reading of the TE ionization chamber:

$$D_t = R \cdot \Pi k_R \cdot k_d \cdot \frac{1}{e.m} \cdot W \cdot r_{m,g} \frac{(\mu_{en}/\rho)t}{(\mu_{en}/\rho)m} \qquad (1)$$

where e is the charge of the electron, m the mass of the gas in the cavity, W the average energy required to produce an ion pair in the gas, $r_{m,g}$ the gas-to-wall absorbed dose conversion factor. The absorbed dose in tissue can subsequently be derived from the ratio of the mass-energy absorption coefficients μ_{en}/ρ in the reference tissue t and the wall material m. For the neutron component, the ratio of the mass-energy absorption coefficients can be replaced by the ratio of kerma in tissue and in wall material K_t/K_m. The mass of the gas in the cavity can be obtained from a calibration of the chamber in a photon field. For calibrating the chamber and for measuring neutron kerma in air, complementary corrections have to be applied including those for wall thickness, stem scatter and angular dependence of response.

For detailed information on the most recent values of basic physical parameters the reader is referred to the ECNEU protocol [9]. The reading of a TE ionization chamber in a mixed field is sometimes interpreted as being proportional to the total absorbed dose D_N+D_G. It has been shown [9] that this approximation can be applied for beams with a relatively low photon contribution. Even when D_G/D_N equals 20% this would result only in an error smaller than 1%. In this section quantitative information is provided on some of the operational characteristics of TE ionization chambers.

2.1. Ion collection characteristics

As the collecting potential of an ionization chamber in a radiation field is increased, the current increases until it approaches the saturation current for the given radiation intensity. The saturation current is reached if all ions formed in the chamber are collected at the electrodes. The recombination of ions decreases with increasing collecting potential. The maximum voltage applicable is limited by electron multiplication. Distinction can be made between volume and initial recombination, for which the following relations are given (see, e.g., Boag [13]):

$$\text{volume recombination:} \quad \frac{Q_s}{Q_V} = 1 + \frac{v}{V^2} \tag{2}$$

$$\text{initial recombination:} \quad \frac{Q_s}{Q_V} = 1 + \frac{w}{V} \tag{3}$$

where Q_s is the saturation charge (charge for infinite voltage), Q_V is the charge when a voltage V is applied and v and w are constants. Volume recombination involves ions from separate tracks and increases with dose rate. Initial recombination is determined only by the ion density along each track and is independent of dose rate. For photons, volume recombination and, for neutrons, initial (columnar) recombination are predominant in commonly used ionization chambers.

In Fig.2 the saturation characteristics for three spherical ionization chambers with cavity radii of 4, 8, and 16 mm are given for ^{137}Cs gamma rays at a dose rate of approximately 13 mGy·min^{-1}. The dependence of recombination on dose rate for photons is included in Fig.2 for measurements with the largest spherical ioniza-

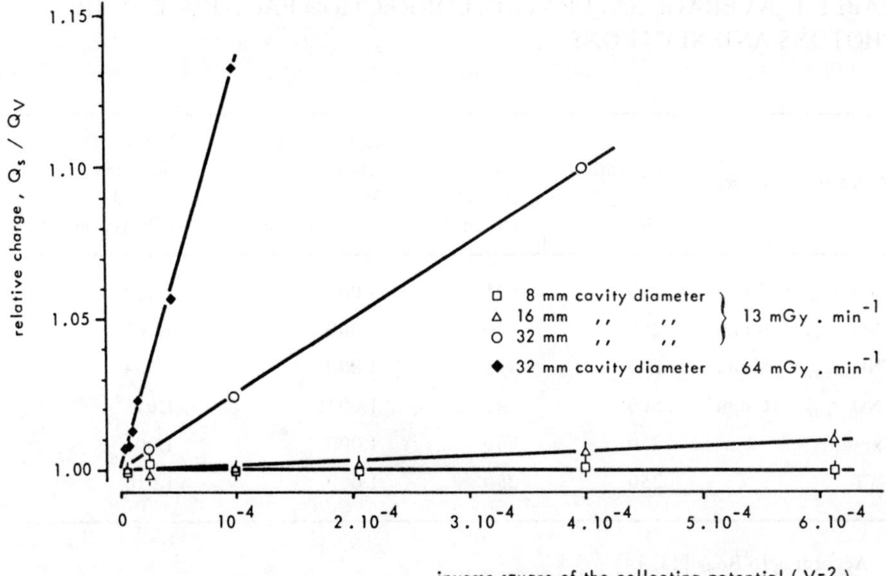

FIG.2. Saturation characteristics of three spherical TNO ion chambers with different cavity radii for ^{137}Cs photons at a dose rate of 13 mGy·min^{-1} for the largest chamber; also at a dose rate of 64 mGy·min^{-1} [6].

tion chamber. The dependence on field strength and on dose rate are both in qualitative agreement with the formulation given by Boag [13] for volume recombination. Applying Mie's constant for air, at a dose rate of 13 mGy·min^{-1} and a collecting potential of 50 V results in measured Q_S/Q_V values somewhat larger than the calculated ones. This is in qualitative agreement with the findings of Boag [14], where also higher recombination values for TE gas compared to air are reported.

The saturation characteristics of ionization chambers in neutron fields are determined by columnar (initial) recombination. Experimental and theoretical treatment of columnar recombination (Jaffé theory) is available for α-particles (see, e.g. Boag [13]). To arrive at the saturation charge the Jaffé function f(x) should be plotted against the collection efficiency. However, the information required to calculate f(x) is not available for all secondaries produced by neutrons and lacking for TE gas. Therefore, Eq.(3) is generally applied to derive the saturation charge or current. This extrapolation is allowed only in the voltage range close to saturation. A rough estimate of the saturation correction has been made by Boag [14] on the basis of the relative contributions to kerma in TE gas and Te plastic from protons, α-particles and heavy recoil nuclei [15]. For neutron

TABLE I. AVERAGE SATURATION CORRECTION FACTORS, k_S, FOR PHOTONS AND NEUTRONS

Ionization chamber	Operation voltage (V)	Field strength[a] ($V \cdot cm^{-1}$)	k_s for ^{137}Cs photons $\dot{K}$ or $\dot{D}$ <60 $mGy \cdot min^{-1}$	k_s for neutrons $\dot{K}$ or $\dot{D}$ <20 $Gy \cdot min^{-1}$
TNO, r_{cav} = 4 mm	250	625	1.000	1.000
TNO, r_{cav} = 6 mm	250	210	1.000	1.005
TNO, r_{cav} = 8 mm	300	125	1.000	1.004
TNO, r_{cav} = 16 mm	500	45	1.002[b]	1.013[c]
Exradin	250	750	1.000	1.001
FWT	250	250	1.000	1.010

[a] According to Boag [13, 14].
[b] Correction at 13 $mGy \cdot min^{-1}$.
[c] Correction determined only for dose or kerma rates up to about 50 $mGy \cdot min^{-1}$.

energies from 0.5 to 15 MeV and field strengths of between 50 and 1000 $V \cdot cm^{-1}$ a loss of charge owing to recombination of less than 3% can be assessed.

A summary of the saturation correction factors, k_S, for neutrons and photons for several ionization chambers is given in Table I for values of the collecting potential commonly used. The correction factors for neutrons are generally higher than those for photons in the dose-rate range investigated. For neutrons, the corrections are all less than 3% in agreement with theoretical estimates.

If differences in response are observed following a change in the polarity of the collecting voltage, the mean value is considered to be the best approximation. However, differences in excess of 2% can be taken as an indication of malfunctioning of the dosimeter system. Some of the possible causes of polarity effects are field inhomogeneities combined with differences in drift velocities of ions of different signs, variation in effective volume of the chamber owing to space charge distortion of the electrical field, collection of charge produced outside the cavity (extracameral charge), and small voltage differences between guard and collector resulting from sources of thermal or electrolytic origin. Cleaning of a chamber sometimes reduces polarity effects. From studies by Schraube et al. [16] it was concluded that there is no need to perform routine measurements at both polarities, since the polarity effect for the chambers was the same for photons and neutrons within the uncertainty limits. However, for

TABLE II. HALF VALUE TIME, $T_{1/2}$ (min), OF THE TWO DECAY COMPONENTS OF THE LEAKAGE CURRENT FOR DIFFERENT CHAMBERS

Chamber	Fast decay	Slow decay
Exradin	0.1	2
FWT	0.3	12
EG & G	2	30
TNO (1 cm^3)	0.8	7

measurements at very low dose rates, the electrometer can show a polarity effect in the low current region which should be investigated.

2.2. Offset and leakage current

The leakage current of a chamber must be small with reference to the minimum current measured in a radiation field (0.5 $mGy \cdot min^{-1}$ for a 1 cm^3 chamber flushed with methane-based TE gas corresponds to 3×10^{-13} A). In this context, the leakage currents from the cable and electrometer must be taken into account and the use of high-quality insulators is of importance. It has been shown [16] that after application of the high voltage, the initial current decreases exponentially with time, demonstrating a fast and a slow component. The decay components of the leakage current of different chambers are given in Table II. The relatively long decay times for the EG&G chamber (essentially of the same design as the FWT chamber) were observed with a chamber which has been used for a period of eight years. The leakage current asymptotically approaches the offset current of the electrometer (about 5×10^{-16} A) when microphonic effects are avoided. The relative correction for leakage current is dependent on the measured ion current. In most practical situations in radiotherapy where ion chambers of about 1 cm^3 gas volume are used, corrections for leakage current are negligible. A comparable bi-exponential decay of the reading has been observed by Parnell [17] for an EG&G chamber exposed to ^{137}Cs gamma rays.

2.3. Radiation sensitivity of cable, connector and stem

During irradiation, extracameral ion currents caused by charge buildup outside the chamber cavity can occur, i.e. in the chamber stem, the connectors or connector

TABLE III. RADIATION SENSITIVITY OF SEVERAL CABLES AND CONNECTORS TO GAMMA AND NEUTRON IRRADIATION COMPARED WITH THE LOWEST CHAMBER SENSITIVITY

Manufacturer	Radiation sensitivity of cable ($C \cdot Gy^{-1}/10$ cm)	Radiation sensitivity of connector ($C \cdot Gy^{-1}$)
Philips triax.	3×10^{-12}	$(4.1 \text{ to } 3\ ) \times 10^{-9}$
Pychlau triax.	1×10^{-11}	1.2×10^{-9}
Amphenol triax.	–	$(3.3 \text{ to } 1.9) \times 10^{-9}$
BNC coax.	–	$(2.1 \text{ to } 0.2) \times 10^{-10}$

Lowest chamber sensitivity $2 \times 10^{-8}\ C \cdot Gy^{-1}$.

blocks at the end of the chamber stem and the connecting cables. The influence of the sensitivity of stem and connectors depends on the irradiation geometry and is appreciably smaller for collimated beams than for uncollimated sources. Special care must be taken in the case of measurements inside a phantom with uncollimated beams where the sensitive volume of the chamber is at a location where relatively low dose rates are measured in comparison with the position of the distal part of the stem, connectors and parts of the cable if these are in the unattenuated neutron field. Massive triaxial connectors show a relatively high sensitivity to radiation in comparison with small coaxial connectors (see Table III). This might be attributed to either the presence of small extracameral cavities or the use of insulating materials of inferior quality in the triaxial connectors.

2.4. Density and composition of TE gas in the cavity

To approximate the homogeneity condition for application of the Bragg-Gray principle, TE ionization chambers are usually flushed with TE gas. The chamber reading depends on the density and composition of the gas in the cavity. Measurements with different gases in a neutron field showed variations up to 1% for ^{252}Cf and negligible variations for d+T neutrons when the readings were normalized to the photon calibration for each gas [18]. The effect of gas flow rate on the reading of the 1 cm^3 TNO, Exradin and FWT chambers relative to the reading obtained with air in the cavity is shown in Fig.3 for ^{137}Cs photons after a pre-flush with 300 cm^3 TE gas. In the region of low flow rates (below 10 cm$^3 \cdot$min^{-1}), relatively low readings are observed for the TNO chamber.

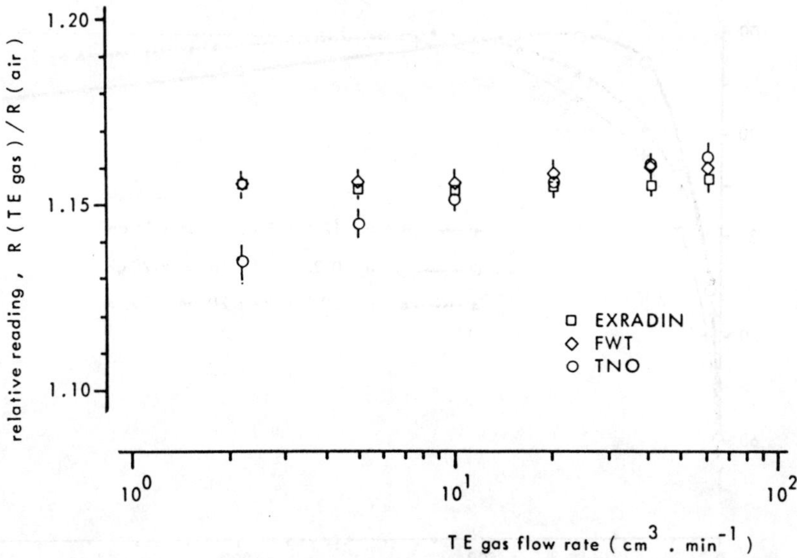

FIG.3. Dependence of chamber reading on TE gas flow rate for three ion chambers irradiated with ^{137}Cs gamma rays. Given are the ratios of the readings when the chambers are flushed with TE gas, R (TE gas), relative to the readings when they are filled with air, R (air), as a function of TE gas flow rate [16].

This might be attributed to differences in the gas outlet constructions of the different chambers which result in differences in diffusion of air constituents into the chambers. For gas flow rates in excess of 10 cm^3·min^{-1}, the responses for all chambers remain nearly constant, if high resistance gas outlet tubes are avoided (see also Mijnheer [19]). The same tendency is observed for flow rates in excess of 10 cm^3·min^{-1} for neutrons. In gas flow systems where gas inlet and outlet pressure are monitored, low flow rates (1 cm^3·min^{-1} or less) are sufficient provided that chamber and gas tubes are airtight. The use of silicone rubber tubing in the gas flow system has to be avoided because of leakage of carbon dioxide through the wall of the tube [20].

The ratio of the readings of an ionization chamber flushed with TE gas to that with air increases with decreasing neutron energy owing to an increasing ionization contribution from protons created in the TE gas rather than in the wall; air-filled chambers should not be used in mixed fields with considerable contributions of low-energy neutrons.

2.5. Wall thickness

The reading of an ionization chamber is determined by the ionization produced in the cavity by charged particles created in the wall, central electrode

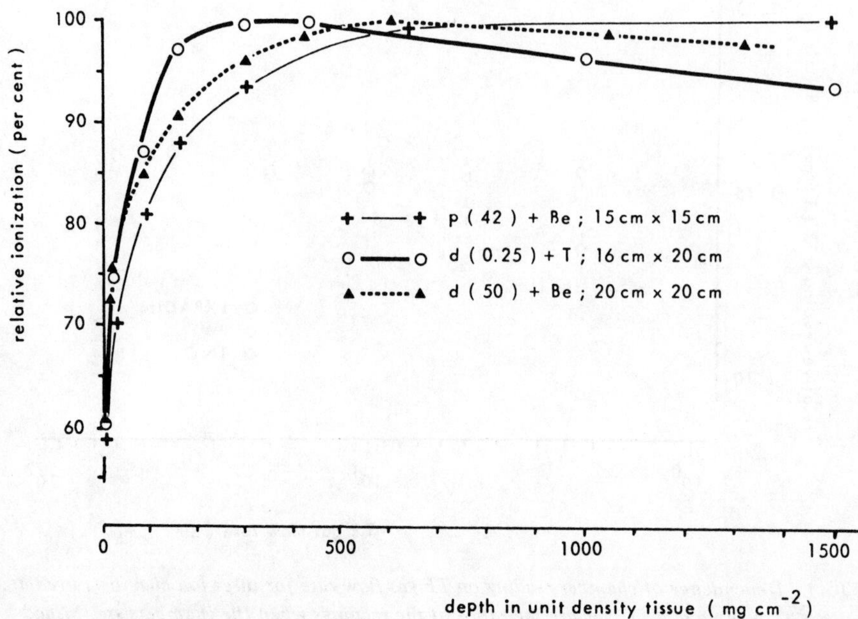

FIG.4. Charged-particle buildup for d(0.25)+T, p(42)+Be and d(50)+Be neutron beams as a function of TE plastic thickness [21].

and gas. The relative contributions from wall material and gas are dependent on the neutron energy spectrum. Starting with zero wall thickness, an increase in the reading with increasing wall thickness is observed owing to the buildup of secondary charged particles. The wall of an ionization chamber must be thick enough to establish secondary charged particle equilibrium. The minimum thickness is determined by the maximum range of the secondaries. However, the wall also causes attenuation of the primary radiation. In addition, the dose will be increased owing to a contribution from scattered primary particles in the wall which otherwise would not have reached the sensitive volume. The loss of response is therefore smaller than derived from the wall attenuation only. Under conditions of charged-particle equilibrium, the absorbed dose will be practically equal to kerma. For the determination of kerma free-in-air, the readings with wall thicknesses in excess of the minimum value required for establishment of charged-particle equilibrium are generally extrapolated to zero wall thickness. It is evident that this method is only an approximation [2].

Figure 4 [21] shows the curves of buildup of secondary charged-particle equilibrium for three neutron beams determined with a disc-type TE ionization chamber. The results for the p(42)+Be beam are given for the situation with a

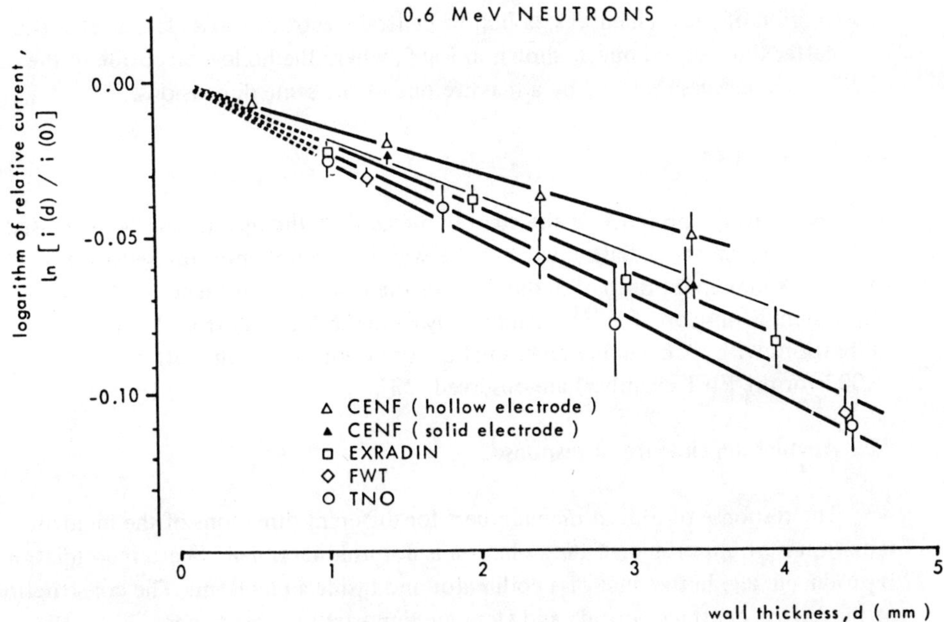

FIG.5. Effects of wall thickness (d) on the ion-chamber reading for four chambers irradiated with 0.6 MeV neutrons. The effects of using a massive or hollow central electrode are shown for the CENF chamber [6].

nylon filter inserted in the collimator duct. For d+T neutrons, the maximum of the buildup curve is reached at a depth of about 2.5 mm in unit density tissue, whereas for the d(50)+Be beam the maximum is reached at a thickness of about 7 mm. The p(42)+Be neutrons have a broad maximum in the buildup curve between 10 and 15 mm, indicating that there are relatively more high energy neutrons in this beam than in the d(50)+Be beam. The depths at which the maxima are reached are in good agreement with the maximum ranges of the recoil protons [2].

The relative change of reading as a function of the thickness of chamber wall and buildup caps for 0.6 MeV neutrons is shown in Fig.5 for the 1 cm^3 TNO, Exradin, FWT and CENF chambers. It can be concluded that the decrease in the reading with wall thickness is more pronounced for the spherical chambers than for the thimble chambers. The introduction of an effective wall thickness as described by Maier [22] will result in a somewhat higher effective thickness for the spherical than for the thimble chambers. Such a type of correction would reduce the differences between the attenuation curves for the chambers of the different shapes. The smaller attenuation for the CENF chamber as compared with the Exradin chamber can be attributed to the difference in central electrode construction of these chambers; a hollow central electrode causes less attenuation and scatter than a solid one as shown in Fig.5, where the hollow electrode in the CENF chamber was replaced by a massive one of the same dimensions.

2.6. Stem scatter

Owing to the presence of the stem of a chamber, the reading can be increased by scattered primary radiation which otherwise would not enter the sensitive volume. A maximum ratio of about 1.02 of the readings in presence and absence of the stem is observed for ^{137}Cs gamma rays and 0.6 MeV neutrons [16]. For calibrations with ^{60}Co gamma rays, smaller corrections for stem scatter (1.002 for the FWT chamber) are observed [23].

2.7. Angular dependence of response

The response of ionization chambers for different directions of the incident radiation is of importance if the radiation is not unidirectional. Scattered radiation is produced, e.g. in the duct of a collimator and inside a phantom. The construction of the cavity, central electrode and stem are important in this respect.

For measurements of neutrons free-in-air, the correction will generally be the same as for the calibration with photons free-in-air. However, for measurements inside a phantom, the angular dependence of the response is more important, since the radiation in-phantom is generally not unidirectional and this dependence has to be incorporated into the correction for the effective point of measurement.

3. EFFECTIVE POINT OF MEASUREMENT

Absorbed dose and kerma are quantities which are defined at points [2], while the ionization chambers used for dose determinations are of finite dimensions. The reading of an ion chamber will be proportional to an average absorbed dose over the sensitive volume and this average value must be related to the absorbed dose at a specific point. The correction for the finite dimensions of an ion chamber can be achieved in two different ways: either a correction factor can be applied when the geometrical centre of the chamber is considered to be the point of measurement, or an effective position of measurement can be adopted.

For photon measurements free-in-air, it is generally accepted that the geometrical centre of the chamber is the effective point of measurement, provided that the distance to the source exceeds five times the diameter of the chamber (for ^{137}Cs and ^{60}Co photons, see Kondo and Randolph [24]; for lower energy X-rays, see Burlin [25]). It is commonly assumed that the principles valid for photons can also be applied to neutrons [2], since the correction factors are essentially geometrical. However, no experimental data which support this assumption for neutrons have been available until recently.

Information on effective points of measurement of cylindrical ionization chambers in-phantom is available for ^{60}Co gamma rays and photons of higher energies (e.g. Ref. [26]). In these studies displacements of between about 0.6 r and 0.8 r (r being the radius of the cavity) are generally observed. It has been suggested [2] that the radial displacement for neutrons is similar to that observed for photons. The experimental corrections for displacement available before 1977 [27, 28] supported this statement. During the European Neutron Dosimetry Intercomparison Project (ENDIP) [4], only 5 of the 12 participants applied displacement corrections for measurements in-phantom. The considerable variation in these corrections could explain part of the differences in the ENDIP results for in-phantom measurements.

3.1. Effective point of measurement free-in-air

For free-in-air conditions, the variation in the radiation field from a point source over the sensitive volume of the ionization chamber is governed by the inverse square law. This implies that larger corrections are to be expected for measurements at relatively short distances from the source.

To investigate the effective point of measurement free-in-air for 15 MeV neutrons, measurements were performed at different distances from the source with tissue-equivalent (TE) spherical ionization chambers with cavity radii of 4 and 16 mm and with a wall thickness of 2.2 mm. The measurements of the chambers (which were flushed with TE gas) were performed with reference to a TE monitor ionization chamber. The readings of the chambers were normalized with reference to the greatest distance from the target. From the results obtained with the smallest chamber, the effective centre of the source was determined. The ratio of the readings of the larger to the smaller chamber, together with the expected ratio according to the volume integral, the surface integral (see Ref. [24]) and a $2/3 \cdot r$ displaced point of measurement are summarized in Table IV. It is concluded from this table that, for free-in-air conditions, a $2/3 \cdot r$ displaced measuring point is not valid for 15 MeV neutrons. A distinction between surface and volume integral cannot be made, since the measurements could not be performed at sufficiently short distances from the source because of the target construction.

The rule of thumb suggested by Kondo and Randolph [24] for photons, that the geometrical centre of a chamber is the point of measurement for distances from the source in excess of five times the chamber diameter, also appears to be valid for neutrons. The correction factors are valid only for a point source. Since neutron sources generally do not fulfil this requirement, an effective centre of the source has to be determined.

TABLE IV. RATIO OF RELATIVE READINGS FROM LARGE AND SMALL SPHERICAL ION CHAMBERS AT DIFFERENT DISTANCES FROM A 15 MeV NEUTRON SOURCE

Distance (cm)	Ratio according to K_v (volume integral)	Ratio according to K_s (surface integral)	Ratio according to $2/3 \cdot r$ displacement	Measured ratio
6.7	1.011	1.018	1.304	1.05 (1 ± 0.03)
11.7	1.004	1.006	1.156	1.00 (1 ± 0.03)
21.7	1.001	1.002	1.079	1.00 (1 ± 0.02)
51.7	1.000	1.000	1.032	1.00 (1 ± 0.02)

3.2. Effective point of measurement in-phantom

Inside a phantom, the effective point of measurement can be displaced relative to the free-in-air condition owing to the replacement of phantom material by the cavity of the ionization chamber. A correction has to be applied for changes in attenuation and scattering of the radiation due to this replacement. This can be achieved in terms of radial displacement, d, which is defined as the displacement of the effective point of measurement from the geometrical centre of the chamber. An alternative method is the use of a displacement correction factor, k_d, when the geometrical centre of the chamber is taken as the point of measurement.

For ^{60}Co gamma rays and 15 MeV neutrons, differences in dose were observed between spherical chambers with different cavity radii (4, 8 and 16 mm) placed with their geometrical centres at the same depth of measurement in the phantom. These differences in dose were converted by using the depth dose curves to differences in the effective point of measurement. When the displacement of the effective point of measurement from the chamber centre was plotted against cavity radius, a linear relationship resulted. The displacements of the effective point of measurement (radial displacements) were therefore extrapolated to zero chamber radius and normalized to this value as shown in Fig.6. For neither ^{60}Co gamma rays nor 15 MeV neutrons was a significant variation of the radial displacement with depth in the phantom observed; therefore the mean results at 5, 10 and 20 cm depth are plotted in Fig.6.

The linear relationship between the radial displacement, d, and the cavity radius, r, can be presented as $d = (0.58 \pm 0.06) \cdot r$ for ^{60}Co gamma rays. The

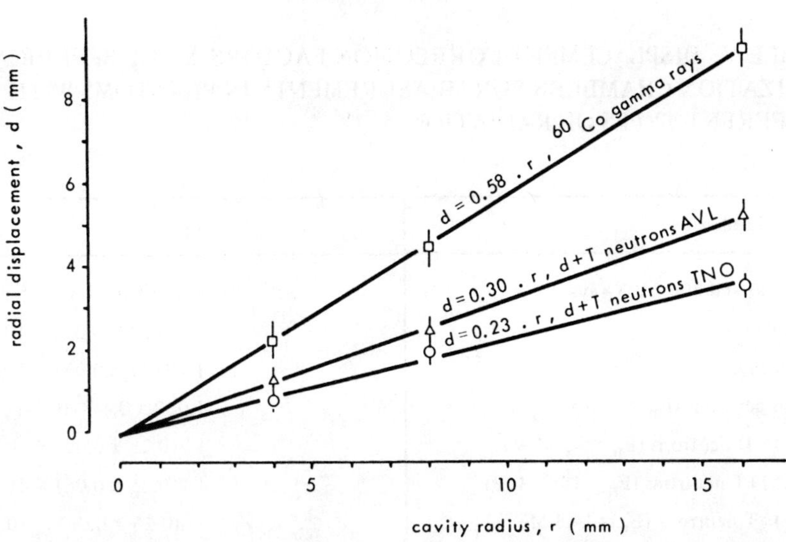

FIG.6. Radial displacement (d) versus cavity radius (r) for spherical ion chambers irradiated with d+T neutrons at Rijswijk: SSD, 42 cm; field area, 6 × 8 cm²; with d+T neutrons at Amsterdam: SSD, 80 cm; field area, 13 × 16 cm²; with ⁶⁰Co gamma rays: SSD, 80 cm; field area, 13 × 16 cm² [29].

derived radial displacements for d+T neutrons at Rijswijk and Amsterdam (see Fig.6) were $(0.23 \pm 0.06) \cdot r$ and $(0.30 \pm 0.06) \cdot r$, respectively, which are significantly smaller than the displacement for ^{60}Co gamma rays and the value of $2/3 \cdot r$ suggested by the ICRU [2]. Displacement correction factors, k_d, can be calculated as the ratio of the actual dose (for an infinitesimal small cavity) to the dose measured. For both d+T neutron beams, the same displacement correction factor of $1-(0.25 \pm 0.06) \times 10^{-2} \cdot r$ is observed where r is expressed in mm. Therefore, further evaluation of the results will be made in terms of displacement correction factors, k_d.

A summary of k_d values for photons and neutrons of different energies is given in Table V. The independence of k_d of field size for neutrons has been confirmed by the measurements with d(50)+Be neutrons with field sizes of 6×8 cm^2 and 13×16 cm^2 at the same SSD. The displacement correction factor for spherical ionization chambers in a water phantom for ^{60}Co gamma rays is smaller than that for ^{137}Cs gamma rays, while for X-rays no displacement is found. For ^{60}Co and ^{137}Cs gamma rays, the displacement correction factors for the thimble-type Baldwin-Farmer chamber of 0.985 ± 0.003 and 0.990 ± 0.003, respectively, are somewhat smaller than that for a spherical chamber of comparable radius, indicating a dependence on the shape of the chambers [29]. The derived displacement correction factors for in-phantom measurements show an increase

TABLE V. DISPLACEMENT CORRECTION FACTORS, k_d, OF SPHERICAL IONIZATION CHAMBERS FOR MEASUREMENTS IN PHANTOMS WITH DIFFERENT TYPES OF RADIATION

Type of radiation	k_d
150, 200 and 300 kV X-rays	$1.000 \pm 0.05 \times 10^{-2} \cdot r$
^{137}Cs γ-rays	$1-(0.22 \pm 0.05) \times 10^{-2} \cdot r$
^{60}Co γ-rays	$1-(0.37 \pm 0.04) \times 10^{-2} \cdot r$
fission neutrons ($\bar{E}_n$ = 1 MeV)	$1.000 \pm 0.1 \times 10^{-2} \cdot r$
d(2.3)+ D neutrons (E_n = 5.3 MeV)	$1-(0.25 \pm 0.09) \times 10^{-2} \cdot r$
d(0.25)+T neutrons (E_n = 14.2 MeV)	$1-(0.25 \pm 0.06) \times 10^{-2} \cdot r$
d(0.5)+T neutrons (E_n = 14.8 MeV)	$1-(0.25 \pm 0.06) \times 10^{-2} \cdot r$
d(50)+Be neutrons ($\bar{E}_n$ = 21 MeV)	$1-(0.21 \pm 0.05) \times 10^{-2} \cdot r$

with decreasing photon energies. Johansson et al. [26] found that k_d increased for X-ray energies in excess of 5 MeV and that the displacement correction factors for ^{60}Co gamma rays and 5 MeV X-rays were similar. This might indicate that displacement correction factors with maximum deviations from unity will occur for photons in the energy region between that of ^{60}Co rays and 5 MeV X-rays.

For neutrons, a dependence of k_d on neutron energy is also observed. For relatively low energy neutrons, no displacement was found, whereas for neutrons with energies in excess of 5.3 MeV, k_d shows an almost constant value, although there seems to be a tendency for an increase in k_d for the highest neutron energy similar to the situation for photons. Shapiro et al. [28] determined displacement correction factors for d(35)+Be neutrons ($\bar{E}_n$ = 14.3 MeV) inside a TE phantom at an SSD of 135 cm. A k_d value of 0.970 was derived for a spherical chamber with a cavity radius of 6.35 mm and filled with air; this is smaller than the value of 0.984 ± 0.004 which can be calculated from Table V. However, the value of 0.970 of Shapiro et al. has an uncertainty of 1% (which can be assessed from their original data); consequently, their result is probably not significantly different from those reported here.

Displacement factors for d(15)+ Be neutrons were derived by Williams et al. [30] from measurements with activation detectors placed at the centre of cylindrical cavities of different diameter. The estimated displacement factor of $1-0.45 \times 10^{-2}$ r is somewhat lower than the values in Table V for 5.3 and 14.2 MeV neutrons. This might be attributed to the use of a differing shape of the cavity and a different experimental approach. The correction factor for the Exradin T2 chamber of 0.991 ± 0.003 is not significantly different from the factor 0.985 ± 0.004 derived by Mijnheer et al. [23] from the data in Table V for comparable neutron energies.

Contrary to all other reports, Hensley and Rassow [31] observed that the displacement correction factor varied considerably with depth in-phantom for positions between 5 and 20 cm.

The introduction of an ionization chamber into a phantom will disturb the dose distribution in and around the cavity. Differences in k_d values for different energies of neutrons and photons might be due to differences in attenuation and scattering processes for the various types of neutron and photon radiation. The introduction of a gas-filled cavity into a phantom disturbs the fluence of primary particles at the position where the secondary charged particles are produced. Central axis depth dose curves were determined with the smallest spherical ionization chamber for the homogeneous water phantom and inside a 32 mm diameter Styrofoam cavity (density, 1.33×10^{-2} g·cm^{-3}) which was immersed in the phantom. The results for the 32 mm diameter cavity are shown schematically in Fig.7 for d+T neutrons at Rijswijk normalized to the value measured at the central position in the homogeneous phantom. The interpretation of the data resulting in the solid line and the discontinuity at the water-Styrofoam interface shown in this figure is comparable to that of analogous measurements performed with high energy photons [32]. At the front interface of the cavity, the reduction in scattered radiation results in a sharp dose decrease, ΔD. The depth dose curve has a reduced slope, μ_{cav}, compared to the slope in water, μ_w. A buildup of the dose occurs at the rear side interface. The results indicate that the correction for displacement of the effective point of measurement is not a geometrical problem depending only on chamber shape as suggested earlier (e.g. Ref. [2]) but results from the complex balance between differences in attenuation and scattering of the various radiation qualities caused by the introduction of a gas-filled cavity into a phantom.

4. DETERMINATION OF RADIATION QUALITY WITH TE IONIZATION CHAMBERS

In neutron dosimetry for biomedical applications, TE ionization chambers with sensitive volumes of about 1 cm^3 are the most commonly applied dosimeters. These dosimeters operated at atmospheric pressure do not provide information on radiation quality. Measurements at different gas pressures and/or collecting voltages can provide information on radiation quality. Zielczynski et al. [33] have shown that the dependence of initial recombination on linear energy transfer can be used to specify radiation quality by introducing the parameter Recombination Index of radiation Quality (RIQ).

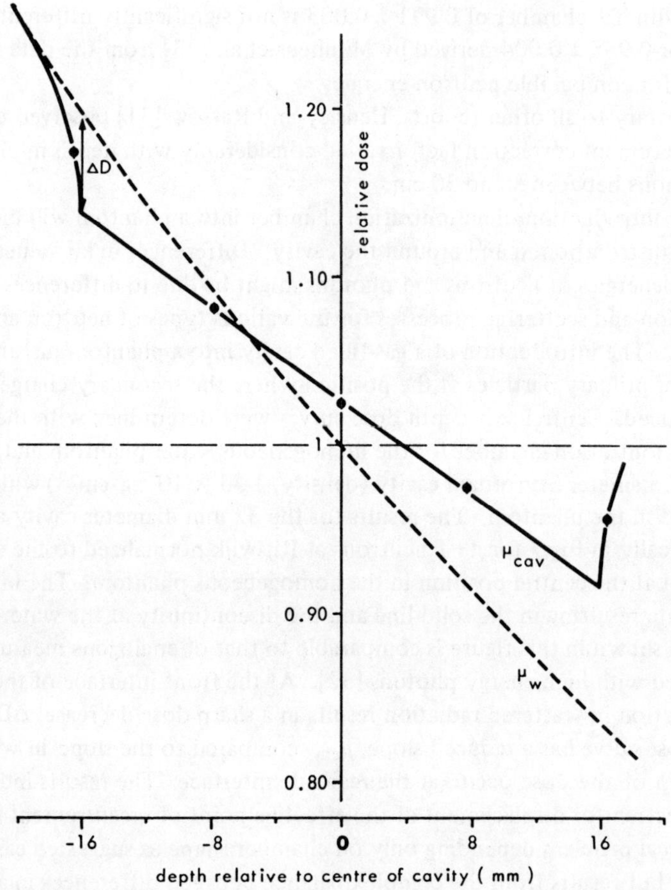

FIG.7. Central axis depth dose distributions in a water phantom for d+T neutrons at Rijswijk measured in the presence and absence of a 32 mm spherical Styrofoam cavity. The depth dose curves for the homogeneous phantom (dashed line) and in the presence of the cavity (solid line) are normalized to the dose at the central position in the homogeneous phantom [29].

A thimble-type TE ionization chamber with a sensitive volume of about 1 cm^3 which can be operated at gas pressures of up to 100 bar (10 MPa) was constructed [34]. The chamber wall is protected against the high gas pressures by an aluminium envelope with a wall thickness of 2 mm. The gas pressure inside the chamber is regulated in a closed system and measured by transducer cells which are based on the application of strain gauges. Measurements have been performed at pressures of up to 80 bar (8 MPa) of methane-based muscle equivalent (TE) gas for ^{137}Cs gamma rays and 0.9 and 14.5 MeV neutrons.

TABLE VI. RATIO OF THE READING AT 500 V TO THE READING AT VOLTAGE V AT A GAS PRESSURE OF ABOUT 6.2 MPa

Collecting potential/V	^{137}Cs γ-rays	14.5 MeV neutrons	0.9 MeV neutrons
50	1.389 ± 0.006	1.876 ± 0.006	2.36 ± 0.01
100	1.278 ± 0.006	1.606 ± 0.006	1.97 ± 0.01
200	1.169 ± 0.006	1.338 ± 0.006	1.53 ± 0.01
300	1.099 ± 0.006	1.185 ± 0.006	1.29 ± 0.01
400	1.045 ± 0.006	1.080 ± 0.006	1.12 ± 0.01
500	1	1	1
600	0.965 ± 0.006	0.938 ± 0.006	0.91 ± 0.01

The effects of dose rate on the saturation characteristics were investigated at several gas pressures for ^{137}Cs gamma rays. For approximately the same pressures, the saturation characteristics are not significantly different. Therefore, only initial recombination is of importance.

The results of measurements at various values of the collecting potential are given in Table VI for TE gas at a pressure of about 62 bar (6.2 MPa) for different types of radiation. The values are given with reference to the reading at 500 V. It can be seen that the influence of initial recombination is dependent on the type of radiation employed and increases with increasing quality.

At a collecting potential of 600 V, the ionization chamber reading relative to the reading at 1 bar as a function of gas pressure is given in Fig.8 for various types of radiation. As a function of gas pressure, an approximately linear increase in the reading is to be expected because of the increase in the mass of the gas in the cavity. However, with increasing pressure, initial recombination also increases and, depending on ion collection field strength and type of radiation, this causes a decrease in the reading as a function of pressure. The results indicate that it is possible to assess radiation quality from the pressure dependence of the reading. It can be seen from Fig.8 that in the investigated pressure region (0.1–8 MPa) the sensitivity of the ionization chamber operated at 600 V can be increased with reference to atmoshperic pressure (1 bar = 0.1 MPa) by factors of about 32, 15 and 6.3 for ^{137}Cs gamma rays, 14.5 and 0.9 MeV neutrons, respectively. Up to about 5 bar (0.5 MPa), the increase in the sensitivity is approximately proportional to the pressure for all types of radiation used. For measurements with TE gas at higher pressures, the saturation characteristics as well as the pressure dependence of the reading are dependent on the quality of the radiation and can be used to assess quality factors.

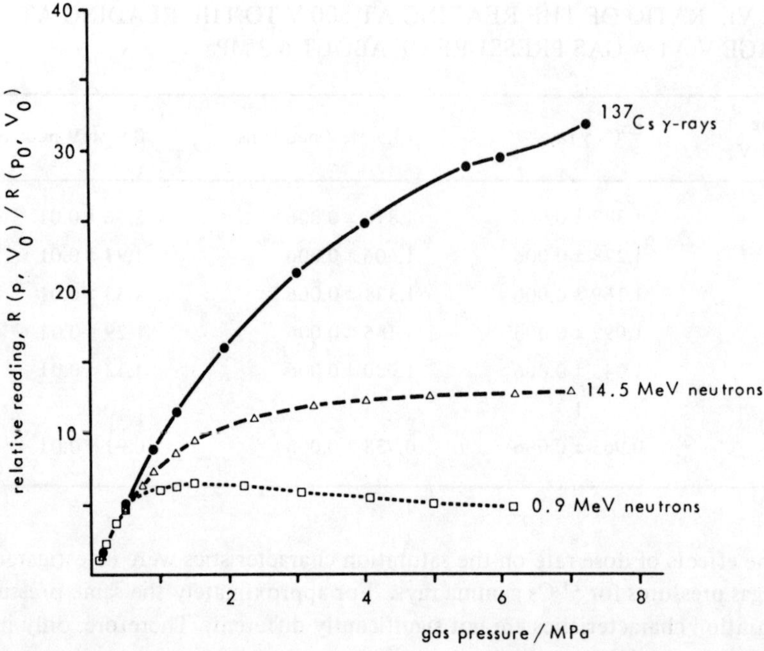

FIG.8. Pressure dependence of the reading relative to that at 1 bar (0.1 MPa) at a collecting potential of 600 V for various types of radiation with a methane-based TE gas ionization chamber [6].

5. CONCLUSIONS

Tissue-equivalent ionization chambers of experimental design have been available for a considerable time [35]. The radiotherapy applications have initiated the commercial production of these devices. An interim inventory of concepts, principles and experimental methods has been compiled in ICRU Report 26 [2]. The results of international [3] and European [4] neutron dosimetry intercomparisons demonstrated relatively large variations in absorbed dose values (up to ±20% from the mean). At present general agreement has been reached on the basic physical parameters to be applied [8, 9]. Investigations on the characteristics of TE ionization chambers (see, e.g. Schraube et al. [16]) and the introduction of common types of chamber resulted in an increased agreement in dose values determined by different groups. Conscientious experimentation and a common data base can provide dosimetry results with variations of less than ±2% (see, e.g. Ref. [36]).

A check of the absolute accuracy of dosimetry with ionization chambers can be inferred from comparisons with other methods, including calorimeters and differential fluence measuring devices. For neutron beams with energies up to 25 MeV, intercomparisons of calorimetric and ionimetric methods showed agreement between the two techniques to within ±2 per cent for total absorbed dose in A-150 plastic [37, 38]. The major uncertainty in deriving the total absorbed dose in soft tissue is introduced by the kerma ratios. Collection of cross-section data, especially for oxygen and carbon in the energy range between 15 and 18 MeV, will be essential to improve the absolute accuracy in neutron dosimetry. For energies up to 15 MeV, the total absorbed dose can be determined with tissue-equivalent ionization chambers with an acceptable degree of accuracy and precision. However, regular calibration of the devices is recommended to maintain and improve the reliability.

REFERENCES

[1] BROERSE, J.J., MIJNHEER, B.J., "Uncertainties in basic physical data for neutron dosimetry in biology and medicine", in Proc. VIIIème Congrès International de la Societé Française de Radioprotection, Saclay, France (1978) 641–59.

[2] INTERNATIONAL COMMISSION ON RADIATION UNITS AND MEASUREMENTS (ICRU), Neutron Dosimetry for Biology and Medicine, ICRU Rep. 26, Washington, DC (1977).

[3] INTERNATIONAL COMMISSION ON RADIATION UNITS AND MEASUREMENTS (ICRU), An International Neutron Dosimetry Intercomparison, ICRU Rep. 27, Washington, DC (1978).

[4] BROERSE, J.J., BURGER, G., COPPOLA, M. (Eds), A European Neutron Dosimetry Intercomparison Project (ENDIP): Results and Evaluation, EUR-6004, Commission of the European Communities, Luxembourg (1978).

[5] BROERSE, J.J. (Ed.), Ion Chambers for Neutron Dosimetry, EUR-6782, Commission of the European Communities, Luxembourg (1980).

[6] ZOETELIEF, J., Dosimetry and Biological Effects of Fast Neutrons. Thesis, Amsterdam (1981).

[7] BROERSE, J.J., MIJNHEER, B.J., "Progress in neutron dosimetry for biomedical applications", Progress in Medical Radiation Physics (ORTON, C.G., Ed.), Plenum Press, New York and London (1982) 1–101.

[8] AMERICAN ASSOCIATION OF PHYSICISTS IN MEDICINE (AAPM), Protocol for Neutron Beam Dosimetry, Task Group No. 18, Fast Neutron Beam Dosimetry Physics Radiation Therapy Committee, AAPM, Rep. 7 (1980).

[9] EUROPEAN CLINICAL NEUTRON DOSIMETRY GROUP (ECNEU), European Protocol for Neutron Dosimetry for External Beam Therapy, European Clinical Neutron Dosimetry Group, Eur. Org. Res. on Treatment of Cancer (EORTC), Br. J. Radiol. **54** (1981) 882–98.

[10] MEEKER, R.D., "Characteristics of thimble type TE ion chamber manufactured by Exradin", Ion Chambers for Neutron Dosimetry (BROERSE, J.J., Ed.), EUR-6782, Commission of the European Communities, Luxembourg (1980) 43–46.

[11] KANTZ, A.D., "Characteristics of spherical ion chamber manufactured by Far West Technology", Ion Chambers for Neutron Dosimetry (BROERSE, J.J., Ed.), EUR-6782, Commission of the European Communities, Luxembourg (1980) 61–68.

[12] RICOURT, A., CHEMTOB, M., PARMENTIER, N., "Characteristics of ion chamber constructed at Fontenay-aux-Roses for the determination of kerma", Ion Chambers for Neutron Dosimetry (BROERSE, J.J., Ed.), EUR-6782, Commission of the European Communities, Luxembourg (1980) 35–42.

[13] BOAG, J.W., "Ionization chambers", Radiation Dosimetry (ATTIX, F.H., ROESCH, W.C., Eds), Vol.2, Academic Press, New York and London (1966) 1–72.

[14] BOAG, J.W., "Cavity theory and its practical implications for neutron dosimetry with ionization chambers", Ion Chambers for Neutron Dosimetry (BROERSE, J.J., Ed.), EUR-6782, Commission of the European Communities, Luxembourg (1980) 151–66.

[15] COYNE, J.J., "Kerma values by particle type", Ion Chambers for Neutron Dosimetry (BROERSE, J.J., Ed.), EUR-6782, Commission of the European Communities, Luxembourg (1980) 195–207.

[16] SCHRAUBE, H., ZOETELIEF, J., BROERSE, J.J., BURGER, G., "Performance tests of four different tissue-equivalent ionization chambers", Ion Chambers for Neutron Dosimetry (BROERSE, J.J., Ed.), EUR-6782, Commission of the European Communities, Luxembourg (1980) 131–49.

[17] PARNELL, C.J., "Ionization chamber dosimetry of fast neutron beams", Proc. 3rd Symp. Neutron Dosimetry in Biology and Medicine (BURGER, G., EBERT, H.G., Eds), EUR-5848, Commission of the European Communities, Luxembourg (1978) 579–88.

[18] BROERSE, J.J., ZOETELIEF, J., BURGER, G., SCHRAUBE, H., RICOURT, A., A Small-Scale Neutron Dosimetry Intercomparison, EUR-6567, CENDOS, Rijswijk (1979).

[19] MIJNHEER, B.J., Influence of the tube length at the gas exhaust port on the reading of a TE-ion chamber flushed with TE-gas", Ion Chambers for Neutron Dosimetry (BROERSE, J.J., Ed.), EUR-6782, Commission of the European Communities, Luxembourg (1980) 125–26.

[20] WILLIAMS, J.R., Problems in neutron dosimetry associated with the incorrect composition of tissue equivalent gas, Phys. Med. Biol. **25** (1980) 501–08.

[21] MIJNHEER, B.J., ZOETELIEF, J., BROERSE, J.J., Build-up and depth-dose characteristics of different fast neutron beams relevant for radiotherapy, Br. J. Radiol. **51** (1978) 122–26.

[22] MAIER, E., Experimentelle Untersuchungen über die Energie-deposition schneller Neutronen in Phantomens, Thesis, GSF (Ges. Strahlen-Umweltforsch.) Ber. S631 (1979).

[23] MIJNHEER, B.J., VAN WIJK, P.C., ZOETELIEF, J., BROERSE, J.J., "Determination of absorbed dose with two types of clinically employed tissue equivalent ionization chambers in fast neutron beams", Proc. 4th Symp. Neutron Dosimetry (BURGER, G., EBERT, H.G., Eds), EUR-7448, Commission of the European Communities, Luxembourg (1981) 361–72.

[24] KONDO, S., RANDOLPH, M.L., Effect of finite size of ionization chambers on measurements of small photon sources, Radiat. Res. **13** (1960) 37–60.

[25] BURLIN, T.E., The effect of inverse square law attenuation on the measurement of grenz-ray exposure with a cavity ionization chamber, Br. J. Radiol. **37** (1964) 693–95.

[26] JOHANSSON, K-A., MATTSON, L.O., LINDBORG, L., SVENSSON, H., "Absorbed-dose determination with ionization chambers in electron and photon beams having energies between 1 and 50 MeV", National and International Standardization of Radiation Dosimetry (Proc. Symp. Atlanta, 1977) Vol.2, IAEA, Vienna (1978) 243–70.

[27] MIJNHEER, B.J., BROERS-CHALLISS, J.E., BROERSE, J.J., "Measurements of radiation components in a phantom for a collimated d-T neutron beam", Proc. 2nd Symp. Neutron Dosimetry in Biology and Medicine (BURGER, G., EBERT, H.G., Eds), EUR-5273, Vol. 1, Commission of the European Communities, Luxembourg (1975) 423–43.

[28] SHAPIRO, P., ATTIX, F.H., AUGUST, L.S., THEUS, R.B., ROGERS, C.C., Displacement correction factor for fast-neutron dosimetry in a tissue-equivalent phantom, Med. Phys. **3** (1976) 87–90.

[29] ZOETELIEF, J., ENGELS, A.C., BROERSE, J.J., MIJNHEER, B.J., Effect of finite size of ion chambers used for neutron dosimetry, Phys. Med. Biol. **25** (1980) 1121–31.

[30] WILLIAMS, J.R., RYALL, R.E., BONNETT, D.E., Measurements of displacement factors in a neutron beam using activation dosimeters, Phys. Med. Biol. **27** (1982) 81–89.

[31] HENSLEY, F., RASSOW, J., "Performance of several types of ion chambers for usual measuring conditions in neutron dosimetry", Proc. 4th Symp. Neutron Dosimetry (BURGER, G., EBERT, H.G., Eds), EUR-7448, Commission of the European Communities, Luxembourg (1981) 543–57.

[32] SAMUELSSON, C., Influence of air cavities on central depth dose curves for 33 MV Roentgen rays, Acta Radiol. Ther. Phys., Biol. **16** (1977) 465–88.

[33] ZIELCZYNSKI, M., GOLNIK, N., MAKAREWICZ, M., SULLIVAN, A.H., "Definition of radiation quality by initial recombination of ions", Proc. 7th Symp. Microdosimetry (BOOZ, J., EBERT, H.G., HARTFIEL, H.D., Eds) Vol.2, EUR-7149, Commission of the European Communities, Luxembourg (1981) 853–62.

[34] ZOETELIEF, J., ENGELS, A.C., BOUTS, C.J., HENNEN, L.A., BROERSE, J.J., "Response of tissue equivalent ionization chambers as a function of gas pressure", Proc. 4th Symp. Neutron Dosimetry (BURGER, G., EBERT, H.G., Eds), EUR-7448, Commission of the European Communities, Luxembourg (1981) 315–26.

[35] ROSSI, H.H., FAILLA, G., Tissue equivalent ionization chambers, Nucleonics **14** (2) (1956) 32–36.

[36] OCTAVE-PRIGNOT, M., PIHET, P., VYNCKIER, S., WAMBERSIE, A., MEULDERS, J.P., ZOETELIEF, J., BROERSE, J.J., MIJNHEER, B.J., VAN WIJK, P.C., Neutron dosimetry intercomparison between Louvain-la-Neuve, Rijswijk and Amsterdam, Strahlenther. **158** (1982) 227–29.

[37] McDONALD, J.C., I-CHANG MA, LIANG, J., EENMAA, J., AWSCHALOM, M., SMATHERS, J.B., GRAVES, R., AUGUST, L.S., SHAPIRO, P., Calorimetric and ionimetric dosimetry intercomparisons I: US neutron radiotherapy centres, Med. Phys. **8** (1981) 39–43.

[38] McDONALD, J.C., I-CHANG MA, MIJNHEER, B.J., ZOETELIEF, J., Calorimetric and ionimetric dosimetry intercomparisons II, d+T neutron source at the Antoni van Leeuwenhoek Hospital, Med. Phys. **8** (1981) 44–48.

IAEA-AG-371/2

NEUTRON DEPTH DOSE CALCULATIONS FOR TREATMENT PLANNING IN FAST NEUTRON RADIOTHERAPY

G. BURGER, A. MORHART
Gesellschaft für Strahlen- und Umweltforschung mbH München,
Institut für Strahlenschutz,
Neuherberg, Munich,
Federal Republic of Germany

P.S. NAGARAJAN
Bhabha Atomic Research Centre,
Division of Radiological Protection,
Trombay, Bombay,
India

Abstract

NEUTRON DEPTH DOSE CALCULATIONS FOR TREATMENT PLANNING IN FAST NEUTRON RADIOTHERAPY.

The central physical goal in treatment planning is to generate optimized isodose distributions. The basic data necessary may be obtained experimentally or by means of coupled neutron-gamma transport calculations. The latter allow investigations far beyond experimental possibilities. The more frequently applied programmes to carry out such calculations and some results and their validity are discussed. The main emphasis is given to the two-dimensional S_n-transport code DOT and a Monte Carlo code SAM-CE, suitable for complicated three-dimensional arrangements. Both codes are implemented at the Gesellschaft für Strahlen- und Umweltforschung (GSF) and used for treating a variety of deep penetration problems in radiation protection and radiation therapy in the energy range of neutrons below 20 MeV. Besides depth doses in homogeneous phantoms, also some results in non-homogeneous phantoms, and the need for corrections in conventional isodose planning are shown. Finally, also discussed are the problem of secondary particle production for investigating interface dose distributions, as well as the means for assessing radiation efficiency, and the possible role of calculable microdosimetric parameters in this context.

1. INTRODUCTION

One of the goals, and at the same time the main physical problem in therapy planning, is the correct administration of dose deposition in the patient. The aim is to deliver the absorbed dose as homogeneously as possible to the target volume, while keeping it as low as reasonably achievable, but in any case below certain thresholds in surrounding critical organs.

The assessment of the depth dose distribution for selected irradiation conditions is usually based upon a limited set of measurements in homogeneous water phantoms irradiated under standard conditions. Generally, such conditions comprise the normal incidence of unfiltered beams for commonly applied field sizes and source to skin distances upon cubic, water-filled phantoms. Also, depending on the emphasis which is given to the physical aspects of treatment planning at a certain installation, non-oblique beam incidence, the effect of filters, or the influence of special irregular phantom arrangements, may be studied experimentally. The measured data are then often interpolated and smoothed or analytically fitted by appropriate algorithms and all distributions suitably documented within a data bank. The generation of new depth dose distributions for arbitrary selected irradiation situations, based upon the available data bank, is the subject of isodose planning. The most straightforward method is the point-wise interpolation and direct combination of measured, spatially distributed dose values. It is called the matrix method. Other methods make use of generating functions which enable the dose to be calculated at any desired point in the phantom. Such functions are normally analytical fit expressions for the central axis depth dose and a few lateral dose profiles.

The main problem is that a true patient irradiation is never comparable with the standard phantom situation. There is often non-oblique incidence with even tangential irradiation of body surfaces and there are tissue inhomogeneities such as cavities, lung, fat, bone and their associated interfaces. All these conditions cause variations of the standard isodose contours, which have to be taken into account to get a complete and sufficiently accurate picture of the exposure situation. It must be admitted that in practice dose administration is often guided by treatment protocols which are based on standardized irradiation schemes only. So far as extensive clinical experience is correlated with such standard conditions, the detailed knowledge of the internal dose distribution may in fact be irrelevant. However, on the other hand it must be stated that even a different patient anatomy may destroy the 'standard condition' and cannot be easily corrected for by contour normalization. Different fat layers, for example, cause considerably different fast neutron dose deposition. The need for correct isodose planning is finally obvious, when individual treatment planning is indicated, that is to say localization of a tumour does not represent a 'standard situation'. There is no doubt that the efficiency of existing radiotherapy can only be increased if radio-oncological experience is more tightly correlated with detailed knowledge about dose distributions inside and around a tumour, rather than with the application of standard irradiation techniques based upon uncorrected isodose plans.

In general, whenever the goals and methods of isodose planning are identical for neutrons and gammas, the problems to be overcome are quite different in both cases. So, even the experimental mixed-field dosimetry is much more complicated than the dosimetry of pure photon beams. It requires the use of more than one detector, and is hence more time-consuming. The separate determination of the neutron and the gamma dose from the multiple detector readings is also less accurate than a single photon dose measurement [1–6]. One of the reasons is the generally unknown local variation of detector response due to the degradation of the neutron spectrum inside the phantom. This degradation may in addition cause a local variation in the biological efficiency, which is difficult to determine experimentally. Finally, the transport as well as the energy deposition of fast neutrons depend on the tissue material and cannot be easily investigated experimentally because of the lack of small integrating dosimeters, which could be used inside solid phantom arrangements.

To overcome all these experimental problems it has proved advantageous to apply radiation transport codes, which allow one to calculate primarily the spectral distributions of neutrons and gammas at any desired point inside phantoms and convert this into kerma or any other dosimetric quantity of interest. It is obvious that the calculations have either to be cross-checked by applying different numerical methods, or to be verified experimentally for easily manageable exposure conditions. They can never totally replace dose measurements, but extend them for investigations and final applications, which are far beyond experimental possibilities.

2. TRANSPORT CODES

Details of transport programs cannot be described in the context of this paper. It seems, however, useful to introduce briefly the physical and mathematical concept of the codes implemented by the authors, since most of the results shown refer to them. This allows one to estimate their operational capabilities.

2.1. The method of discrete ordinates

The discrete ordinates, or S_n method, has been reviewed by several authors [7–12]. It is a numerical technique to solve the steady-state Boltzmann transport equation

$$\nabla\vec{\Omega}\phi(\vec{r},E,\vec{\Omega}) = S(\vec{r},E,\vec{\Omega}) - \sum_t (\vec{r},E)\,\phi(\vec{r},E,\vec{\Omega})$$

$$+\iint\sum_s (\vec{r},E'-E,\vec{\Omega}'-\vec{\Omega})\,\phi(\vec{r},E',\vec{\Omega}')\,dE'\,d\vec{\Omega}' \qquad (1)$$

which can be interpreted in terms of physics as the balance equation for the angular flux density $\phi(\vec{r}, E, \vec{\Omega})$ in differential phase-space elements $dP = d\vec{r}dEd\vec{\Omega}$, with $\vec{r}$ being the position vector, E the particle energy and $\vec{\Omega}$ the direction vector. Σ_t and $\Sigma_s = d\Sigma/d\Omega dE$ are the total reaction cross-section and the scattering cross-section, respectively. The term on the left-hand side of the equation is the net particle flow in the phase-space element. The first term on the right represents the uncollided particle gain from external sources, the second one counts for particle loss due to collision (outscattering and absorption), the third term finally describes the particle gain from in-scattering.

The net particle flow in a phase-space element is essentially controlled by in- and out-scattering, according to the double differential cross-sections $d\Sigma/d\Omega dE$. Hence, these cross-sections contain the complete information for simulating the radiation transport. It is now the principle of the S_n method to discretize the cross-sections and approximate the integral term in the Boltzmann transport equation by a Gauss quadrature formula. For this purpose one approximates the differential cross-section by a truncated expansion into Legendre polynomials P_m in the cosine of the scattering angle and evaluates them for a limited set of n directions. Replacing the integrals over all incident energies and all incident angles by sums of integrals over the phase-space cells, $P = V_i E_k \vec{\Omega}_1$, one finally obtains a set of coupled finite difference equations. The V_i are volume elements as defined by the spatial mesh chosen, to derive the desired local resolution for the resulting spectral fluences, the E_k are discrete energies and the $\vec{\Omega}_1$ discrete directions.

The solid angle associated with each direction is expressed by a corresponding normalized weight W_n. The selection of appropriate sets of angles and weights for performing an accurate numerical integration to derive average cross-sections depends on the method of quadratur, and is a delicate task in applying the S_n method. For the computer solution, sets of discrete angles are available in the literature, developed by several authors [13, 14]. The energy discretization is practically done by using group averaged cross-sections.

During the numerical solution of the set of difference equations, problems may arise from the relationship between the values of the directional fluences at the edges and midpoints of the phase-space cells to provide correct extrapolation from one mesh point (or angular interval) to the other. Several models may be applied and influence the results. The reason for this is that the program starts at given conditions at the outer boundaries and goes through all spatial and angular intervals. The fluences in each energy group are calculated, starting with an initial assumption from the highest energy group downward. The whole process is called inner iteration. It is finished when all boundary conditions are satisfied and the flux density has converged. According to the order of the Legendre polynomials chosen (m), and the number of discrete angular segments (n), the approximation is described by P_m-S_n. In many shielding and phantom calculations it has been proved that any approximation higher than P_3-S_8 does not increase the accuracy of the results.

At the Radiation Shielding Information Center at Oak Ridge two implementations of the S_n method are available, namely the one-dimensional code ANISN [15] and the two-dimensional code DOT [16–19]. With the first it is possible to calculate the radiation transport in spheres with a source located at the centre as well as the transport in an infinite slab or even in an infinite upright cylinder. With the latter it is possible to calculate the transport in cylinders with the radiation incidence parallel to the long axis. The codes are regularly revised. The most recent version of DOT, specially designed to allow large problems to be solved on a wide range of computers, even with limited memory capacity, is called DOT-IV, available for users in the United States of America only.

A known handicap of the DOT code is the so-called ray effect, due to the discretization of the directions. It happens if the source or the detectors of interest are small compared with the total arrangement, or the spatial meshes are too narrow compared with the average free path lengths of the particles. It can be overcome by introducing an analytically determined first-collision source based upon the uncollided (UNCL) fluence. It is important to note that the existing revised versions of DOT have different capabilities for making use of such UNCL routines. Whereas the older versions, DOT-II and DOT-III, could be coupled with the powerful external GRTUNCLE code, this is for example not the case for the DOT 3.5 version, which is therefore less valuable for phantom calculations despite its improved programming and the existence of an 'internal' UNCLE routine.

2.2. Monte Carlo methods

More complicated three-dimensional geometrical configurations, such as anthropoid phantoms, can only be treated by means of Monte Carlo codes. The most widely used code for this purpose seems to be MORSE [20], which was originally planned for rather low energy neutron-photon transport calculations up to 20 MeV neutron energy at maximum. For higher energies up to 50 MeV, occurring at some of the bigger cyclotrons used for neutron radiotherapy in the USA, the code HETC [21] is often applied. It allows one to determine not only the spectral neutron fluences at the points of interest, but also the spectral distributions of charged secondary particles. From this the doses due to separate secondaries can be determined.

We implemented SAM-CE, a comprehensive all-Fortran code for time-dependent neutron, photon and secondary electron problems. The code was developed by the Mathematical Applications Group, Inc. (MAGI) in Elmsford, USA. It is described in detail by Lichtenstein et al. [22].

In the most recently updated version of the program package the main parts are: SAM-X, for the cross-section preparation, and SAM-R, for the normal forward and adjoint transport calculations. In the present work only the forward option was used, with ENDF/B-IV as the primary cross-section data input. All data were processed into point energy sets, with each nuclide tabulated in its own unique and appropriate distribution of energies. The code provides a variety of sampling options regarding, for example, space, direction, energy and time. Quantities to be output may be the spectral neutron fluence, energy fluence, absorption or fission rates. Detectors inserted can be point or volume detectors. There is a post-editor program with which the stored results of the MC runs can be easily re-used by applying, for example, various response functions to the spectral fluences.

We have used only one capability of the code up to now, namely chord length scoring of the neutron fluence in phantoms of non-multiplying material. This will be described in a little more detail to give some insight into the possible geometrical realization of therapeutically relevant phantom arrangements.

SAM-CE employs a scheme of combinatorial geometry to define scoring regions which differs from most other Monte Carlo practices. The user can visualize each geometry as consisting of a set of combinations of regular geometric bodies, the definition and arrangement of which are part of the input. Thus, when a particle traverses an extended combined body, the code computes first the points of intersection of sub-bodies and from this the corresponding track or chord length s in the sub-body until the particle's history is terminated. For each sub-body or scoring region the contribution of the particle to the spectral fluence is then determined by the equation

$$\Phi = \langle s \rangle / V \tag{2}$$

with $\langle s \rangle = \int \exp(-\mu s) ds = [1 - \exp(-\mu s)]/\mu$

$\langle s \rangle$ is the expected total chord length in the scoring region, V is the corresponding volume, and μ is the total cross-section of the material in the region. Individual particle fluence contributions are accumulated to derive the total average fluence in the scoring region. This capability is of great importance in radiation protection, where the average doses in individual organs inside three-dimensional man-like modelled phantoms are of special interest. However, to determine isodose distributions the code also provides the possibility of computing fluxes at specified points within the geometry. It is this option which in principal could make SAM-CE an extremely powerful tool in refined isodose planning. It should allow one to calculate the combined effect of tissue inhomogeneities for typical exposure situations on the basis of an existing detailed three-dimensional model of a human body. However, there are up to now some severe limitations regarding the number of points and, finally, the whole point-dose option is not yet completely tested and running well.

3. CROSS-SECTIONS AND KERMA FACTORS

The basic information on cross-sections consists of original experimental and theoretical data. These data need to be intercompared, properly selected, in some cases averaged and interpolated. Parametrization and reduction of the data to a suitable tabular form finally produce an evaluated reference data set. At energies below E_n = 20 MeV extensive evaluated nuclear data files (ENDF/B) are available [23, 24]. From time to time the files are revised. At the moment, we use version ENDF/B IV. ENDF/B V is already released, however, but only for use in the USA. Different transport codes require different data libraries. Therefore, a series of processing codes are available, generating suitable secondary libraries from ENDF/B. Examples are SAMX, which prepares cross-sections for the SAMR Monte Carlo code as described in Section 2.2., or SUPERTOG or AMPX code systems, which prepare neutron, photon production, and photon interaction cross-sections for a variety of transport codes in the group-averaged, the so-called ANISN format, with generally less than 100 groups. For all important and more frequently occurring elements the ENDF/B libraries provide in any case the gamma-ray production information needed to carry out the coupled neutron/gamma-ray transport problem.

Fine-group cross-sections are available in special data library collections from Oak Ridge Nat. Lab. (RSIC 1980), as in the DLC-2 file (100 neutron groups, P_8-approximation), or the DLC-37 file (coupled 100 neutron, 21 gamma groups, P_8-approximation). These data libraries may again be used as input into a spectrum-generating code as, for example, ANISN or APRFIX, to become compressed further into group-averaged, flux-weighted data sets.

The libraries do not contain sufficient information on secondary particle production to obtain the dose directly. The latter has to be approximated by the kerma, derived from fluence-to-kerma factors, as given for example by ICRU 26 [25]. There is also a processing code available, MACK [26], which produces kerma factors as a function of energy, for all materials in the ENDF libraries.

4. RESULTS OF CALCULATIONS

4.1. Neutron and coupled gamma transport in homogeneous media

One of the first phantom studies, by means of Monte Carlo codes, concerned the free in-air irradiation of infinite slabs or upright cylinders for the analysis of

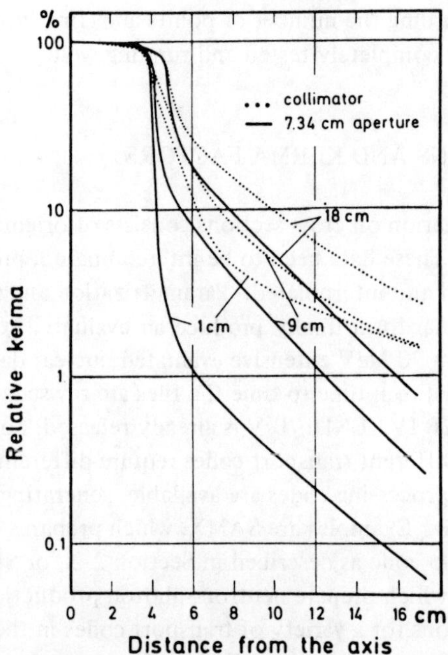

FIG.1. Calculated lateral kerma profiles in a water phantom at several depths for an ideal and a real collimated neutron beam of 15 MeV.

central axis depth doses relevant in radiation protection [27]. As could be shown by Grünauer [28], DOT results for the central axis depth dose in the case of the narrow beam exposure of a horizontal cylinder for 14 MeV neutrons agree well with the corresponding Snyder data. In both cases the beams were ideal, geometrically defined ones. However, when concentrating upon neutron radiation therapy one has to deal with real collimated beams. The question then arises as to what extent such beams are altered by scattered neutrons from the collimator duct.

Grünauer [29] has investigated this problem in great detail and found that real collimation practically does not influence the axial depth doses, compared with an ideal narrow beam; however, it drastically influences the beam penumbra (Fig.1). Similar results were reported by Helm and Becker [30], and Schlegel and Dietze [31]. Therefore, calculations of complete isodose distributions have to be performed for the whole arrangement of source, collimator head and phantom, which makes them time-consuming. Such calculations were performed for several collimator systems and sources. It could be shown that the total doses calculated axially and laterally in a 30 cm^3 water phantom for a specially designed test collimator at 15 MeV show good agreement with TE chamber

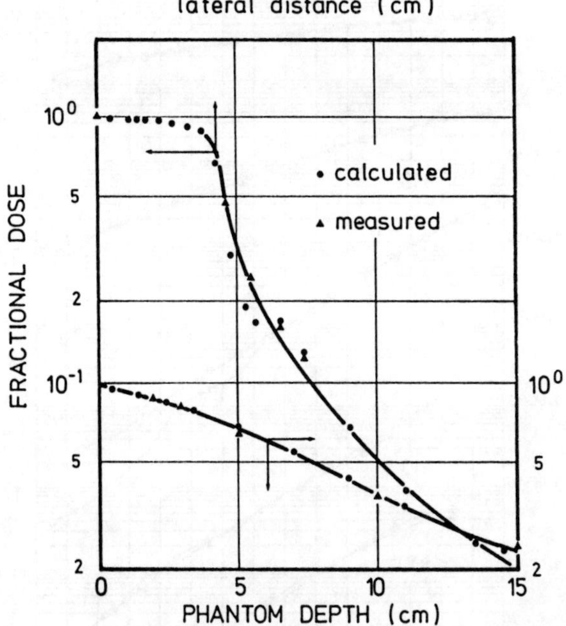

FIG. 2. Comparison of calculated and measured axial depth doses and of a lateral profile dose in a water phantom for the Neuherberg test collimator for a 15 MeV neutron beam.

measurements. This holds also for calculations of a collimator phantom arrangement at Rijswijk, where irradiations were performed for therapeutical pilot studies with 15 MeV neutrons and later on with 15 MeV and 5.3 MeV neutrons for a European Dosimetry Intercomparison project [32, 33]. There is excellent agreement, especially between the measured and calculated neutron dose components (Fig.2). The gamma contribution in the measured fields along the axis in a water phantom is normally much smaller than the neutron dose contribution and more difficult to determine. It depends further, to an often unknown extent, on external gammas stemming from the collimator-source arrangement. The agreement in the gamma doses is therefore generally less good, which is, however, of no significance for verifying the transport codes.

On several occasions the Stuttgart group reported on the intercomparison of calculations with central axis dose measurements for 15 MeV beams, as was also done at the Louvain cyclotron [34, 35]. Alsmiller [36] gave a broad review on depth dose calculations with the high energy MC code HETC for d(35)-Be neutron beams in water phantoms and intercomparisons with measurements. Brenner et al. [37] reported on calculations with an MC code and measurements at p(65)-Be neutron beams at the Fermilab/UC Davis. Morstin and Kawecka [38]

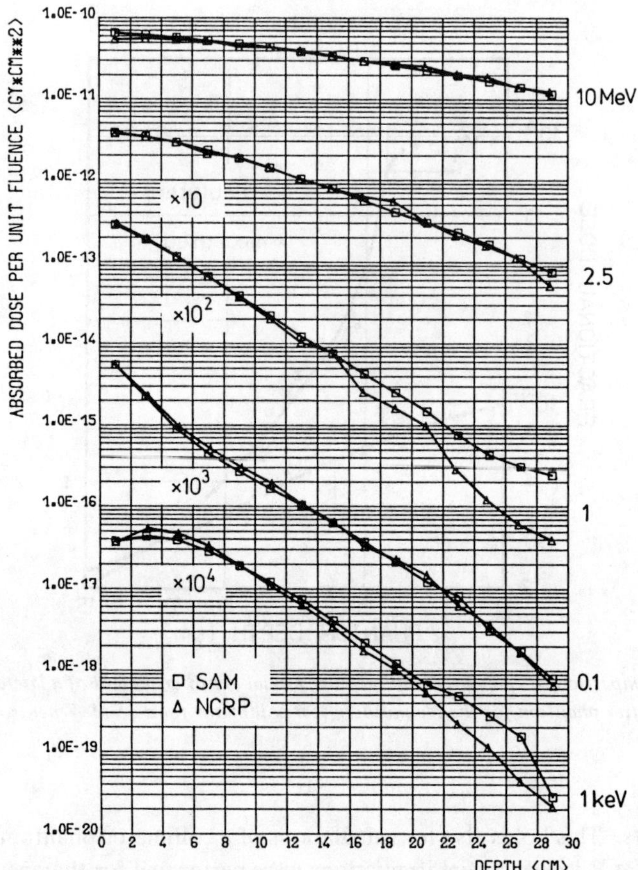

FIG.3(a). Calculated axial depth doses from heavy recoils in cylinder phantoms irradiated by broad beams of monoenergetic neutrons (Intercomparison of Rep. ICRP-38 with SAM-CE results).

applied the ANISN code to an infinite slab phantom and intercompared the axial depth dose results with experimental data for d(12.5)-Be neutrons. In all these cases the experimental and numerical data were normalized to each other, since an absolute measure of the neutron fluence is normally not possible. However, the shapes of the depth dose curves agreed excellently.

To verify one of the most important data sets in radiation protection we repeated calculations for the exposure of circular cylinders by monoenergetic parallel broad neutron beams, as described in NCRP 38 [39]. The calculations were performed by means of SAM-CE for neutron energies from 14.1 MeV down to 1 keV. As a result it came out that the central axis heavy particle recoil doses were practically identical in both cases for all energies, aside from one

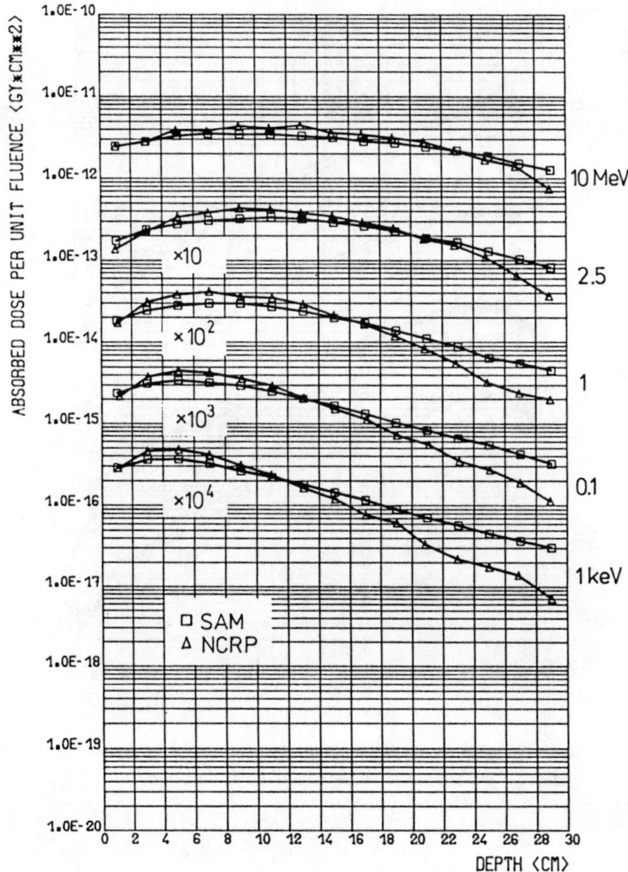

FIG.3(b). Calculated axial depth doses from internally produced gamma rays in cylinder phantoms irradiated by broad beams of monoenergetic neutrons (Intercomparison of Rep. ICRP-38 with SAM-CE results).

small irregularity at 1 MeV (Fig.3a). Also the H(n, γ) dose distributions showed reasonable agreement down to 2.5 MeV incident neutron energy (Fig.3b). At lower energies the SAM curves became systematically flatter than the NCRP curves, differing finally at 1 keV by factors of 3–4 at the greatest phantom depths of 29 cm. Also, at that lowest calculated energy, the heavy particle recoil dose, determined at the greater depths essentially by the N(n, p) reaction, becomes now slightly flatter.

Both effects signalize a difference either in the thermal neutron distribution or in the kerma calculations based upon these distributions, depending critically on the group structure and hence the average kerma factors. To investigate this phenomenon further we performed broad beam calculations for ICRU's

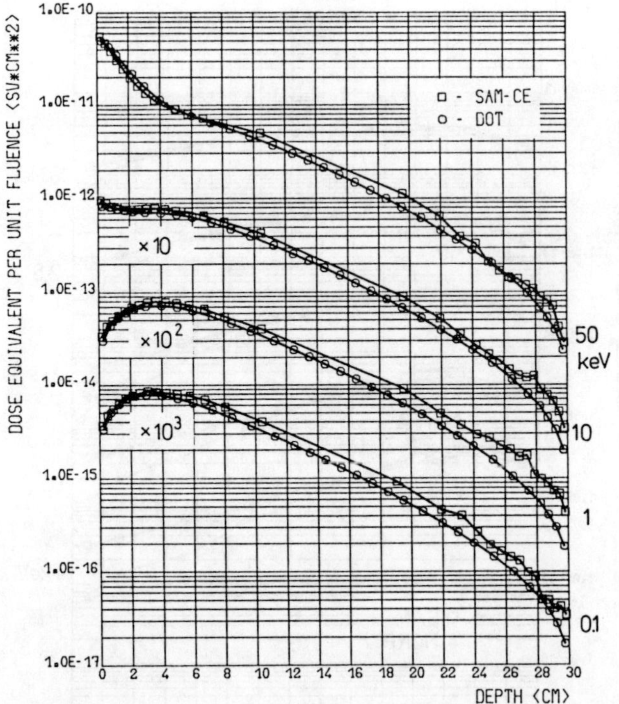

FIG.4(a). Calculated axial depth doses from heavy recoils in a 30-cm-dia. spherical phantom irradiated by broad beams of monoenergetic neutrons (Intercomparison of DOT and SAM-CE results).

30-cm-diameter spherical phantom for neutron energies between 14.1 MeV and 1 eV by means of both DOT and SAM-CE. It could be shown that the final results for the thermal flux density distribution in the SAM calculations depend critically on the energy limit for the MC histories, below which all neutrons were considered to have Maxwellian distribution. For DOT the final choice of width of the thermal group and its lower energy is critical. In Figs 4a and 4b, the depth dose equivalent distributions are shown for an MC energy limit in SAM of 10^{-3} eV, and a thermal group interval in DOT from 0.0001 eV to 0.414 eV. The results shown are in terms of dose equivalent (correctly kerma-equivalent) for the sake of application in radiation protection. This is, however, of no relevance with respect to the inter-comparison of both cases as long as the conversion functions used are the same. The latter task is not easy to achieve because of the different group structure. Figure 4c shows the functions used. Their generation is explained in Burger et al. (1980). The value of 2.93×10^{-12} $Sv \cdot n^{-1}$ cm^{-2} for the broad thermal DOT group was chosen to be identical with the value at 0.025 eV for the SAM calculations. This is a

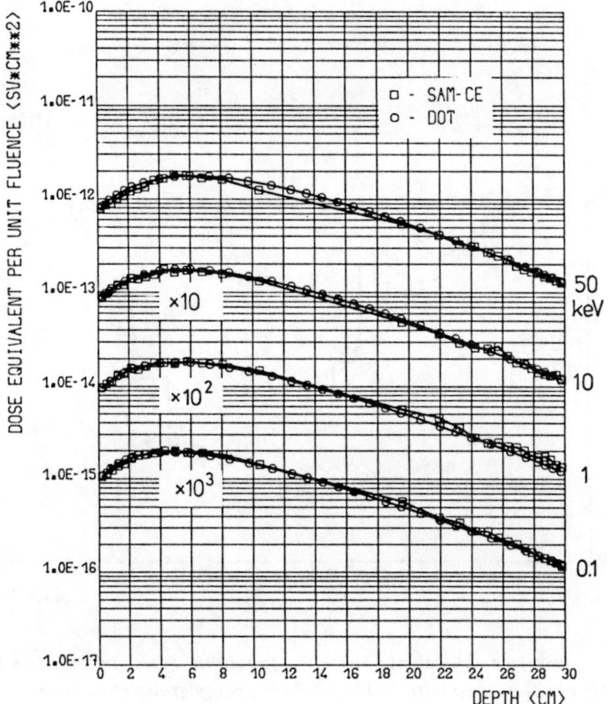

FIG.4(b). Calculated axial depth doses from internally produced gamma rays in a 30-cm-dia. spherical phantom irradiated by broad beams of monoenergetic neutrons (Intercomparison of DOT and SAM-CE results).

somewhat arbitrary procedure, which will have to be revised in the future. Nevertheless the results are surprising.

The thermal flux density in the case of the DOT calculations for neutron energies below 10 MeV defined between 0.0001 eV and 0.414 eV is 40–50% higher than that derived by SAM-CE, defined between 10^{-3} eV and 0.46 eV. The dose-equivalent components of the heavy recoils are however slightly higher for SAM-CE than for DOT, whereas the gamma dose distributions agree perfectly. The reason for this is not quite understood, as both dose distributions depend mainly on the thermal neutron source distribution. On the other hand, the total axial depth doses and depth dose equivalents as derived by SAM for higher energy neutrons ($E_n \geqslant 100$ keV) agree perfectly [40] with results reported by Chen and Chilton [41], also calculated by means of a Monte Carlo code.

From all the intercomparisons it can be concluded that generally the transport and MC codes available in combination with the cross-section data based upon the ENDF libraries guarantee sufficiently accurate calculations of depth kerma distributions in regular homogeneous phantoms.

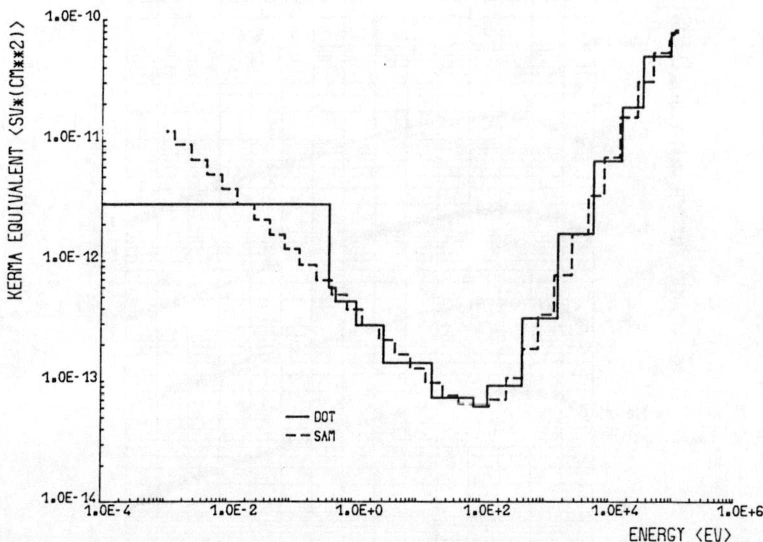

FIG.4(c). Fluence-to-kerma-equivalent conversion functions, based on Rep. ICRU-26 [25] and Burger and Maier [2] as used with DOT and SAM energy group structures for incident neutron energies below 100 keV.

4.2. Radiation transport in non-homogeneous media

The dose at any point within a non-homogeneous body may be determined either by direct three-dimensional Monte Carlo calculation, or be derived from the known calculated or measured dose distribution in homogeneous bodies by the application of corrections. In this case the first step is the description of the dose distribution for the homogeneous case by analytical expressions, which only depend on the SSD, field size and phantom depth. There are several so-called beam models, which calculate the dose distributions by means of 'generating functions', based normally upon a reasonable set of measurements.

If, in a non-homogeneous situation, the point of interest is within a medium different from the homogeneous case, the dose correction can then be performed in the following manner. First an effective depth of homogeneous medium for this point and the corresponding dose is determined and finally this dose is corrected by means of an appropriate kerma ratio.

This may be illustrated by the following figure and formula:

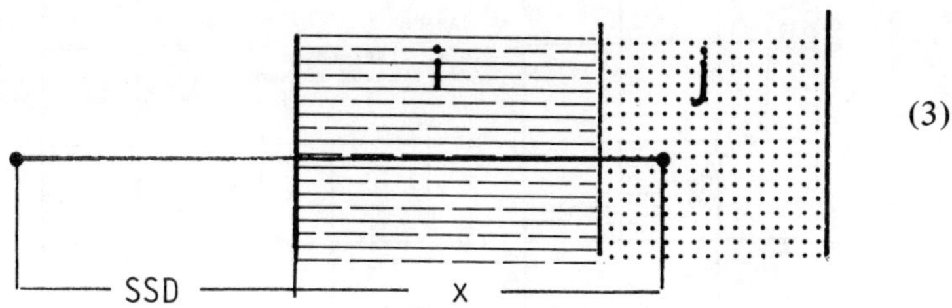

(3)

$$D_j(x) = (SSD/(SSD + x_{i,eff}))^2 \cdot D_{o,i} \cdot \exp(-\mu_{att} \cdot x_{eff}) \cdot B(x,D) \cdot K_j/K_i$$

$D_{0,i}$ = known dose at the surface of homogeneous body with medium (i)
SSD = source-to-surface distance
μ_{att} = effective attenuation coefficient
B(x,D) = dose buildup function (depending on depth x and field size D)
K_j/K_i = kerma ratio of neutron fields at point of interest for media (j) and (i)

Formulas of this kind have been applied by several authors [42–45]. The corrections may be called the effective-tissue-layer or effective-chord-length correction. It can be understood as a dose correction along straight lines drawn from the source to each point of the field, depending only on the material intersected or hit, but being independent of the neighbourhood of each line. In this sense the correction is ‘one-dimensional’.

In Figs 5 and 6 results are shown for the broad beam exposure of heterogeneous trunk phantoms representing two different arrangements, the first consisting of layers of fat, soft tissue, lung, soft tissue, and the second consisting of fat, bone and soft tissue. Calculations were performed for 14 MeV neutrons by DOT, SAM-CE and finally by a ‘beam model’ according to Eq.(2) with effective-tissue-layer correction, based upon DOT depth doses for the homogeneous phantom. The results of the transport calculations agree excellently for the first phantom situation. The effective-tissue-layer method, however, delivers somewhat too high values, especially in the lung region. The effect is clearly due to the lack of backscatter in the lung material, and is visible even in the surface fat and soft tissue layers.

Quite a different situation occurs at the second arrangement. Here the bone and large soft tissue region following the fat establishes a backscatter component, increasing the fat dose absolutely by roughly 4% in comparison with the first arrangement. In both cases the fat dose is, of course, higher than the corresponding soft tissue dose by roughly 11%, owing to the increased hydrogen content. The small differences between DOT and SAM-CE results, especially in the bone, are not quite understood, but are likely due to the not yet adequate and well corresponding scoring regions in both cases.

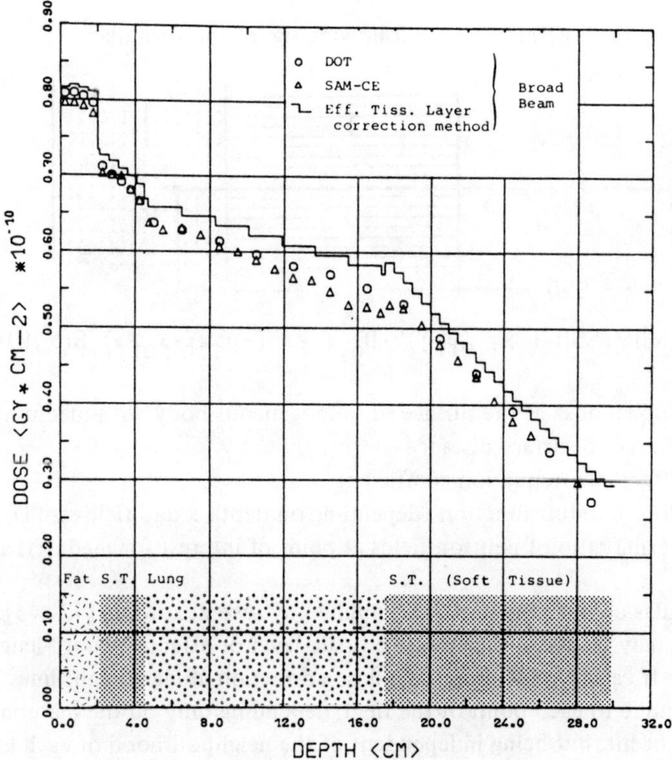

FIG.5. Calculated broad beam depth dose distributions for normal incidence of neutrons on the flat face of a heterogeneous cylindrical phantom, composed of fat, soft tissue, lung tissue and soft tissue layers. Comparison of DOT and SAM-CE results and those derived by the effective-tissue-layer (ETL) correction method, applying the DOT-depth dose curve.

In real radiotherapy the geometric conditions are, however, more complicated. Generally, narrow beams are used and the inhomogeneities are irregularly arranged, representing spatially finite volumes, as special bones, lung, air cavities, subcutaneous fat layers etc. Figure 7a shows the depth dose distribution for a small beam irradiation (8 cm field diameter) of a phantom with a cylindrical air cavity 4 cm dia. × 2 cm, located between depths of 3 and 5 cm. The cavity may simulate, for example, the trachea in the case of a neck irradiation. Figure 7b shows the depth doses along the axis and parallel to it out to the edge of the cavity in comparison with the effective-tissue-layer corrected depth dose. It can be seen that the latter is independent of the lateral distance, i.e. the same for all rays as long as they represent the same geometrical situation along their path. The true depth doses, however, show a continuous transition from the disturbed situation in the middle of the cavity to the undisturbed situation outside the cavity.

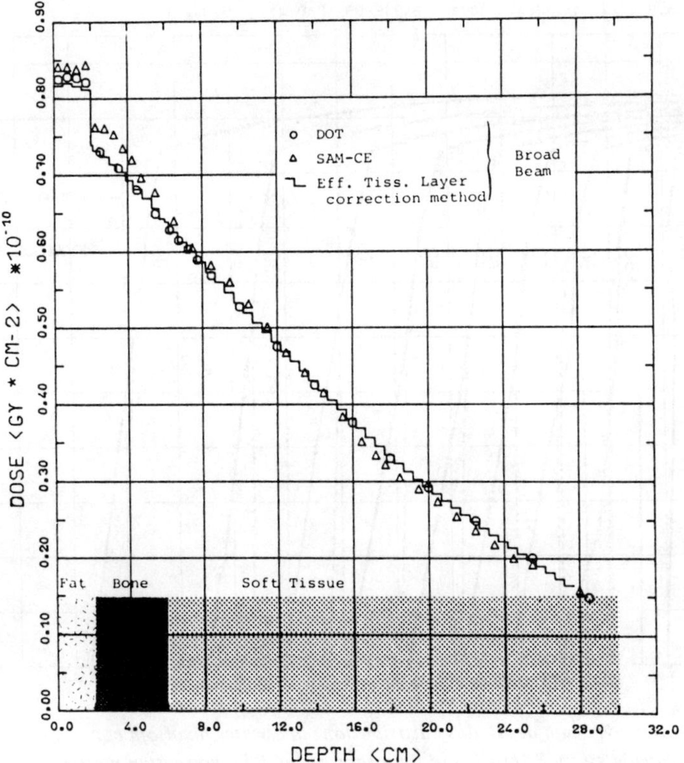

FIG. 6. Calculated broad beam depth dose distributions for a similar irradiation situation as that shown in Fig. 5. The phantom in this case is composed of fat, bone and soft tissue layers.

From the results shown it can be clearly concluded that the omission of corrections for lung density, missing tissue in air cavities, as well as for the increased fat kerma, would lead to highly incorrect depth doses. The effective-tissue-layer correction, based upon matched patient and organ contours as derived from CT scans or probably, even better, NMR scans in the future, provides a reasonable first-order correction. It cannot, however, account for the change of scatter conditions in different materials and will lead to wrong results at the edges of spatially finite inhomogeneities.

4.3. Secondary particle production

Another problem is the calculation of charged secondary particle production and their slowing-down distribution. Several codes are available and applied for different purposes. One purpose is the calculation of true dose distributions rather than kermas at interfaces and within microscopical structures of organs.

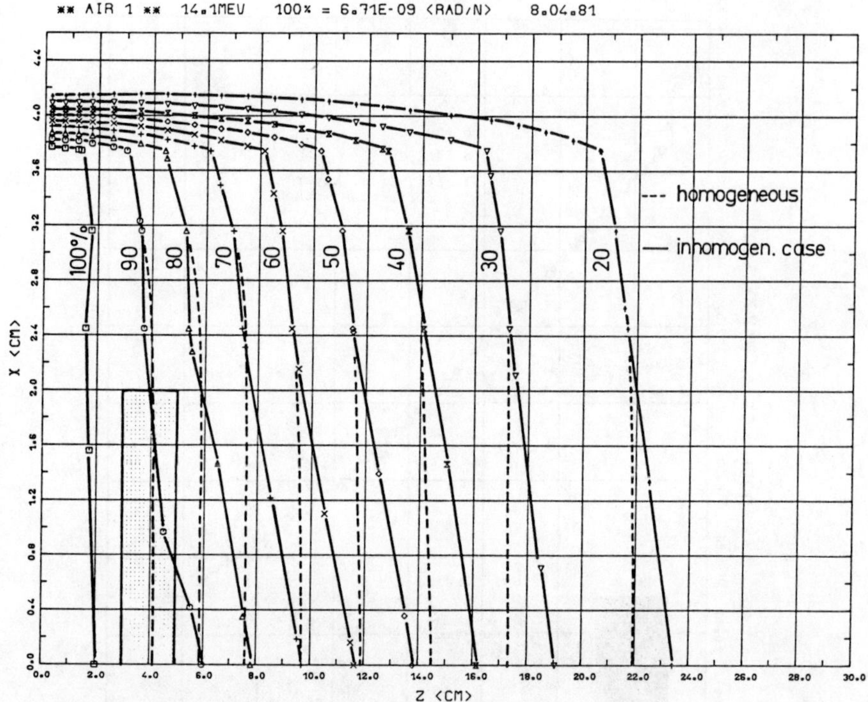

FIG. 7(a). Narrow beam depth dose distributions in a water phantom containing a small air cavity, calculated by SAM-CE and by means of the ETL correction method.

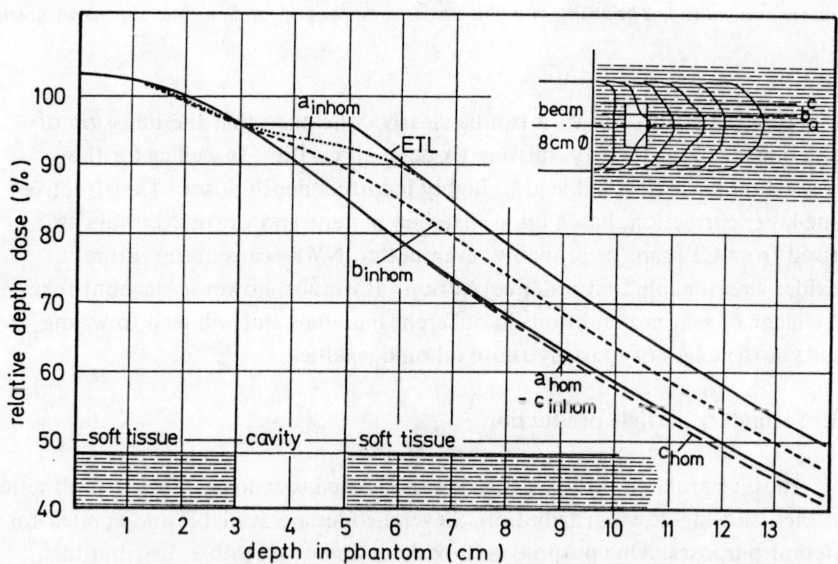

FIG. 7(b). Depth doses from Fig. 7(a) along rays parallel to the beam axis for various lateral distances.

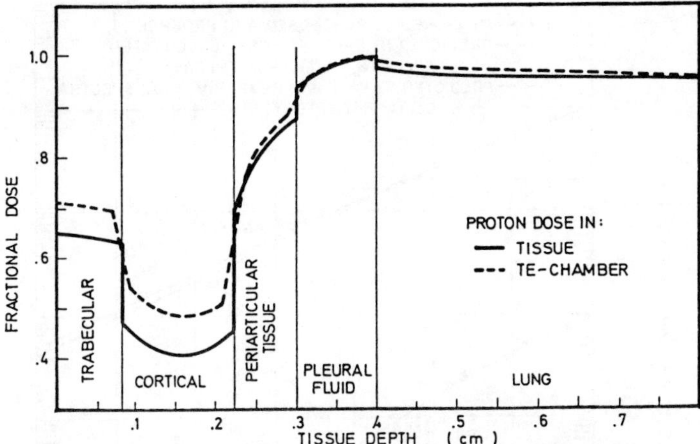

FIG.8. Calculated proton dose curve for 15 MeV neutron irradiation through an inhomogeneously composed rib in front of the lung in the material and in a TE detector at the points of interest (after Pfister et al. [46]*).*

Pfister et al. [46] modified the S_n code ANISN and adapted it for the simulation of secondary particle transport. For that purpose the scattering integral in the transport equation is replaced by a term describing the continuous slowing down of the protons. The energy transfer matrix for the proton generation is determined by the programme RICE [47], using neutron reaction cross-sections from ENDF-B versions. With this program the authors could not only study the skin sparing effect, taking into account the real morphology of the skin, but also other inhomogeneity situations, as for example a rib in front of the lung. In this case they distinguished trabecular, cortical and periarticular bone layers, followed by 1 mm pleural fluid and then lung tissue (Fig.8).

Another purpose for computing the secondary particles is to investigate the local variation of effective dose. There are in principle two different methods of assessing radiation efficiency inside a phantom. The most direct approach calculates the recoil particle spectra and their slowing-down distributions, and applies a model of cellular radiation response based upon track segment measurements and theoretical ion track models. An example for this procedure is given by Alsmiller [36], who used the code HETC for the neutron transport. This code allows one to record at each neutron collision the type and energy of all of the charged particles produced for later analysis, which may then be used as input for the cell inactivation model of Katz et al. [48] to produce cell-survival probabilities, oxygen enhancement ratios (OERs) and relative biological effectiveness factors (RBEs) for each point of interest. A detailed description of how HETC results and the Katz model are combined is given

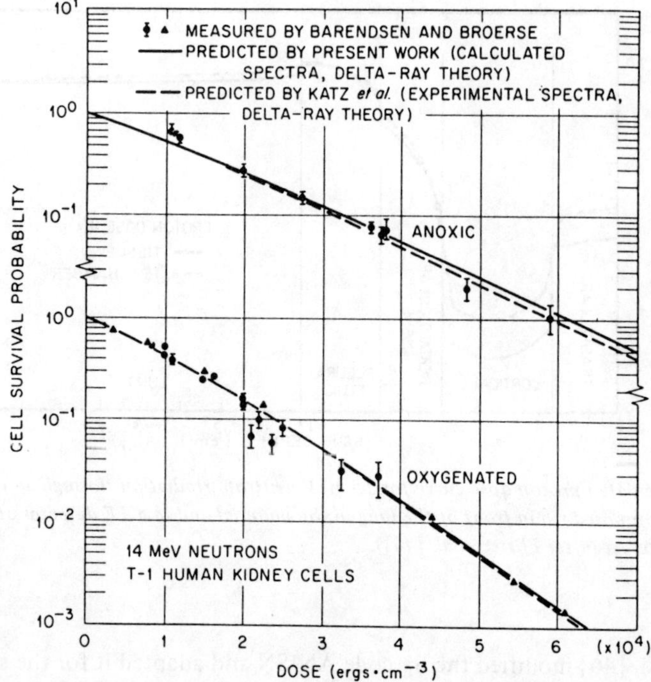

FIG. 9. Calculated cell survival probability versus absorbed dose for T-1 kidney cells irradiated by 14 MeV neutrons (after Alsmiller [36]).

by Armstrong and Chandler [49]. Figure 9 shows cell survival probabilities of aerobic and anoxic T-1 kidney cells as a function of absorbed dose for various cyclotron-produced neutron spectra and for photons, calculated in the described manner from the average recoil spectra within a 6 X 6 X 10 cm^3 cube on the axis, between 10 and 20 cm depth inside a phantom. From such curves the relevant dose-dependent RBEs and radiobiologically effective doses can be directly derived.

A similar approach, however, based upon a more phenomenological survival model rather than a true track efficiency model, was reported by Morstin and Kawecka [38].

Another, more classical approach is the determination of LET spectra from the secondary particle distribution by applying suitable stopping power functions. These LET spectra can then be folded with RBE values as a function of LET. The procedure is time-consuming and in fact unnecessary, as a few cellular RBEs are already available as a function of neutron energy [50]. These functions can be folded directly with the neutron spectrum at any point of interest [1].

A somewhat more general approach in this respect is the microdosimetric one. In this case one relies upon the hypothesis that RBEs for the survival of mammalian cells are more uniquely linked with the energy deposition within small assumed target volumes, rather than with LET, neutron energy or any other physical parameter characterizing the radiation field. Complete microdosimetric distributions may be used in the same manner for the quantitative specification of radiation efficiency as complete LET spectra. This seems, however, too complex a procedure. It is obviously desirable to derive one single parameter from the distributions, which permits an adequate specification of radiation quality. One proposed, radiobiologically matched quantity is y^*, the so-called saturation corrected dose-mean lineal energy. The dependence of y^* on neutron energy [51] reflects sufficiently well a kind of average RBE for cell killing of mammalian cells. Its use is therefore justified for assessing expected local variations of radiation efficiency within a phantom, irrespective of any underlying microdosimetric model. The great advantage of this approach is the experimental verification of y^* by means of proportional counter measurements.

It is obvious that the degradation of the neutron spectrum, and hence the axial variation of y^* in a phantom, depends not only on the neutron energy, but also on the field size. It could be shown [52] that, for an ideal narrow beam of 100 cm^2 field size, the axial variation of y^* for the neutron component only is about 15%. This is to a great extent compensated for by the gamma component in the phantom, as measured at several neutron therapy installations. According to this the calculated y_t^*, including both radiation components, generally varies only by a few per cent (Fig.10). A similar situation holds for the penumbra. Most radiobiological experiments performed inside phantoms do in fact confirm this finding. It is hence rather doubtful whether the combination of neutron and gamma dose by means of applying a constant RBE to the neutron component is generally correct. It must, however, be stated that the real situation in a beam depends on the specific target-collimator arrangement, and that it is further uncertain as to what extent the clinically relevant tumour or healthy tissue response follows such simple RBE considerations. This may be the main reason that, at the moment, an indication of the source spectrum and a separation of neutron and gamma-ray absorbed dose is considered to be sufficient in practical radiotherapy in describing a possible variation of radiation quality inside a patient [25].

However, if the possibility exists of more sophisticated treatment planning, it is recommended that the radiation quality of the beam, and its local variation inside the phantom, be better determined and eventually taken into consideration. One possible way is the direct conversion of the spectral neutron fluence, resulting from transport calculations at any point of interest into a radiobiologically effective dose, which must further be suitably combined with the gamma component. A possible way how this can be done in an integrated treatment planning system has been shown by Hehn et al. [43].

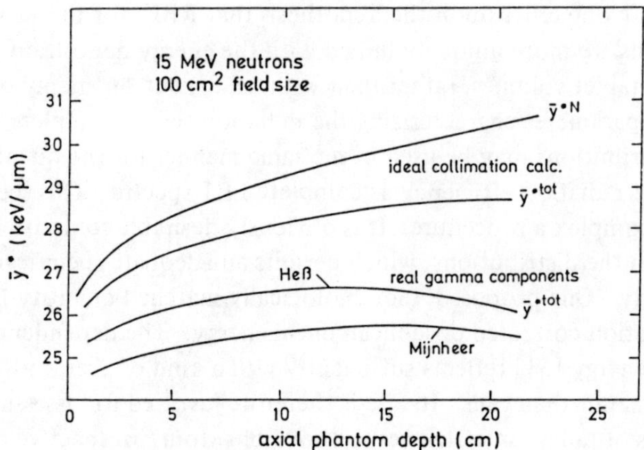

FIG.10. Variation of the saturated dose mean lineal energy densities $y^*_{neutron}$ *and* y^*_{total}, *calculated for an ideal beam of 15 MeV neutrons at 5 cm depth in a water phantom for a field size of 100 cm², taking into consideration phantom produced and real gamma radiation dose components.*

5. SUMMARY

The derivation of perfectly agreeing depth dose results from two totally different neutron transport codes for the broad and ideal narrow beam irradiation of homogeneous phantoms was demonstrated. It was further shown that by means of such codes not only ideal exposure situations but also real target-collimator-phantom arrangements, even with non-homogeneous phantoms, can be treated. In many cases numerical results could be verified experimentally.

Some of the more frequently used codes and the necessary cross-section libraries were discussed. They are available from international distribution centres, such as RSIC in Oak Ridge or the NEA Data Bank in Saclay. The primary quantities to be calculated are spectral neutron and coupled gamma-ray fluxes within any desired scoring element in the material of interest. Additional codes allow the generation of secondary charged particle distributions.

From this information all dosimetrically relevant quantities, such as kerma, absorbed dose at interfaces or within microscopic organ structures, micro-dosimetric parameters and, finally, the expected cellular responses, can be derived. Based upon this, optimized treatment planning in fast neutron radio-therapy is possible.

REFERENCES

[1] BURGER, G., GRÜNAUER, F., MAIER, E., "Mixed field phantom dosimetry", Biomedical Dosimetry (Proc. Symp. Vienna, 1975), IAEA, Vienna (1975) 71.

[2] BURGER, G., MAIER, E., "Data for uncertainty analysis", Basic Physical Data for Neutron Dosimetry (BROERSE, J.J., Ed.), EUR 5629 (1976) 239.

[3] BURGER, G., GRÜNAUER, F., "Neutron degradation inside a phantom and its influence on dosimetric quantities", ibid., p. 53.

[4] BICHSEL, H., RUBACH, A., "Uncertainty of the determination of absolute neutron dose with ionization chambers", Proc. 3rd Symp. Neutron Dosimetry in Biology and Medicine (BURGER, G., EBERT, H.G., Eds), EUR 5848 (1977) 549–63.

[5] BROERSE, J.J., BURGER, G., COPPOLA, M. (Eds), A European Neutron Dosimetry Intercomparison Project (ENDIP), EUR 6004 (1978).

[6] GOODMAN, L.J., "Uncertainty analysis for dosimetry in a mixed field of neutrons and photons", Proc. 2nd Symp. Neutron Dosimetry in Biology and Medicine (BURGER, G., EBERT, H.G., Eds), EU 5273 (1974) 227–36.

[7] CARLSON, B.G., "The numerical theory of neutron transport", Methods in Computational Physics (ADLER, B., et al., Eds) Vol.1, Academic Press (1963) 1–42.

[8] LEE, C.E., The Discrete S_n Approximation to Transport Theory, Los Alamos Lab. Rep. LA-2595, Los Alamos (1962).

[9] MYNATT, F.R., "The discrete ordinates method in problems involving deep penetrations", A Review of the Discrete Ordinates S_n Method for Radiation Transport Calculations, Oak Ridge Nat. Lab. Rep. ORNL-RSIC-19 (1968) 25–51.

[10] TRUBEY, D.K., MASKEWITZ, B.F. (Eds), A Review of the Discrete Ordinates S_n Method for Radiation Transport Calculations, ORNL-RSIC-19, Oak Ridge Nat. Lab. (1968).

[11] LATHROP, K.D., Discrete-ordinates methods for the numerical solution of the transport equation, Reactor Tech. **15** (1972) 107–35.

[12] STEVENS, P.N., Use of the discrete ordinates S_n method in radiation shielding calculations, Nucl. Eng. Des. **13** (1970) 395–408.

[13] LATHROP, K.D., CARLSON, B.G., Discrete Ordinates Angular Quadrature of the Neutron Transport Equation, LA-3186, Los Alamos Scient. Lab. (1964).

[14] CARLSON, B.G., Tables of Equal Weight Quadrature EQ_n over the Unit Sphere, LA-4734, Los Alamos Scient. Lab. (1971).

[15] ENGLE, W.W., Jr., A User's Manual for ANISN, a One-Dimensional Discrete Ordinates Transport Code with Anisotropic Scattering, K-1693, Comp. Techn. Center, Union Carbide Corp. (1967).

[16] MYNATT, F.R., A User's Manual for 'DOT', K 1694, Union Carbide Corp. (Jan. 1967).

[17] MYNATT, F.R., MUCKENTHALER, F.J., STEVENS, P.N., Development of Two Dimensional Discrete Ordinates Transport Theory for Radiation Shielding, CTC-INF-952, Comp. Techn. Center, Union Carbide Corp. (1969).

[18] RHOADES, W.A., MYNATT, F.R., The DOT III Two-Dimensional Discrete Ordinates Transport Code, ORNL-TM-4280, Oak Ridge Nat. Lab. (1973).

[19] RHOADES, W.A., SIMPSON, D.B., CHILDS, R.L., ENGLE, W.W., Jr., The DOT-IV Two-Dimensional Discrete Ordinates Transport Code with Space-Dependent Mesh and Quadrature, ORNL-TM-6529, Oak Ridge Nat. Lab. (1979).

[20] STRAKER, E.A., STEVENS, P.N., IRVING, D.C., CAIN, V.R., The MORSE-Code – A Multigroup Neutron and Gamma-Ray Monte Carlo Transport Code, ORNL-4585, Oak Ridge Nat. Lab. (1970).

[21] CHANDLER, K.C., ARMSTRONG, T.W., Operating Instructions for the High-Energy Nucleon-Meson Transport Code HETC, ORNL-4744, Oak Ridge Nat. Lab. (1972).

[22] LICHTENSTEIN, H., COHEN, M.O., STEINBERG, H.A., TRUBETZKOY, E.S., BEER, M., The SAM-CE Monte Carlo System for Radiation Transport and Criticality Calculations in Complex Configurations (Revision 7), Rep. EPR/CCM-8 (1979).

[23] OZER, O., GARBER, D., ENDF/B Summary Documentation, BNL-17541, Nat. Cross Section Center, Brookhaven Nat. Lab. (1973).

[24] GARBER, D., DUNFORD, C., PERLSTEIN, S. (Eds), Data Formats and Procedures for the Evaluated Nuclear Data File, ENDF, BNL-NCS-50496, Nuclear Data Center, Brookhaven Nat. Lab. (1975).

[25] INTERNATIONAL COMMISSION ON RADIATION UNITS AND MEASUREMENTS (ICRU), ICRU 26, Neutron Dosimetry for Biology and Medicine, ICRU 26, Washington DC (1977).

[26] ABDOU, M.A., MAYNARD, C.W., WRIGHT, R.A., MACK, a Computer Program to Calculate Neutron Energy Release Parameters (Fluence-to-kerma-factors) and Multigroup Neutron Reaction Cross Sections from Nuclear Data in ENDF Format, ORNL-TM-3994, Oak Ridge Nat. Lab. (1973).

[27] SNYDER, W.S., JONES, T.D., "Depth dose due to neutrons as calculated for a tissue phantom and man", Proc. 1st. Symp. Neutron Dosimetry (BURGER, G., SHRAUBE, H., EBERT, H.G., Eds), EUR 4896 (1972) 597–623.

[28] GRÜNAUER, F., Design of a Neutron Collimator for a Medical-Biological Irradiation Set-Up, Dissertation, Technical University, Munich (1975).

[29] GRÜNAUER, F., "Calculation of a neutron collimator for therapeutical purposes", Proc. 1st Symp. Neutron Dosimetry (BURGER, G., SCHRAUBE, H., EBERT, H.G., Eds), EUR 4896 (1972) 511.

[30] HEHN, G., BECKER, R., "Optimization of a collimator for 14-MeV-neutrons", ibid., p. 529.

[31] SCHLEGEL, D., DIETZE, G., "Monte Carlo calculations of a collimator for fast neutron therapy", ibid., p.783.

[32] GRÜNAUER, F., "Neutron collimator calculations", Proc. 2nd Symp. Neutron Dosimetry in Biology and Medicine (BURGER, G., EBERT, H.G., Eds), EUR 5273 (1974) 757.

[33] BURGER, G., MORHART, A., SCHRAUBE, H., "Recent results in neutron depth dose calculations", Proc. 3rd Symp. Neutron Dosimetry in Biology and Medicine (BURGER, G., EBERT, H.G., Eds), EUR 5848 (1977) 125.

[34] BÖHM, J.K., EISSA, H.M., HEHN, G., PFISTER, G., STILLER, P., "Transport calculations, analytical representation of isoeffect curves for therapy with fast neutrons", ibid., p.145.

[35] PFISTER, G., HEHN, G., FRIEDLEIN, H.P., "Calculations of neutron and gamma components, spectra and inhomogeneity effects in phantoms for neutron therapy facilities up to 50 MeV", Proc. 4th Symp. Neutron Dosimetry (BURGER, G., EBERT, H.G., Eds) Vol.2, EUR 7448 (1981) 79–89.

[36] ALSMILLER, R.G., Jr, "Calculation of the transport of fast neutrons (⩽50 MeV) through matter", Proc. 2nd Symp. Neutron Dosimetry in Biology and Medicine (BURGER, G., EBERT, H.G., Eds), EUR 5273 (1974) 681.

[37] BRENNER, D.J., PRAEL, R.E., DICELLO, J.F., ZAIDER, M., "Improved calculations of energy deposition from fast neutrons", Proc. 3rd Symp. Neutron Dosimetry in Biology and Medicine (BURGER, G., EBERT, H.G., Eds), EUR 5848 (1977) 103.

[38] MORSTIN, K., KAWECKA, B., "Distribution of radiobiological parameters in the tissue phantom irradiated with fast neutrons", Proc. 4th Symp. Neutron Dosimetry (BURGER, G., EBERT, H.G., Eds), EUR 7448 (1981) 115–27.

[39] NATIONAL COUNCIL ON RADIATION PROTECTION, NCRP-Rep. No.38, Protection Against Neutron Radiation: Recommendations of the Nat. Council on Radiation Protection and Measurements (1971).

[40] BURGER, G., MORHART, A., NAGARAJAN, P.S., WITTMANN, A., "Effective dose equivalent and its relationship to operational quantities for neutrons", Proc. European Seminar on Radiation Protection Quantities for External Exposure (BURGER, G., EBERT, H.G., HARDER, D., KRAMER, R., WAGNER, S., Eds), EUR 7101 EN, Harwood Academic Publishers (1980) 117–35.

[41] CHEN, S.-Y., CHILTON, A.B., Calculation of fast neutron depth-dose in the ICRU standard tissue phantom and the derivation of neutron fluence-to-dose-index conversion factors, Radiat. Res. **78** (1979) 335–70.

[42] BURGER, G., MORHART, A., NAGARAJAN, P.S., WITTMANN, A., "Determination of depth dose distributions by means of transport calculations", Treatment Planning for External Beam Therapy with Neutrons (BURGER, G., BREIT, A., BROERSE, J.J., Eds), Urban & Schwarzenberg (1981) 83–92.

[43] HEHN, G., PFISTER, G., FRIEDLEIN, H.P., KICHERER, G., "Recent development of the Stuttgart Program System for treatment planning in neutron therapy", ibid., pp.72–82.

[44] HÖVER, K.H., LORENZ, W.J., SCHARFENBERG, H., SCHLEGEL, W., "Treatment planning at the Heidelberg Neutron Therapy Facility (with special considerations of the inhomogeneity problem)", ibid., pp.116–22.

[45] EENMAA, J., WOOTTON, P., KALET, I., "Treatment planning procedures at the University of Washington)". See Ref. [42], pp 199–207.

[46] PFISTER, G., PRILLINGER, G., HEHN, G., KRASS, C., STILLER, P., "Absorbed dose and recoil spectra at critical tissue boundaries characterized by the absence of recoil equilibrium", Proc. 4th Symp. Neutron Dosimetry (BURGER, G., EBERT, H.G., Eds) Vol.2, EUR 7448 (1981) 91–101.

[47] JENKINS, J.D., 'RICE', A Program to Calculate Primary Recoil Atom Spectra from ENDF/B Data, ORNL-TM-2706 (1970).

[48] KATZ, R., SHARMA, S.C., Response of cells to fast neutrons, stopped pions, and heavy ion beams, Nucl. Instrum. Methods **111** (1973) 93–116.

[49] ARMSTRONG, T.W., CHANDLER, K.C., Calculations related to the application of negatively charged pions in radiotherapy: Absorbed dose, LET-spectra, and cell survival, Radiat. Res. **58** (1974) 293–328.

[50] HALL, E.J., ROSSI, H.H., KELLERER, A.U., GOODMAN, L., MARINO, S., Radio-biological studies with monoenergetic neutrons, Radiat. Res. **54** (1973) 431.

[51] CASWELL, R.S., COYNE, J., "Energy deposition spectra for neutrons based on recent cross section evaluations", Proc. 6th Symp. Microdosimetry (BOOZ, J., EBERT, H.G., Eds), EUR 6064 (1978) 1159.

[52] BURGER, G., MAIER, E., MORHART, A., "Radiation quality and its relevancy in neutron radiotherapy", ibid., p.451.

MEASUREMENT OF PHOTON DOSE FRACTION IN THE FAST NEUTRON THERAPY BEAM

A. ITO
Cyclotron Unit,
The Institute of Medical Science,
The University of Tokyo,
Tokyo, Japan

Abstract

MEASUREMENT OF PHOTON DOSE FRACTION IN THE FAST NEUTRON THERAPY BEAM.

Photons are always present in fast neutron irradiations. As the relative biological effectiveness (RBE) of neutrons and photons is different the fractional photon dose should be measured and its significance be clarified. Paired dosimetry is the most common method for this purpose, and here the relative neutron dose sensitivity of the neutron insensitive dosimeters, k_U value, is the key parameter. The k_U values were determined for C-CO_2 and Mg-Ar ionization chambers and a micro-Geiger-Müller counter (Philips 18529) by independent techniques. These include the tissue equivalent (TE) proportional counter, the lead attenuation and the time-of-flight techniques. The k_U values of the d(14) + Be neutron beam from the Institute of Medical Science (IMS) Cyclotron are estimated to be 0.150 ± 0.004, 0.062 ± 0.001 and 0.011 ± 0.001. The photon dose fraction, (D_G/D_T), of this neutron beam in air amounted to 1.1 to 1.8% at SSD = 150 cm for a field size of 17 to 600 cm^2. All measurements in air by the independent techniques showed consistent results. The neutron and photon dose distributions in the phantom were measured as the basic data for treatment planning. The paired dosimeters of TE-TE and Mg-Ar are preferably used. A computer-controlled paired dosimetry system was developed, which precisely measures the depth dose curve, the transverse dose profile and the two-dimensional dose map. The fractional photon dose in a phantom is the function of both field size and depth, ranging from 2 to 20% along the central axis. Measurements with different paired dosimeters sometimes showed inconsistent results. Because the radiation fields in a phantom are different from those in free air, the k_U value must be carefully examined.

1. INTRODUCTION

Photons are always present in fast neutron irradiations. As the relative biological effectiveness (RBE) is different for neutrons and photons, the neutron dose and the photon dose should be separately measured and recorded for fast neutron therapy. It is generally recognized that an accuracy of ± 5% is required to determine the absorbed dose in the tumour in radiation therapy. To achieve this goal, a precision of ± 2% is required for the practical dosimetry in a homogeneous phantom.

The photon in the fast neutron therapy beam is produced in the neutron target and collimator system, and through the neutron capture process within the patient or phantom material. Hence, measurements of the photon dose fraction in the mixed neutron and photon field must be made carefully.

Paired dosimetry is the most common method for measuring the photon dose and the neutron dose separately, as is precisely described in ICRU Rep. 26 [1], AAPM Rep. 7 [2] or ECNEU protocol [3]. One of the central problems in paired dosimetry is the accurate determination of the relative neutron dose sensitivity of the neutron insensitive dosimeters to the calibration photon dose, the k_U value[1] (ICRU 26). Extensive efforts have been made to improve knowledge of $k_U(E_n)$ of the various dosimeters for monoenergetic neutrons as well as broad spectral neutron beams. Non-hydrogeneous ionization chambers such as C-CO_2, Al-Ar and Mg-Ar have been utilized. Also, more recently, the miniature Geiger-Müller (GM) counters of various designs were used. The $k_U(E_n)$ values of these dosimeters are usually determined at the calibrated neutron field whose photon contamination is kept as low as possible.

We have investigated the $k_U(E_n)$ values of C-CO_2 and Mg-Ar ionization chambers and a miniature GM counter in the fast neutron therapy beam (d(14) + Be) at the Cyclotron Unit, the Institute of Medical Science (IMS), Tokyo, employing the different techniques. These include the tissue equivalent (TE) proportional counter technique, the lead attenuation and time-of-flight techniques. A brief review of these results of $k_U(E_n)$ determinations are given and some problems discussed.

We measured the neutron and photon dose distributions in the phantom as the basic data for the fast neutron treatment planning. The paired TE-TE and Mg-Ar ionization chamber is preferably used for this purpose. A computer-controlled paired dosimetry system was developed for the precise neutron and photon dose distribution measurements.

When the photon dose in a phantom is measured, it is important to remember that the quality of the radiation field is different from that in air. The neutron spectrum changes. Especially the proportion of slower neutrons increases. The photon spectrum is not identical, and in some cases the energetic secondary recoil protons may enter the dosimeters. Thus, in certain cases, the k_U values measured in air may not be applicable to the photon dosimetry in a phantom.

2. DETERMINATION OF k_U VALUES

To determine the k_U values of any neutron insensitive dosimeter, a calibrated neutron field is required. Such a neutron field, either monoenergetic or spectral,

---

[1] k_U = the relative neutron sensitivity of dosimeters used for photon-fraction determinations.

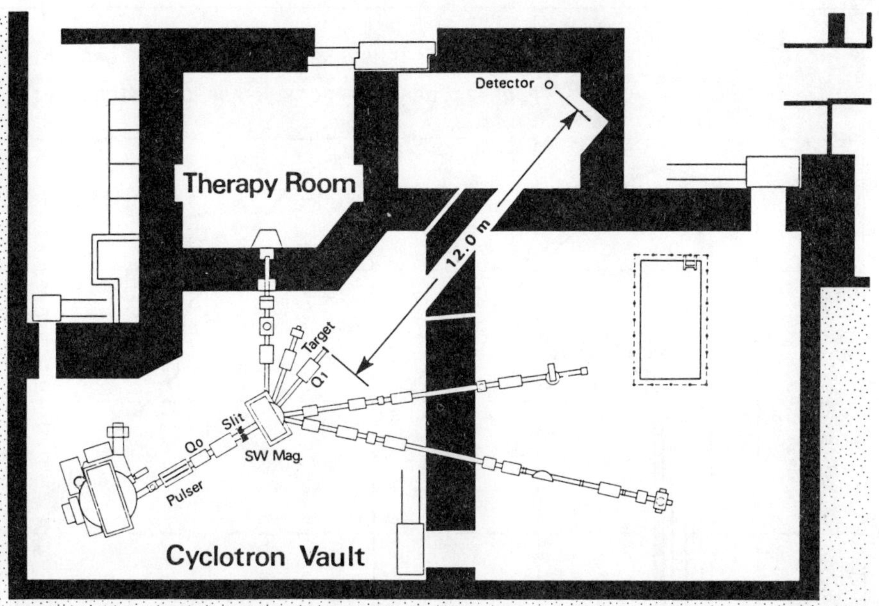

FIG.1. Plan view of the Fast Neutron Therapy Facility at Institute of Medical Science (IMS), Cyclotron Unit, Tokyo. Most of the photon dosimetry was carried out with the therapy beam. Time-of-flight experiments were also done.

should have the minimum photon contamination. Various techniques have been developed to separate the neutron response of the given dosimeter from the photon response [1]. Depending on the nature of the fast neutron beam and the available methods and instrumentation, the applicable techniques are limited. We have employed as many techniques as possible to determine the $k_U(E_n)$ values accurately.

Figure 1 shows the plan view of our medical cyclotron facility. In addition to the conventional d(14) + Be fast neutron beam in the therapy room [4], we have an external beam-pulsing system and enough space to carry out the time-of-flight (TOF) experiments [5].

In the following, the techniques are reviewed that we have used to determine the k_U values for our neutron beam.

2.1. Proportional counter technique

The spherical TE proportional counter was originally designed by Rossi and Rosenzweig [6] to measure the event-size spectrum in a microscopic region and to deduce the LET distribution. Apart from microdosimetry, this proportional

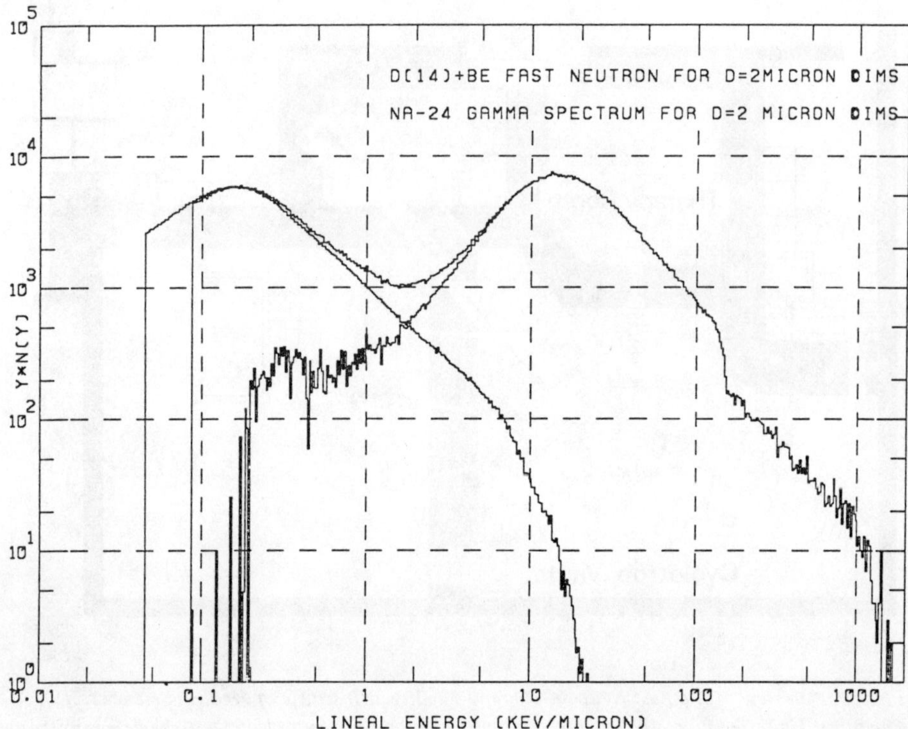

FIG.2. Y-spectra for the d(14)+Be neutron measured in free air and ^{24}Na photons. The conventional TE proportional counter with an equivalent diameter of 2 μm was used. The photon fraction can be subtracted by "peeling off" the y-spectrum of ^{24}Na gamma rays.

counter technique was found very useful for characterizing the given mixed neutron and photon field. Especially, from the analysis of the y-spectrum, one can deduce the fractional photon dose contribution in the mixed field [7].

Figure 2 shows an example of the y-spectrum of the d(14) + Be neutron beam measured in air. The methods employed are described elsewhere [8, 9]. The fraction of the photon dose can be estimated by "peeling off" the y-spectrum due to the photon component. In Fig.2, the photon component is assumed to be identical with that from the ^{24}Na gamma rays. Both spectra fit fairly well, and the photon dose fraction of 1.4% (D_G/D_N) is obtained.

This calibrated field could be used as the test field to measure the k_U values of any neutron insensitive dosimeters. Paired dosimetry with TE-TE and C-CO_2 (EG&G 2 cm^3) ionization chambers as well as with TE-TE and Mg-Ar (laboratory-built 2.6 cm^3 cylindrical) ionization chambers was carried out. Then, the k_U values were deduced from these two different experiments. The resulting k_U values of our d(14) + Be neutron beam were 0.150 for the C-CO_2 chamber and 0.062 for the Mg-Ar chamber [8].

Using the same technique, the photon dose fractions in various fast neutron therapy beams were measured, namely, d(16) + Be neutrons at the Texas A&M Variable Energy Cyclotron (TAMVEC), College Station, Texas, d(35) + Be neutron at the Naval Research Laboratory (NRL), Washington, D.C., and d(50) + Be neutrons at TAMVEC. Thus, a group of k_U values for C-CO_2 and Mg-Ar chambers was obtained as a function of the deuteron energy used to produce fast neutron beams [8].

2.2. Lead attenuation technique

Lead is known as a good filtering material for photons in a mixed neutron and photon field. Attix et al. [10] have proposed a technique to determine the photon dose fraction and the k_U value of any neutron-insensitive dosimeter. This technique, called the lead attenuation technique, requires only a narrow neutron beam, a plug collimator and a lead filter. Should a narrow neutron beam not be available, the modified lead attenuation method by Waterman et al. [11] or by Hough [12] can be used. These techniques are very simple and useful for the practical determination of the k_U values of the fast neutron therapy beam. We have, therefore, employed this technique to supplement the other techniques. Figure 3 shows the result of the measurements of our d(14) + Be neutron beam with the TE-TE and the Mg-Ar ionization chamber pair. The right side of Fig.3 gives the measurements with a TE-TE chamber. On the left are the results with the Mg-Ar chamber. In solving the six equations [10], we found the k_U value of the Mg-Ar chamber to be 0.062. The k_U value for the C-CO_2 chamber was determined to be 0.147 [8].

2.3. Time-of-flight technique

As our facility has the capability of producing a pulsed beam of any intervals [5], the neutron time-of-flight (TOF) technique was used to measure the energy spectrum of the d(14) + Be neutron [13]. By this TOF technique we determined the $k_U(E_n)$ values of a miniature GM counter (Philips 18529 with the Sn-Pb filter in the acrylic resin sleeve as suggested by the manufacturer) for neutrons with energy between 1 and 18 MeV. As a result, the average k_U value obtained for the d(14) + Be neutron was found to be 0.011 ± 0.001 [14]. This value, however, did not agree well with the k_U of the similar MX 163 GM counter with the Perspex sleeve of the MRC cyclotron neutron (0.0073 ± 0.0007) [12]. The difference may be due to the differences in the GM counters and also in the energy compensation filters.

We are combining this TOF technique with the proportional counter technique to study the response of TE carbon, magnesium or aluminium walled chambers.

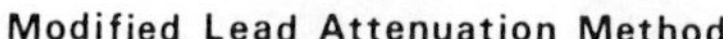

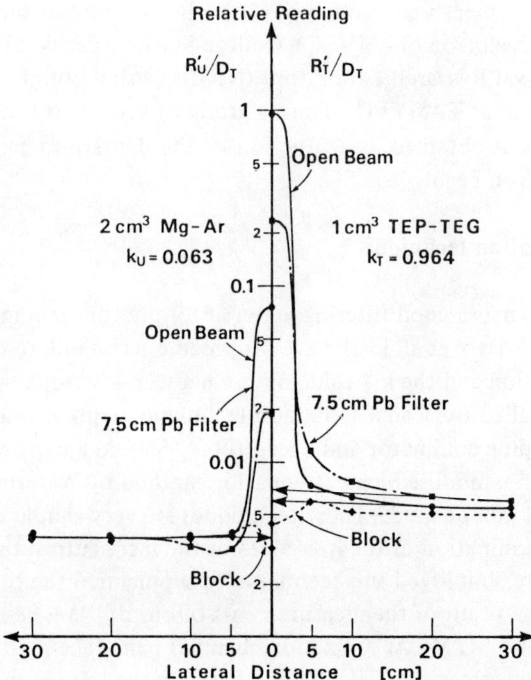

FIG.3. Results of the modified lead attenuation measurements with the TE-TE (right) and the Mg-Ar ionization chamber. From the lateral profile measurements the scattered photon component is determined.

2.4. Comparison of neutron insensitive dosimeters

We have been investigating the k_U values of the neutron insensitive dosimeters, using the different techniques as described above. Table I summarizes the results of the intercomparison of the three types of neutron insensitive dosimeter, namely the C-CO_2 chamber, the Mg-Ar chamber and a GM counter. The agreement of the k_U values is quite satisfactory between the independent methods. Thus, the uncertainty of k_U for each dosimeter is assigned as the maximum deviation of the data, which is also listed in Table I. Then, from Eqs (2.7–7) and (2.7–8) in ICRU Rep. 26 [1], one can deduce the uncertainty in photon dose determination. Both the Mg-Ar chamber and the GM counter could measure the photon dose as precisely as 0.1%. Even with the C-CO_2 chamber the estimated error is only 0.5%.

These k_U values were determined for specific dosimeters. If these values are applicable for similar dosimeters of the same kind, it is quite useful. For C-CO_2

TABLE I. COMPARISON OF NEUTRON INSENSITIVE DOSIMETERS

Dosimeter model	$C\text{-}CO_2$ EG&G 2 mL	Mg-Ar 2.6 mL thimble	GM counter Philips 18529[a]
k_U for d(14) + Be	0.150 (PC)[b] 0.147 (LAM)[c] 0.147 (MLAM)[d] 0.146 (Calc.)[e]	0.062 (PC)[b] 0.063 (LAM)[c] 0.062 (MLAM)[d]	0.011 (PC)[b] 0.011 (TOF)[f]
$k_U \pm \Delta k_U$	0.150 ± 0.004	0.062 ± 0.001	0.011 ± 0.001
$\Delta D_G / D_G$	± 5%	± 0.1%	± 0.1%
$k_U(n_{th})$	Small	Small	Large (9)
Applicability	Chamber-size dependent	Less chamber-size dependent	Tube & filter dependent
Z_{Wall}	6	12	High (~ 26)
Δh_U	± 2%	± 5%	± 20%
^{60}Co γ-ray calib.	1.135×10^9 R/C	8.39×10^8 R/C	8.76×10^{-7} R/count
Dose rate	0.001 – 10 Gy/min	0.001 – 10 Gy/min	< 2 mGy/min
Saturation correction	Columnar (~ 10%)	Volume (< 1%)	Dead time (< 30 μs)
Gas	CO_2 flow	Air flow	Halogen-filled
Measurement instrument	Charge electrometer	Charge electrometer	Pulse counter

[a] With the photon energy compensation filter of Sn and Pb in an acrylic resin sleeve, as suggested by the manufacturer.
[b] (PC) Proportional counter technique by Ito [8].
[c] (LAM) Lead attenuation method by Attix et al. [10].
[d] (MLAM) Modified lead attenuation method by Waterman et al. [11].
[e] (Calc.) Theoretical calculation by Kawashima et al. [20].
[f] (TOF) Time-of-flight technique by Ito [14].

chambers there is some chamber-size dependency on the dose conversion factor [15]. But, the chamber-size dependency is within a few per cent for the conventional chamber size (0.1 through 10 ml). There are no data on the chamber-size dependency of the Mg-Ar chamber. For the 18529 GM counter, there are some similar GM tube and different photon energy compensation filters, such as ZP1300, ZP1100, 18550 (Philips) or MX163 (Mullard), commercially available. The published values of k_U of these GM counters compared well with the $k_U(E_n)$ data obtained with the TOF method [14]. Except for an extreme case, most k_U values agreed within ± 30% at neutron energies around 5 and 15 MeV.

If the k_U values are to be determined accurately enough to give an accurate estimation of the photon dose, we must pay more attention to the energy dependence of the photon dose sensitivity, $h_U(E_\gamma)$. In this connection, the C-CO_2 has the best advantage over the Mg-Ar chamber and the GM counters. Although GM counters are usually covered with photon energy compensation filters, the $h_U(E_\gamma)$ is not constant compared with ionization chambers [16].

For practical use, some characteristics of the photon dosimeters are compared in Table I. The C-CO_2 chamber has columnar recombination in the neutron beam. The amount of recombination loss could be measured and extrapolated by using the inverse of the applied voltage. This amounts to several per cent in the d(14) + Be neutron field in air. The Mg-Ar chamber has the least problem in the recombination loss. As far as the GM counters are concerned, however, they have limitations in the dose rate (up to 2 mGy/min) and necessitate correction for the dead time. For practical paired dosimetry, the TE-TE ionization chamber is exclusively used at high dose rate, at least 100 mGy/min. The GM counters cannot be used simultaneously with the TE-TE chamber.

Considering all the points listed in Table I, we are using the Mg-Ar chamber for routine dosimetry.

3. PHOTON DOSE FRACTION IN AIR

The photon dose fraction in air has been repeatedly measured by different methods. Figure 8 shows the photon dose fraction relative to the total (neutron plus photon) dose (D_G/D_T) as a function of the field size in the "equivalent circular radius", as measured with the TE-TE and Mg-Ar pair dosimetry at the source-to-surface distance (SSD) of 150 cm. The lowest curve (open circles (○)) in Fig.8 shows the D_G/D_N in air. As the field size increases, the photon fraction increases slightly from 1.1 to 1.8%. This increase can be attributed to the photons originating from the internal surface of the collimator. And the field size independent portion of about 1.0% may be divided into two parts. One is the

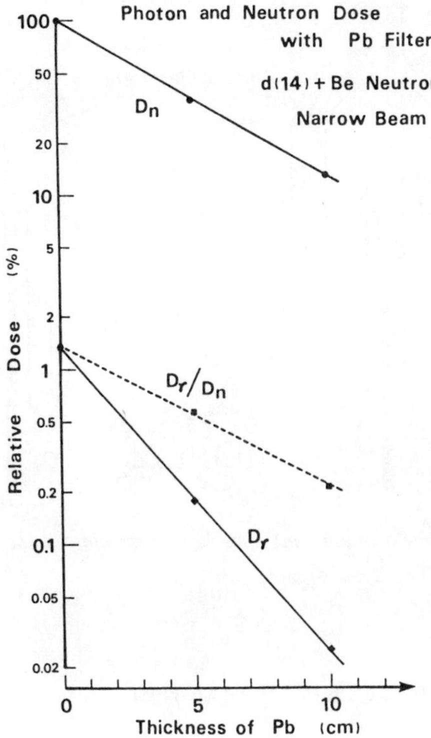

FIG.4. Filtering of the photon component in the mixed neutron and photon field. By increasing the thickness of the lead filter, the fraction of photon dose decreases.

scattered photons from the collimator body. This portion is estimated to be about 0.5% from the measurements with the collimator plug as described in Section 2.2. And the other part is attributed to the photons generated in the Be target (prompt gammas).

For calibration purposes, it is preferable to have the test neutron field that has the very low photon dose contamination using the lead filter. The photon contamination will be filtered out owing to the difference in attenuation coefficients between neutrons and photons. Figure 4 shows the fractional photon dose as a function of the thickness of the lead filter measured with the proportional counter technique. The photon dose fraction decreases down to 0.5% with a 5-cm-thick lead filter. To achieve this low photon contamination, the narrow neutron collimator and long SSD are needed to avoid the scattered photons from the neutron collimator.

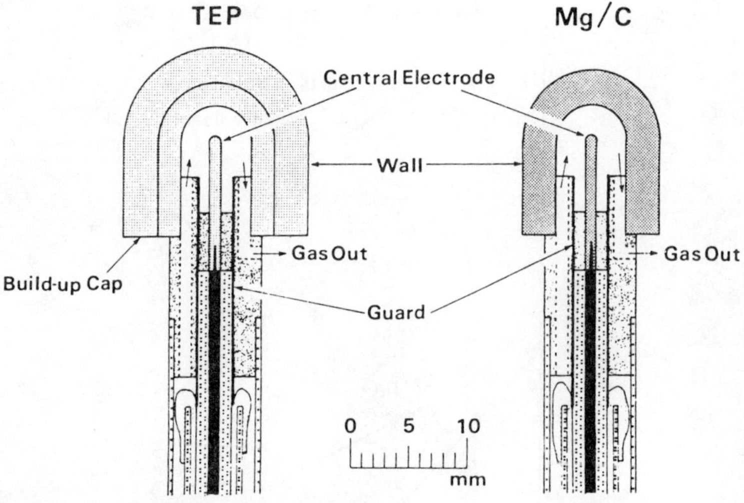

FIG.5. Cross-sectional view of the 0.1 mL spherical ionization chambers used for the pair dosimetry in the phantom.

4. PHOTON DOSE FRACTION IN PHANTOM

We measured the neutron and photon dose distributions in the Frigerio solution (density 1.07 g/cm^3) [17] as the basic data for the treatment planning of the fast neutron therapy [18]. As has already been discussed in Section 2.4, we are using the Mg-Ar ionization chamber as the neutron insensitive dosimeter. To improve the spatial resolution of dosimetry in phantom, we have recently adopted the 0.1 mL TE-TE and Mg-Ar paired dosimetry system. Figure 5 shows the cross-sectional view of the laboratory-made 0.1 mL spherical ionization chambers (6 mm in dia.). Care was taken in the design of the central electrode insulator and the guard electrode to reduce the leakage current. Also, to reduce the extracameral ionizations, the central electrode is not exposed to air at all, except for the sensitive chamber volume. Precise measurements of the collected charge, especially for the Mg-Ar chamber, are required. The leakage current from both chambers is about 2×10^{-14} A in normal operating conditions. If we are to measure the dose as precisely as 0.5% of the peak dose rate of about 0.2 Gy/min, the ionization charge must be measured as precisely as possible. Also, as the measurements of the neutron and photon dose distributions in the phantom are frequently performed for clinical purposes, we have developed a computer-controlled paired dosimetry system. Figure 6 illustrates the block diagram of the system. Owing to the recent progress in micro-computer technology and also to the standardized system for data acquisition and interfaces, a complicated

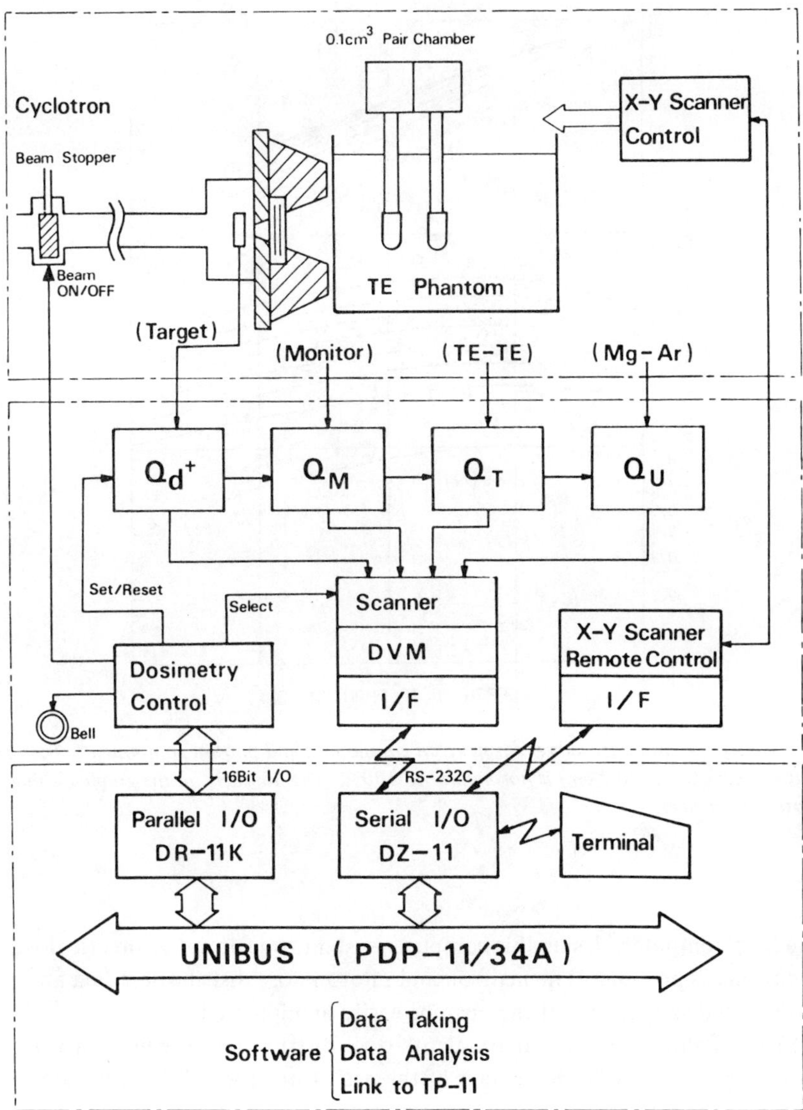

FIG.6. A block diagram for the automated paired dosimetry system, used for clinical dosimetry.

dosimetry system can be configured easily. The ionized charges from the TE-TE and Mg-Ar chambers, together with charges in the monitor chamber and Be target, are fed into four electrometers. The output voltages are read off by the digital voltmeter by the host computer, through the GP-IB interface. To increase the precision of the charge measurements, the neutron beam on and off is controlled

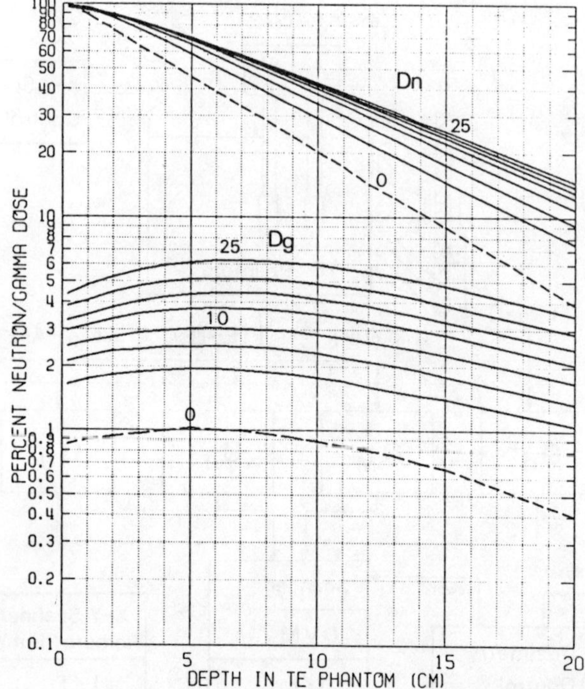

FIG.7. Per cent depth dose curves for neutron component and photon component. The total dose at the peak depth (0.2 cm) is normalized to 100%. The numbers on the graph are the irradiation field sizes.

by the host computer. Using this computer system, we can get as precise dose distributions as possible. The neutron and photon dose distribution data are stored in the disk file, so that they can be easily manipulated.

Figure 7 shows the central axis depth dose distributions for neutrons and photons, separately, as the percentage of the peak total dose. The curves are measured at the different field sizes ranging from 5 cm diameter circular to 25 × 25 cm^2 square field at SSD = 150 cm. The photon dose fraction increases as the field size increases. The photon dose has the peak at around 5 cm depth. This means that the photons in the phantom are mainly generated through the capture of the slowed-down neutron by the hydrogen ($H(n,\gamma)$) reaction. Separate measurements of the thermal neutron flux distribution in a phantom by Au activation foils support this assumption.

Figure 8 shows the plot of photon dose fraction of the total dose at each depth of measurement. In this plot it is clearly seen that the relative photon dose fraction increases almost linearly as the equivalent field radius increases. For

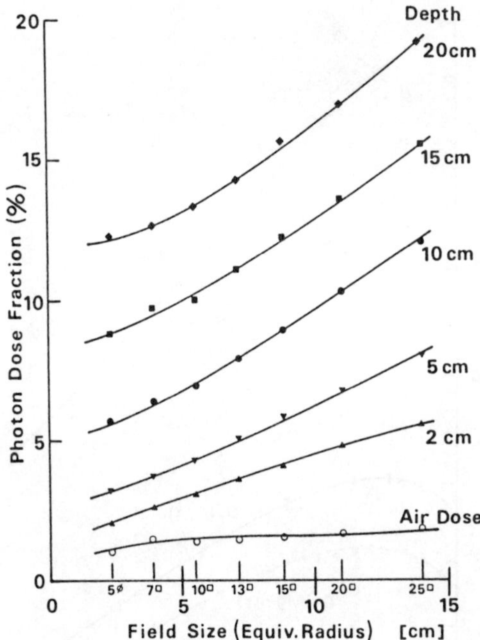

FIG.8. Fractional photon dose, D_G/D_T, as a function of the field size as the "equivalent radius". Graphs are shown for the measurements in air (○) and at the depths in the TE phantom.

neutron therapy of small superficial tumours, i.e. tumours in the head and neck region or melanoma, the photon contribution is minimal (around 3%). However, for deep-seated tumours with a bigger field, i.e. lung cancer, the photon contribution will not be negligible.

Mijnheer and Broerse compared the photon dose fraction data among fast neutron therapy centres [19]. There are some discrepancies in photon dose fractions among the similar fast neutron centres, namely, Tokyo, Essen, Edinburgh and London, having a similar d(~ 15) + Be neutron source. Differences in the neutron collimator design would yield some differences in photon dose fraction in the entrance region, because the photon contributions from the collimator are different. Hence, the direct photon dose from the collimator will influence the phantom dose at the surface. Other causes for the differences are attributed to the differences in the phantom materials used and also to the differences in the technique employed for the photon dosimetry. Thus, we measured the photon dose fraction in the water phantom and compared it with the data measured in the Frigerio solution. The differences are found to be small (within 0.5%). Also, to check the difference in the technique in photon dosimetry, the photon dose fraction was measured using the GM counter as described in

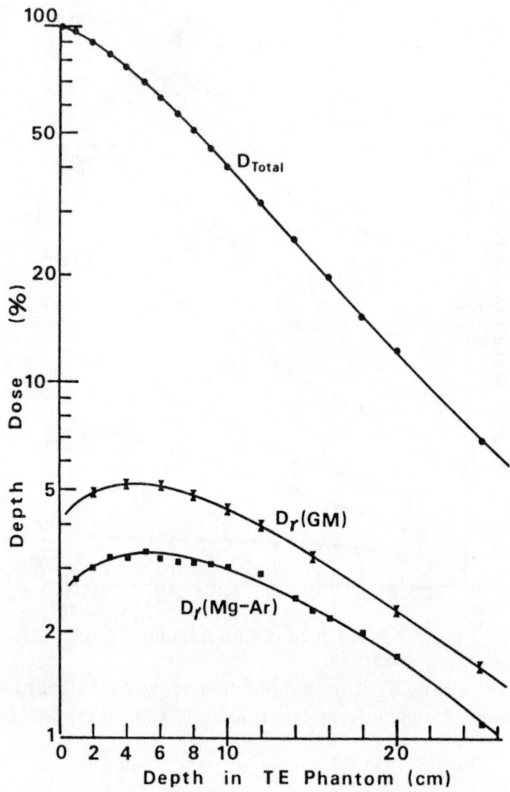

FIG.9. Difference in photon dose fraction between the Mg-Ar chamber and the GM counter as the neutron insensitive dosimeter. Overestimation by the GM counter is attributed to the excessive response of the GM counter to the thermal neutron.

Section 2.2. The measurements were made without any thermal neutron shield. Figure 9 compares the results for both the Mg-Ar chamber and the GM counter measurements. The higher photon dose fraction is estimated with the GM counter. The overestimation of the photon dose measured with the GM counter could be attributed partly to the excessive response of the GM counter to the thermal neutrons. Another possibility is the difference in the photon energy dependence on the photon dose sensitivity $h_U(E_\gamma)$ of the GM counter [16].

5. CONCLUSION

We have determined the k_U values for the three types of neutron insensitive dosimeter, namely the C-CO_2 and Mg-Ar ionization chambers as well as a

GM counter. The k_U values were determined by the three independent methods. The proportional counter technique was used in all cases. For the ionization chambers, the lead attenuation method was applied. The TOF technique was successfully applied for the GM counter. The k_U values thus determined for the d(14) + Be neutron beam are 0.150 ± 0.004, 0.062 ± 0.001 and 0.011 ± 0.001 for the C-CO_2 and the Mg-Ar chambers, and the GM counter, respectively. In addition, the $k_U(E_n)$ values for the GM counter for neutrons with energies between 1 and 18 MeV were determined. These $k_U(E_n)$ compared well with the published data for a similar GM counter and photon energy compensation filter. The characteristics of these three dosimeters are compared. From the practical point of view, the Mg-Ar ionization chamber is found to be most useful.

The photon dose fractions in the d(14) + Be therapy beam have been extensively studied. The photon fraction in air increases from 1.1 to 1.8% as irradiation field size increases. The photon dose fraction in a TE phantom is a function of field size and the depth in the phantom, ranging from 2% at the shallow depth with small field size to 20% at the deeper region and with the large field size. Attention should be paid to the possible change of k_U value in the phantom.

ACKNOWLEDGEMENTS

The fast neutron therapy programme at IMS was supported by the Ministry of Education of the Japanese Government. The continuous encouragement and valuable discussions of Dr. Y. Iino, head of the Cyclotron Unit, are gratefully acknowledged. The excellent operation of the IMS cyclotron by Mr. Y. Kashiwa and H. Takasaki is also acknowledged. Also, gratitude is expressed for the assistance in practical dosimetry by Mr. S. Ohtsu.

REFERENCES

[1] INTERNATIONAL COMMISSION ON RADIATION UNITS AND MEASUREMENTS, ICRU Rep. 26, Neutron Dosimetry for Biology and Medicine, ICRU, Washington, D.C. (1977).

[2] AMERICAN ASSOCIATION OF PHYSICISTS IN MEDICINE, Protocol for Neutron Beam Dosimetry, AAPM Report No. 7, Task Group 18, Radiation Therapy Committee, (1980).

[3] ECNEU, European Protocol for Neutron Dosimetry for External Beam Therapy, European Clinical Neutron Dosimetry Group, EORTC (1981).

[4] ITO, A., IMS cyclotron, Jpn. J. Cancer Clin. **23** (1977) 258.

[5] YOSHIDA, Y., NAKAYAMA, H., NAKAI, K., YAMAZAKI, T., Beam-pulsing system for the IMS cyclotron, Nucl. Instrum. Methods **138** (1976) 579.

[6] ROSSI, H.H., ROSENZWEIG, W., A device for the measurement of dose as a function of specific ionization, Radiology **64** (1955) 404.

[7] WEAVER, K., BICHSEL, H., EENMAA, J., WOOTTON, P., Measurement of photon dose fraction in a neutron radiotherapy beam, Med. Phys. **4** (1977) 379.

[8] ITO, A., "Neutron sensitivity of $C\text{-}CO_2$ and Mg-Ar ionization chamber", Proc. Third Symp. Neutron Dosimetry in Biology and Medicine, Munich, 1977, EUR 5848 (1978) 113.

[9] ITO, A., HENKELMAN, R.M., Microdosimetry of the Pion Beam at TRIUMF, Radiat. Res. **82** (1980) 413.

[10] ATTIX, F.H., THEUS, R.B., ROGERS, C.C., Measurement of Dose Components in an n-γ Field, NRL Progress Rep. (Dec. 1974).

[11] WATERMAN, F.M., KUCHNIR, F.T., SKAGGS, L.S., Comparison of two independent methods for determining the n/γ sensitivity of a dosimeter, Phys. Med. Biol. **22** (1977) 880.

[12] HOUGH, J.H., A modified lead attenuation method to determine the fast neutron sensitivity k_U of a photon dosimeter, Phys. Med. Biol. **25** (1979) 734.

[13] KOYAMA, H., ITO, A., Measurements of the Neutron Energy Spectra from the IMS Cyclotron, IMS Progress Rep. (1977).

[14] ITO, A., The determination of the neutron dose sensitivity of a Geiger-Müller counter between 1 and 18 MeV, Phys. Med. Biol. (to be published).

[15] RUBACH, A., BICHSEL, H., ITO, A., Neutron dosimetry with spherical ionization chambers. Part V: Experimental results and comparison with calculations, Phys. Med. Biol. (to be published).

[16] ZOETELIEF, J., Dosimetry and Biological Effects of Fast Neutrons, Thesis, 1981, p.67.

[17] FRIGERIO, N.A., COLEY, R.F., SAMPSON, M.J., Depth dose determination I. Tissue-equivalent liquids for standard man and muscle, Phys. Med. Biol. **17** (1972) 792.

[18] ITO, A., KUMASAWA, A., IINO, Y., "Treatment planning system for the fast neutron therapy at the Institute of Medical Science, University of Tokyo", Treatment Planning for External Beam Therapy with Neutrons (BURGER, G., Ed.), Urban & Schwarzenberg, Munich (1981) 227.

[19] MIJNHEER, B.J., BROERSE, J.J., "Dose distributions of clinical fast neutron beams", High LET Radiation in Clinical Radiotherapy (BARENDSEN, G.W., BROERSE, J.J., BREUR, K., Eds), Suppl. Eur. J. Cancer, Pergamon Press, Oxford (1979) 109.

[20] KAWASHIMA, K., HOSHINO, K., HIRAOKA, T., MATSUZAWA, H., HASHIZUME, T., ITO, A., ALMOND, P.R., SMATHERS, J.B., BICHSEL, H., The second neutron dosimetry intercomparison between Japan and USA, Jpn Radiol. Phys. **1** (1981) 31.

DEMONSTRATION OF A FAST METHOD FOR EVALUATING THE GAS-TO-WALL ABSORBED DOSE CONVERSION FACTOR VERSUS CAVITY SIZE

M. MAKAREWICZ, S. PSZONA
Radiation Protection Department,
Institute of Nuclear Research,
Swierk, Poland

Abstract

DEMONSTRATION OF A FAST METHOD FOR EVALUATING THE GAS-TO-WALL ABSORBED DOSE CONVERSION FACTOR VERSUS CAVITY SIZE.

On the basis of data for an infinitesimal cavity and for infinite cavities, the conversion factor dependence on cavity size has been derived. Using this derivation the conversion factor was evaluated versus cavity size of a TE ionization chamber exposed to 5, 9 and 14.5 MeV neutrons. The stopping-power ratios and the partial kerma by particle type data have been used as input data. Comparison of the results obtained with those obtained by means of another method as well as with results by another author shows good agreement. It has been shown that TE cavities of 0.1 to 15 cm^3 in volume, in 5 to 14.5 MeV neutron fields, cannot be assumed to be infinitesimal or infinite, otherwise up to 3% uncertainties in the value of the conversion factor may be introduced. It was confirmed that in this range of cavity size the conversion factor changes its values by less than 0.5, 2 and 2.5% for 5, 9 and 14.5 MeV neutrons, respectively.

1. INTRODUCTION

There is no doubt that neither the Bragg-Gray theory nor the Fano theorem can be applied without reservations when the sensitivity of ionization chambers, commonly used for determining the absorbed dose in a biological medium exposed to fast neutrons, is being evaluated. The TE-TE, C-CO_2 chambers are not fully homogeneous and are in general not infinitesimal. It is necessary to replace the Bragg-Gray relation by [1]

$$D_w = D_g/r \tag{1}$$

where $1/r$ is the gas-to-wall absorbed dose conversion factor, which relates the absorbed dose D_w in the wall to the absorbed dose D_g in a gas-filled non-infinitesimal cavity of an ionization chamber.

The influence of cavity size on the response to neutrons has been clearly demonstrated for C/CO_2 [2], TE/Air [3] and TE [3, 4] ionization chambers by Monte Carlo methods. Dependence of the 1/r factor on TE cavity size in the fast neutron energy range up to 20 MeV is discussed in Ref. [5], where a very simple estimation of r is proposed. This estimation is based on the result that for some charged secondaries a cavity is infinitesimal whereas for another particle it is infinite and gives constant r values within definite size limits of a parallel-plate ionization chamber (p-p ionization chamber).

In the method employed for this work and reported in Ref. [6] it is assumed that each kind of charged secondary has its "gas" and "wall" component of energy deposited in a cavity and the proportion of the components changes with the cavity size. Thus, the method gives the conversion factor as a continuous function of cavity size.

The main aim of this paper is to demonstrate the method and to check its validity for an example of a TE ionization chamber, employing commonly available average stopping-power ratios and the best available partial kerma by particle type data.

2. METHOD

The method of calculating the gas-to-wall absorbed dose conversion factor versus cavity size of a p-p ionization chamber with an infinite surface has been reported in Ref. [6]. The reciprocal of the conversion factor, r, is given by the following expression

$$r = \frac{1}{K_w} \sum_{i,j} \left[K_{i,j}\, w_i^g\, r_{i,j} + K_{i,j}\, w_i^w\, (1 - r_{i,j}) \frac{1}{\rho_{w,g}^{i,j}} \right] \qquad (2)$$

where

w_i^g w_i^w – is the weight content of the nuclide of type i in the gas and in the wall, respectively.

$K_{i,j}$ – is the partial kerma factor by particle type, i.e. the energy released by charged particles of type j in interaction of neutrons with nuclides of type i, per unit mass of the nuclide and per unit neutron fluence (it should be noted that, for example, alpha particles coming from the (n, α) reaction at two different excitation levels are distinguished and they are characterized by two different values of j).

$\rho_{w,g}^{i,j}$ – is the average wall-to-gas stopping power ratio for an ij particle.

K_w – is the kerma in the wall material.

$r_{i,j}$ – is the ratio of energy deposited in a gas cavity to the sum of the initial energy of ij particles originating in a cavity.

The following expression has been assumed for an estimation of $r_{i,j}$

$$r_{i,j} = \frac{1}{1 + c\, R_{i,j}/d} \qquad (3)$$

where $R_{i,j}$ is the range in a gas of the most penetrating ij particle, and d is the distance between the walls of a p-p ionization chamber. Such an estimation of $r_{i,j}$ seems to be justified by the dependence of D/K on R/d (Figs 1 and 2). The symbol D/K is the $r_{i,j}$ratio defined above and appears as "gas contribution" in the figures. The "wall distribution" has been obtained in a similar way: D/K means here the ratio of energy deposited in a gas cavity by ij particles originating in a chamber wall to the sum of its initial kinetic energy. It is worth noting that the results were obtained by means of the Monte Carlo method for a p-p ionization chamber of infinite surface, and that only elastic and inelastic scattering of neutrons have been assumed as non-isotropic reactions in the centre-of-mass system. The last assumption implies rectangular or semi-rectangular primary particle spectra. The dotted lines in Figs 1 and 2 are the fits to the points: $R_{i,j}$ for the gas contributions and $(1 - r_{i,j})$ for the wall contributions according to Eq.(3): c = 0.376 for 1 MeV neutrons and c = 0.310 for 15 MeV.

3. INPUT DATA

Two sets of primary data have been employed, which may be recognized as the most reliable at present: the average stopping-power ratios of Bichsel and Rubach [7] and the kerma values by particle types of Coyne [8]. "Miscellaneous", which appears in Table I of Ref. [8] has been added to the partial kerma of carbon, nitrogen and oxygen recoils in elastic scattering reactions proportionally to the contribution of these elements in ICRU tissue kerma. The absolute values of kerma for elements have been taken from ICRU Report 26 [9]. These sets of stopping-power ratio data and those of partial kerma values determine the neutron energies to be considered here to be 5, 9 and 14.5 MeV. The reactions considered were assumed to take place in the ground state with the exception of inelastic scattering at oxygen (excitation energy of 6.13 MeV) and of the $^{16}O(n, \alpha)$ reaction for 14.5 MeV neutrons (7.19 MeV). The ranges of charged secondaries were calculated by means of the SPAR code [10] without correction of the stopping powers for phase effect (as done previously [6]). For ^{13}C and ^{15}N recoils the range values in TE gas for ^{12}C and ^{14}N have been used. The numerical values for the composition of a TE chamber made of A-150 plastic and filled with a methane based gas mixture are the same as those used by Bichsel and Rubach [7]. The calculations were performed at a value of 0.310 for the parameter c which appears in $r_{i,j}$ (Eq. (3)).

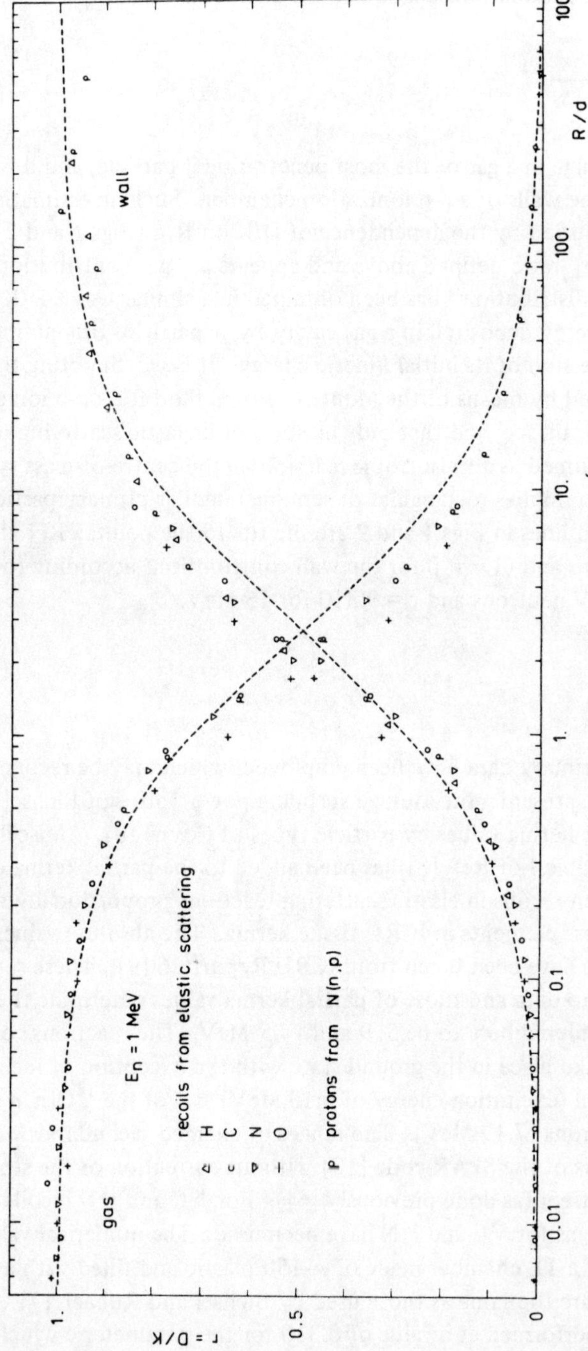

FIG.1. The ratio of energy deposited in a gas cavity to the sum of initial kinetic energy of charged particles originating in a gas, or in walls of a TE p-p ionization chamber, for 1 MeV neutrons (see Section 2).

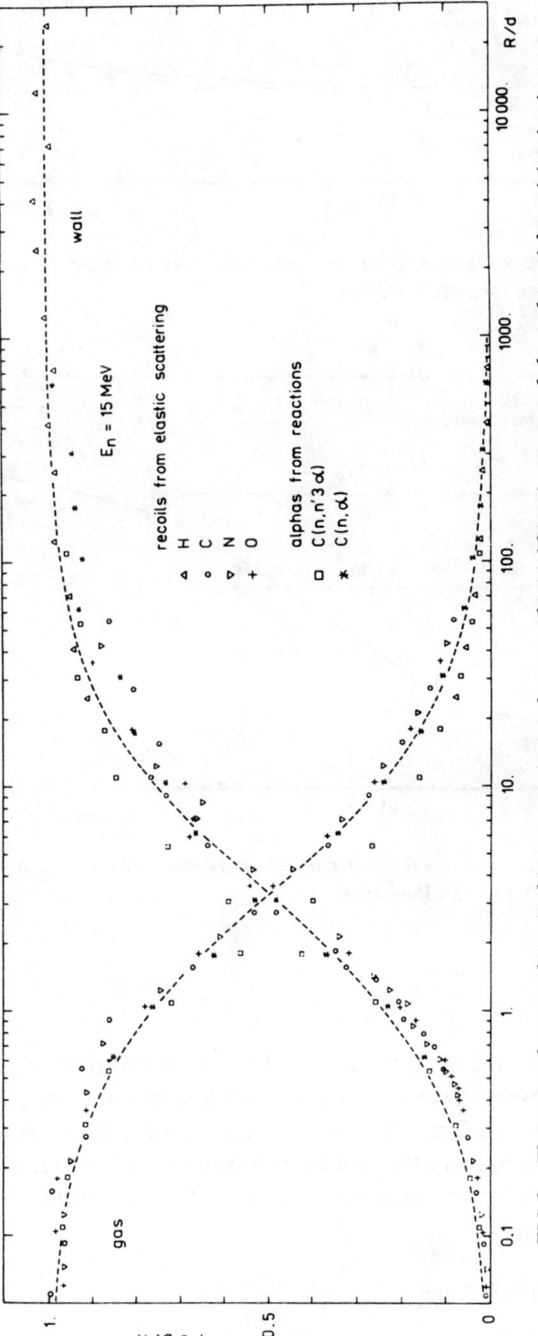

FIG.2. The ratio of energy deposited in a gas cavity to the sum of initial kinetic energy of charged particles originating in a gas, or in walls of a TE p-p ionization chamber, for 15 MeV neutrons (see Section 2).

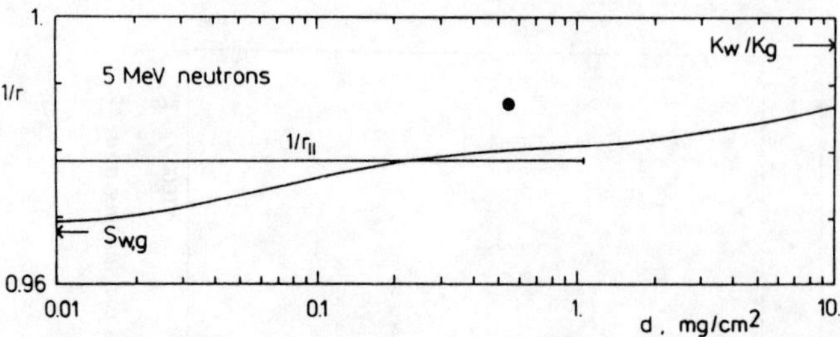

FIG.3. The gas-to-wall conversion factor 1/r versus cavity size d of a p-p TE ionization chamber (see Section 4), for 5 MeV neutrons.

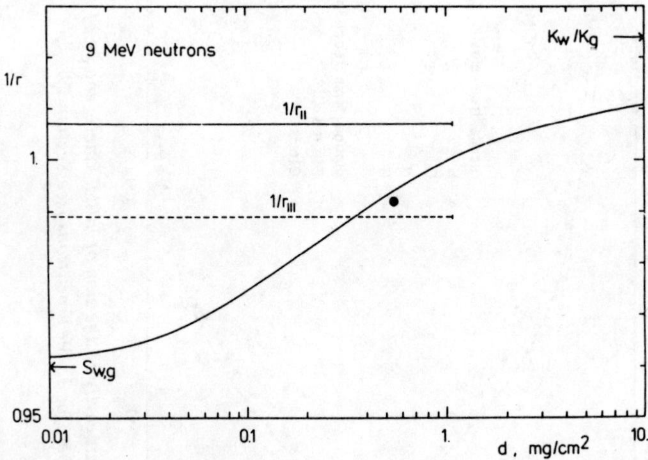

FIG.4. The gas-to-wall conversion factor 1/r versus cavity size d of a p-p TE ionization chamber (see Section 4), for 9 MeV neutrons.

4. RESULTS

The results obtained for 5, 9 and 14.5 MeV neutrons are shown as solid curves in Figs 3–5 respectively. They are compared with results of calculations done by Bichsel and Rubach for a 2 cm^3 TE spherical chamber [7] (black points) and with those performed according to the procedure recommended in Ref. [5] (sectors). According to the recommendation the reciprocal of the conversion factor r was determined as follows:

(a) $E_n < 1$ MeV, $d > 0.2$ cm

$$r_I = K_g/K_w \tag{4a}$$

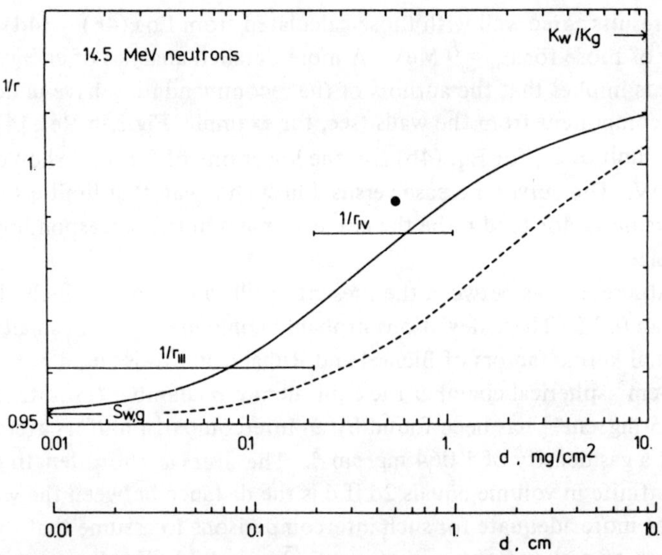

FIG.5. The gas-to-wall conversion factor 1/r versus cavity size d of a p-p TE ionization chamber (see Section 4), for 14.5 MeV neutrons.

(b) $1\ \text{MeV} \leqslant E_n < 10\ \text{MeV},\ d < 1\ \text{cm}$

$$r_{II} = \frac{K_w^p}{K_w}\ \frac{1}{\rho_{w,g}^p} + \frac{K_g^\alpha}{K_w} + \frac{K_g^{CNO}}{K_w} \tag{4b}$$

(c) $10\ \text{MeV} \leqslant E_n < 20\ \text{MeV},\ d < 0.2\ \text{cm}$

$$r_{III} = \frac{K_w^p}{K_w}\ \frac{1}{\rho_{w,g}^p} + \frac{K_w^\alpha}{K_w} + \frac{K_g^{CNO}}{K_w} \tag{4c}$$

(d) $10\ \text{MeV} \leqslant E_n < 20\ \text{MeV},\ 0.2\ \text{cm} < d < 1\ \text{cm}$

$$r_{IV} = \frac{r_{II} + r_{III}}{2} \tag{4d}$$

where p, α and CNO indexes refer to the protons, alphas and heavy recoils, and d is the distance between the walls of a p-p TE ionization chamber, E_n is the neutron energy.

The results agree well with those calculated from Eqs (4b) – (4d) with the exception of those for E_n = 9 MeV. A more detailed analysis of energy-deposited components implies that the authors of the recommendation have underestimated the alpha component from the walls (see, for example, Fig.3 in Ref. [4]). Thus, the higher limit of E_n for Eq. (4b) and the lower one of Eq. (4c) should be shifted to 8–9 MeV. The curves increase versus d in such a way that limit estimations of the conversion factor tend to be the curve averages in the corresponding limits of cavity size.

The discrepancies between the present results and those of Bichsel and Rubach are less than 0.7%. These deviations probably come mainly from small differences in the partial kerma factors of Bichsel and Rubach and those used in this work. For the 2 cm^3 spherical chamber the equivalent p-p chamber (infinite in volume) of d = 0.55 $mg \cdot cm^{-2}$ has been found by an intercomparison of its average chord lengths, at a gas density of 1.064 $mg \cdot cm^{-3}$. The average chord length of a p-p chamber infinite in volume equals 2d if d is the distance between the walls. It seems to be more adequate for such intercomparisons to assume that the average chord length of such cavities lies between $\sqrt{2}d$ and 2d. For these limits the discrepancy in the values of the conversion factor is less than 0.5% for 14.5 MeV and 0.4% for 9 MeV neutrons.

The influence of the shape of the initial charged particle spectra on the results is shown in Fig. 5. Two curves have been obtained by different treatments of the spectrum of alphas coming from the $^{12}C(n, n')3\alpha$ reaction. The dashed curve was obtained in the standard procedure for this method, whereas the solid one takes account of a correction for the shape of the initial alpha spectrum, which differs greatly from that of a rectangle (see Section 2). The primary alpha spectrum for the reaction [11] was divided into semi-rectangular segments [11] and then each of these was treated by the standard method.

The sensitivity of the conversion factor to changes in the value of c (Eq. (3)) has been investigated over a wide c range: c = 0.310 ± 0.100. This change in c value corresponds to +50% for c = 0.210 and −25% for c = 0.410 and has the same effect as a change of cavity size d, or systematic change of gas stopping powers for all charged particles considered (with conservation of $\rho_{w,g}^{i,j}$ ratios) by the same percentages. The uncertainty of the conversion factor in terms of "square root of the sum of squares" was calculated at these c limits and the following dispersion of 1/r values, in the cavity size range of 0.2 $mg \cdot cm^{-2}$ to 1 $mg \cdot cm^{-2}$, was obtained:

from ±0.1% to ~ 0% for 5 MeV neutrons,

from ±0.5% to ±0.3% for 9 MeV, and

from ±0.5% to ±0.6% for 14.5 MeV.

The main sources of the uncertainties come from: alphas from $^{12}C(n, \alpha)^9Be$ reaction for 9 MeV neutrons, alphas from $^{12}C(n, n')3\alpha$ for 14.5 MeV, as well as carbon recoils from elastic scattering for 5, 9 and 14.5 MeV neutrons.

If we suppose that most ionization chambers used for dosimetry in biology and medicine have volumes between 0.1 cm^3 and 15 cm^3, then the most interesting region of the conversion factor lies in the range of d values from 0.20 $mg \cdot cm^{-2}$ to 1.1 $mg \cdot cm^{-2}$ (spherical cavity to parallel-plate cavity conversion was employed – as above). In this range of cavity sizes the conversion factor changes its values by less than 0.3% for 5 MeV, 1.8% for 9 MeV and 2.3% for 14.5 MeV neutrons. The differences between 1/r values (in terms of the first derivative, for example) seem to be more reliable than the values of the conversion factor. This makes it possible to use the known gas-to-wall conversion factor of a defined chamber for the determination of the 1/r value for another one by applying a small correction. Thus, if we want to determine the conversion factor, for example, for a 0.1 cm^3 spherical cavity on the basis of that for a 2 cm^3 one, then the 1/r value should be decreased by 1.1% for 9 MeV neutrons or by 1.4% for 14.5 MeV. If the value of the conversion factor is considered, for a TE ionization chamber in 9 or 14.5 MeV neutron fields, it is not possible either to approximate a 0.1 cm^3 cavity by the average stopping-power ratio, or a 15 cm^3 one by the kerma ratio, since a 2–3% error in the conversion factor may be introduced.

In view of the present results the values of conversion factors for cavities of 0.1 cm^3 to 15 cm^3 occupy the middle part of the conversion factor range and these are still "medium size" cavities in 5 MeV to 14.5 MeV neutron fields.

5. CONCLUSIONS

(i) In view of the results shown, the method presented gives a satisfactory approximation of the gas-to-wall conversion factor as a function of cavity size.

(ii) The method is weakly sensitive to uncertainties in the constant parameter c. The errors in estimated values of the conversion factor arise mainly from uncertainties in partial kerma values and those in average gas-to-wall stopping-power ratios for different kinds of charged secondaries. The (n, n), (n, d) and (n, n')3α reactions in carbon should be considered especially carefully.

(iii) After deciding on a common measure of size for cavities, the data shown make it possible to convert the well-known gas-to-wall conversion factor for a defined cavity to that for the cavity of interest.

(iv) TE cavities of 0.1 cm^3 to 15 cm^3 in volume, in 5 MeV to 14.5 MeV neutron fields, cannot be assumed to be infinitesimal or infinite, otherwise up to 3% uncertainty in the value of the conversion factor may be introduced. It has been confirmed that in this range of cavity size the conversion factor changes its value by less than 0.5%, 2% and 2.5% for 5 MeV, 9 MeV and 14.5 MeV neutrons, respectively.

(v) For the precise evaluation of the gas-to-wall conversion factor with respect to cavity size, further stopping-power ratios and partial kerma data are needed for other neutron energies. More details on "miscellaneous" (Table I [8]) are desirable, especially when a C/CO_2 cavity is considered.

(vi) The method needs further verification. The results of the paper are preliminary and are essentially to be considered as an aid for intercomparing the two methods. To obtain more correct results for the absolute value of the conversion factor, the partial kerma of pair production in the $^{16}O(n, n')$ reaction should be taken into account and a more appropriate partition of "miscellaneous" should be made. In particular, that part of "miscellaneous" connected with the $^{12}C(n, n')$ reaction may considerably influence the results.

(vii) The method is easy to use, simple, fast (saves computer time) and allows new conversion factor values to be obtained rapidly from new values of partial kerma and/or stopping-power ratios.

REFERENCES

[1] BICHSEL, H., RUBACH, A., Basic Physical Data for Neutron Dosimetry, EUR 5692 (1976) 121.

[2] MAKAREWICZ, M., PSZONA, S., Nucl. Instrum. Methods **153** (1978) 423.

[3] MAKAREWICZ, M., PSZONA, S., in Proc. 6th Symp. Microdosimetry, 1978, Vol. 1, CEC, Harwood Academic Ltd. (1978) 549.

[4] PSZONA, S., MAKAREWICZ, M., "Influence of size of cavity on the response of partially homogeneous tissue-equivalent ionization chambers for fast neutrons", Biomedical Dosimetry: Physical Aspects, Instrumentation, Calibration (Proc. Symp. Paris, 1980), IAEA, Vienna (1981) 139.

[5] PSZONA, S., MAKAREWICZ, M., Phys. Med. Biol., to be published.

[6] MAKAREWICZ, M., PSZONA, S., in Proc. 4th Symp. Neutron Dosimetry (BURGER, G., EBERT, H.G., Eds) Vol. 2, EUR-7448, Commission of the European Communities (1981) 307.

[7] BICHSEL, H., RUBACH, A., in Proc. 3rd Symp. Neutron Dosimetry in Biology and Medicine (BURGER, G., EBERT, H.G., Eds), EUR-5848, Commission of the European Communities (1978) 549.

[8] COYNE, J.J., Ion Chambers for Neutron Dosimetry, EUR 6782 (1980) 195.

[9] INTERNATIONAL COMMISSION ON RADIATION UNITS AND MEASUREMENTS, Neutron Dosimetry for Biology and Medicine, ICRU Rep. 26 (1977) 74.

[10] ARMSTRONG, T.W., CHANDLER, K.C., Oak Ridge National Laboratory, ORNL-4869 (1973).

[11] CASWELL, R.S., COYNE, J.J., Radiat. Res. **52** (1972) 448.

PROPORTIONAL COUNTER MEASUREMENTS IN NEUTRON THERAPY BEAMS

H.G. MENZEL
Fachrichtung Biophysik und Physikalische Grundlagen der Medizin,
Boris Rajewsky Institute,
Saarlande University,
Homburg, Saar,
Federal Republic of Germany

Abstract

PROPORTIONAL COUNTER MEASUREMENTS IN NEUTRON THERAPY BEAMS.

Dosimetry for clinical neutron therapy requires a characterization of radiation quality in addition to the specification of absorbed dose. Generally, a very simple approach has been adopted which consists in separating total absorbed dose into neutron and photon fractions. This is explained by the requirement of clinical dosimetry to apply methods suitable for routine measurements, by the lack of generally accepted improved alternatives, and by the fact that radiation quality is only one of several problems in neutron therapy not sufficiently solved. Spectra measured with low-pressure tissue-equivalent proportional counters (experimental microdosimetry) provide a detailed description of the physical properties of the radiation field at neutron therapy facilities. These descriptions are suitable for explaining the influence of different parameters (collimation, field size, phantom) on radiation quality. Although the physical properties of the radiation field as described by the measured microdosimetric distributions and quantities are not the only properties relevant for radiation effects, in general there are reasons to believe that they provide a suitable radiation quality characterization for the limited range of applications in neutron therapy.

1. INTRODUCTION

In clinical neutron therapy a characterization of radiation quality in terms of the dose fractions of neutrons and photons is given in addition to the specifications of absorbed dose in one or more reference points. The fact that this very simple treatment of the problem of radiation quality is considered to be adequate for radiation therapy by many therapists and physicists involved is explained by several reasons, which include:

(1) Radiation quality is only one of many problems which are of clinical relevance and are not yet sufficiently solved. Others are, for example, the temporal distributions of administering the dose and the differences in the Relative

Biological Effectiveness (RBE) for different types of tissue and biological endpoints, for instance early versus late effects;

(2) Despite almost four decades of research to understand radiation mechanisms, and thus radiation quality, our knowledge still does not allow us to predict induced biological effects only on the basis of measured physical parameters of a radiation field;

(3) At most neutron therapy facilities changes of radiation quality within phantoms, which would be of clinical significance, have not been observed [1, 2];

(4) The practical requirements of clinical neutron dosimetry do not allow the routine performance of sophisticated and laborious physical measurements.

To exploit fully the potential benefits of fast neutron therapy, however, the related complex problems, including that of radiation quality, have to be solved. This paper discusses which role experimental microdosimetry, i.e. measurements with low-pressure tissue-equivalent (TE) proportional counters, can play in this context.

Phenomenologically, radiation quality is related to the type and energy of the radiation. More precisely, radiation quality is closely linked to the microscopic pattern of energy depositions which result from the discontinuous energy transfers in discrete interactions of the primary radiation and its secondary particles with the biological structures. To date, however, no manageable description of these microscopic distributions has been established which could be generally and quantitatively correlated to the effectiveness of different radiations. An understanding of the relation between the physical properties of radiations and the induced biological mechanisms and effects is part of current research and therefore an area of controversial views. In practical situations, therefore, practical solutions are required. For instance, in radiation protection a pragmatic method to account for radiation quality is applied by using the quantities quality factor and dose equivalent [3]. In neutron therapy, radiobiological intercomparisons between different neutron irradiation facilities [4] and so-called biological dosimetry [5] have been carried out.

The microdosimetric proportional counter measures the spectra of ionizations, which are produced in the gas cavity by the secondary charged particles released in the interactions of incoming single neutrons and photons with primarily the counter wall. The chosen pressure of the tissue-equivalent gas in the proportional counters can be so low that the energy loss of a charged particle traversing the counter along its diameter is equal to that in a tissue volume of dimensions of around 1 μm. The spectra are thus closely correlated to the energy deposition distributions in small tissue-like volumes and represent therefore a particular description of the above-mentioned microscopic pattern of energy distributions. Certain radiobiological indications had led to the belief that the physical properties of radiations, as described by these distributions, are closely related to the physical properties of radiations which are relevant for radiation quality [6, 7]. Although the general validity of this assumed link has not been found [8] and research in this field is still continuing, this paper shows that the purely physical information obtained with TE proportional counters could be of practical importance for neutron therapy and that, for the limited range of application in neutron therapy, microdosimetric quantities are suitable for specifying radiation quality.

After describing the experimental method and the measured quantity, the results of microdosimetric measurements at several neutron therapy facilities are presented. They are used to discuss systematically the influence of different parameters such as collimation, field size and phantom on radiation quality. An important aspect is the intercomparison of results for different irradiation facilities. The paper ends with a discussion on microdosimetric radiation quality parameters which may be applied in neutron therapy.

2. EXPERIMENTAL METHOD

2.1. The tissue-equivalent (TE) proportional counter

Most proportional counters used in experimental microdosimetry are similar in construction to the counter originally introduced by Rossi and Rosenzweig [9]. Such Rossi counters are spherical, have walls made of A-150 TE plastic and the central collecting wires are usually surrounded by a helix to provide homogeneity of the electrical field. The counting gases are either a methane- or propane-based TE gas mixture. The gas pressure is chosen so that the product of gas density and the geometrical diameter of the counter is equal to the equivalent product for a tissue sphere of unit density and diameter of the order of 1 μm. Thus, a charged particle traversing the gas cavity experiences an energy loss equal to that of an identical particle traversing a tissue sphere along an equivalent trajectory. In a neutron or mixed neutron-photon radiation field, the majority of charged particles originate from interactions of the primary radiation with the wall material which usually has a thickness to provide charged particle equilibrium. Depending on the energy of the primary radiation, and on the simulated diameter, there is a certain fraction of the charged particles which starts or stops inside the cavity.

The ionizations produced by the secondaries of a single primary particle in the gas cavity are detected as an electrical pulse with a pulse height proportional to the number of primary ionizations. In principle, the physical processes induced by external radiation in TE ionization chambers used in neutron dosimetry and in TE proportional counters are identical. The proportional counter, however, enables one to measure not only the total electric charge produced by a large number of primary particles, but also to register and measure the ionization produced by each interacting primary particle (single event spectra).

There are commercially available Rossi-type counters.[1] A particular model (LET-1/2), which is suitable for measurements inside phantoms filled with liquid, was used in the investigation reported here. It has an inner diameter of 12.7 mm (0.5 in.) and a built-in alpha particle source which can be used for calibrating the counter. The gas pressure in all measurements was chosen so that a diameter of 2 μm was simulated. To calibrate with the ^{244}Cm alpha source an energy loss of 176 keV was assumed for this diameter. Further details can be found in earlier publications [2, 10].

2.2. Lineal energy spectra

The number of ionizations produced by a charged particle traversing the gas cavity and the energy imparted to the gas are proportional (W value). Thus, the calibration of the proportional counter with the help of the known energy deposition of alpha particles coming from a collimated source, enables the measurement of energy imparted and lineal energy spectra in small volumes of tissue-like material. The microdosimetric quantity lineal energy, y, is defined [3] as the quotient of the energy imparted by a single primary particle and its associated secondaries to the considered small volume and the mean chord length in that volume, and is measured in $keV \cdot \mu m^{-1}$. The mean chord length in a sphere is two thirds of the diameter.

However, the conversion of the measured ionization yield spectra into lineal energy spectra is distorted by the fact that the W value is dependent on the type and energy of the charged particles. Although this dependence is principally quite well known, it is in practice extremely difficult to deconvolute this dependence from measured spectra. Following a generally adopted convention the results in this report are nevertheless presented in terms of lineal energy. This is equivalent to the assumption that the W value of the alpha particles used in the calibration is a good approximation for the W values of the secondary charged particles released by the primary radiation. For neutrons of energies used for therapy, and for the average W value, this assumption only introduces a small error [11]. For the spectra, however, the average W value is only of limited significance and the dependence of W on the energy of charged particles introduces distortions. The significance of the results presented here, and the conclusions, are not significantly affected by this systematic uncertainty.

The measured spectra are closely related to the slowing-down spectra of the charged particles released in the counter wall, and thus also to the composition and energy spectra of the primary radiation. It is therefore possible to understand the spectra in terms of other physical properties of the radiation such as neutron energy and photon dose fraction [12, 13]. To illustrate this in Fig. 1 four micro-

---

[1] Far West Technology, Goleta, California, USA.

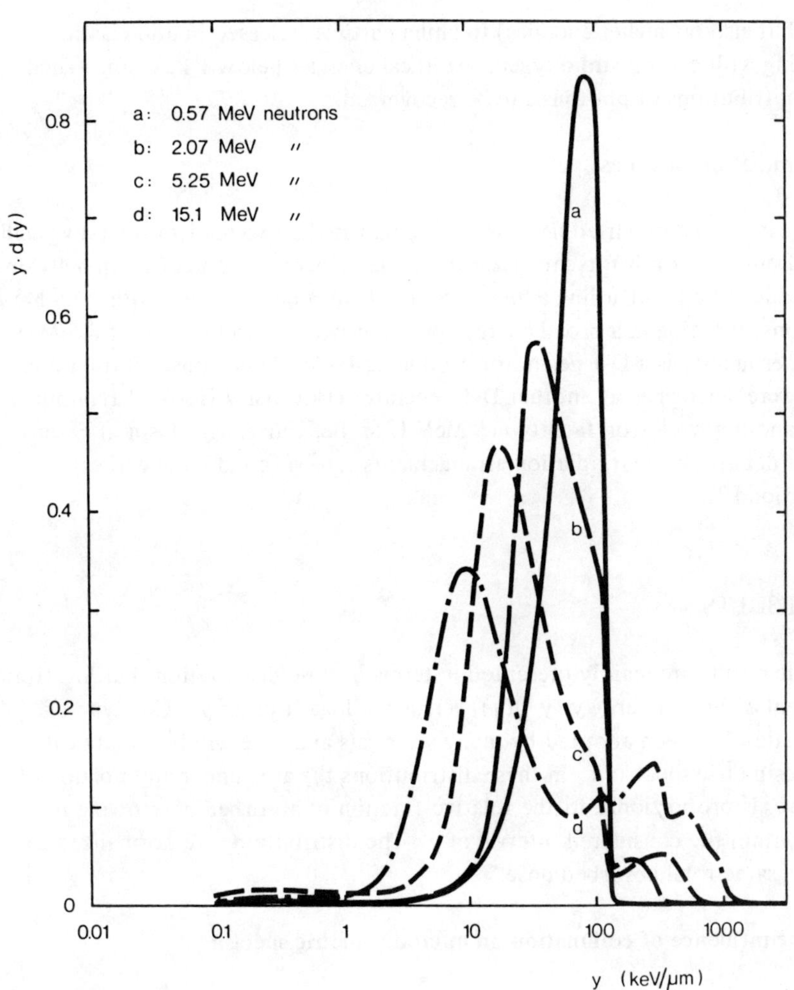

FIG.1. Microdosimetric spectra for monoenergetic neutrons of 0.57, 2.07, 5.25 and 15.1 MeV measured in the low scattering environment of the neutron irradiation facility of Gesellschaft für Strahlen- und Umweltforschung (GSF), München-Neuherberg (simulated diameter: 2 μm) [13].

dosimetric spectra, measured with monoenergetic neutrons between 0.57 and 15.1 MeV, are set out. The increase of the mean energy of the recoil protons and the corresponding decrease in stopping power with increasing neutron energy are obviously related to a shift of the peak in the proton distribution from about 100 keV·μm^{-1} to about 10 keV·μm^{-1}. The dose contribution above a lineal energy of 140 keV·μm^{-1}, which corresponds to the maximum energy deposition by protons in the gas sphere ("proton edge"), increases markedly with neutron energy. This is related to the increasing energy of elastically scattered heavy nuclei but also (at higher energies) to alpha particles released in non-elastic scattering with carbon and oxygen. At lineal energies below 1 keV$\cdot \mu m^{-1}$ small dose contributions of photons can be recognized.

2.3. Irradiation facilities

Most results presented here were obtained at the two neutron therapy facilities of the Deutsches Krebsforschungszentrum, Heidelberg. At one of them neutrons are produced by bombarding a high-pressure deuterium gas target with 10.5 MeV deuterons, resulting in a broad energy spectrum with a mean energy of 8.5 MeV. The other facility is a D-T generator producing 14 MeV neutrons. Further measurements were performed at another D-T generator (University Hospital Hamburg) and at another cyclotron facility (15 MeV D on Be, University Hospital Essen). Further details on the irradiation arrangements can be found in an earlier publication [2].

3. RESULTS

The results are mainly presented in terms of dose distributions per logarithmic increment of lineal energy y (y·d(y)) versus the logarithm of y. This type of presentation has been adopted by many scientists and here will be simply called microdosimetric spectrum. In these distributions the area under any portion of the curves is proportional to the relative fraction of absorbed dose owing to events within the considered interval of y. The distributions are normalized to have the same total absorbed dose.

3.1. The influence of collimation on microdosimetric spectra

Scattering of neutrons in the collimator and the surrounding wall material will introduce changes to the energy spectrum of the primary neutrons. This is reflected in microdosimetric spectra, an example of which is given in Fig. 2, where two spectra measured at D-T neutron generators are compared. At one facility (GSF, Munich, dotted line) no collimation was used and the irradiation arrangement has a low scattering environment. The other was the collimated beam at the Heidelberg generator, KARIN (solid line), with walls of the irradiation room nearby. Although a difference also exists in the primary neutron energy (GSF: 15.1 MeV; Heidelberg: 14 MeV) it is recognizable that there is a dose contribution of low energy neutrons owing to scattering in collimator and walls.

Microdosimetric spectra measured free in air at different lateral distances to the axis of the collimated cyclotron neutron beam (field size: 6×8 cm^2) in Heidelberg are shown in Fig. 3. At 5 cm distance to the axis the counter is still

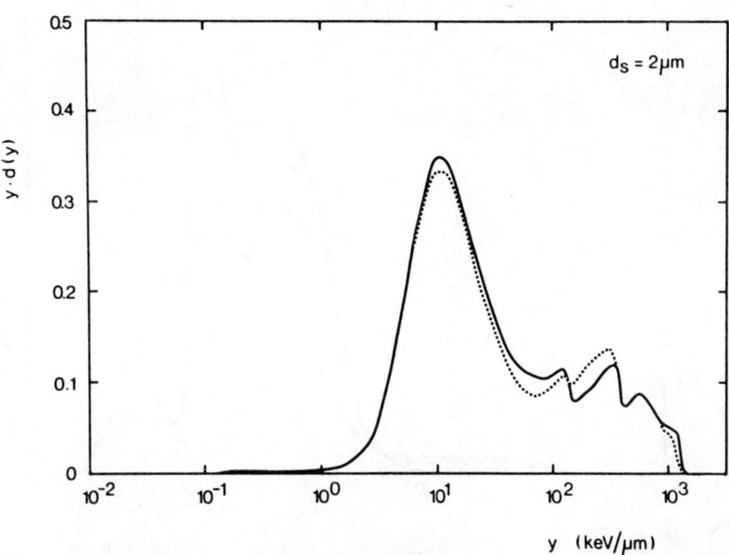

FIG.2. Microdosimetric spectra for 15.1 MeV non-collimated neutrons (dotted line) and collimated 14 MeV neutrons (D-T neutron generator KARIN at Deutsches Krebsforschungs-zentrum (DKFZ), Heidelberg; solid line) (simulated diameter d_s = 2 μm).

partially inside the direct beam and the resulting spectrum (b) is not markedly different from that obtained on the beam axis. An increase of the distance by a further centimetre results in a substantial change of the spectrum (c). A significant fraction of the dose is due to protons with lineal energies just below the proton edge, and the dose contribution with y values above the proton edge is considerably reduced. Both effects must be explained by a decrease in the mean neutron energy. At 12 cm distance to the beam axis (d) the dose is almost entirely due to low-energy neutrons. Below 1 keV$\cdot\mu m^{-1}$ a small increase of the photon dose fraction with increasing lateral distance can be observed.

The dependence on lateral distance shown in Fig. 3, however, is not representative for all neutron therapy collimators [2]. The influence of the collimator on both the photon dose fraction and the mean neutron energy depends strongly on the material used in the collimator. This was shown in an intercomparison between different therapy facilities [2] and can be seen directly in Fig.4. At the D-T neutron generator in Heidelberg the spectra were measured free in air at 12 cm distance to the beam axis (i.e. outside the direct beam) for a collimator with a wood insert (dashed line) and a steel insert (solid line) to define the field size. The distinct difference in the spectra has to be understood as a large difference in mean neutron energy. It should be mentioned that the lateral profiles of total absorbed dose are very similar for both types of collimator insert.

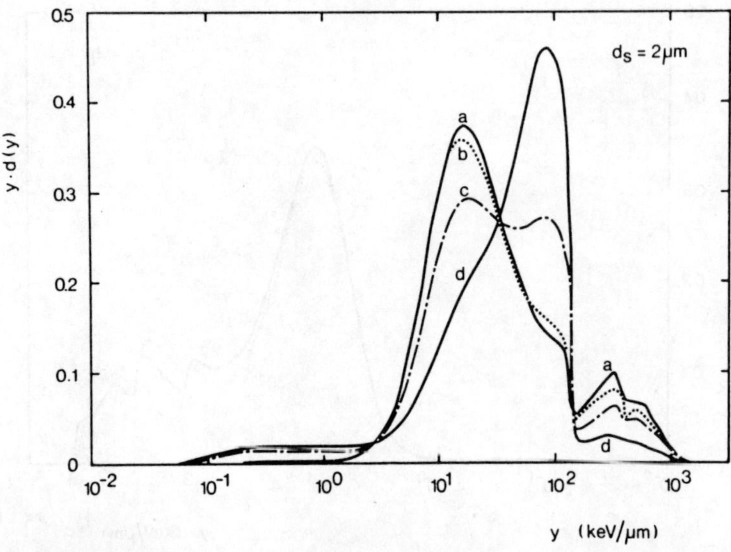

FIG.3. Microdosimetric spectra measured free in air at (a) 0; (b) 5; (c) 6; and (d) 12 cm lateral distance to the collimated beam (field size: 6 × 8 cm²) of cyclotron-produced neutrons at DKFZ, Heidelberg (d_s = 2 μm).

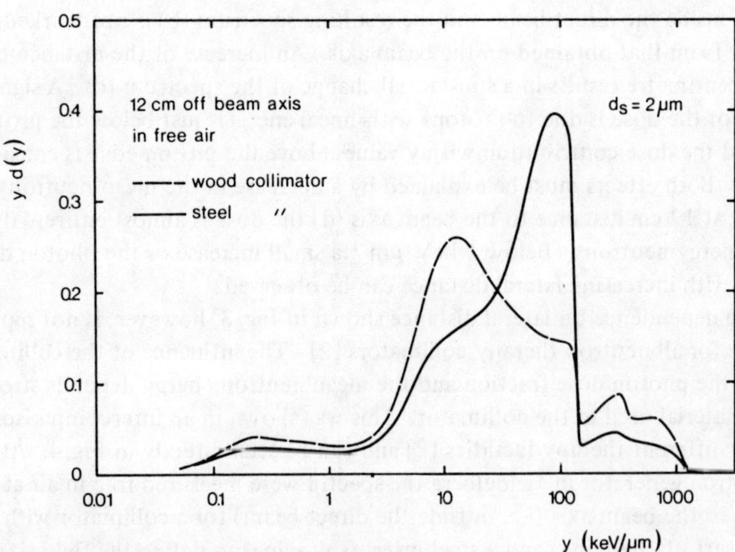

FIG.4. Microdosimetric spectra measured at the D-T generator of DKFZ, in free air outside the direct beam in identical positions with a wooden collimator insert (dashed line) and a steel collimator insert (solid line) (d_s = 2 μm).

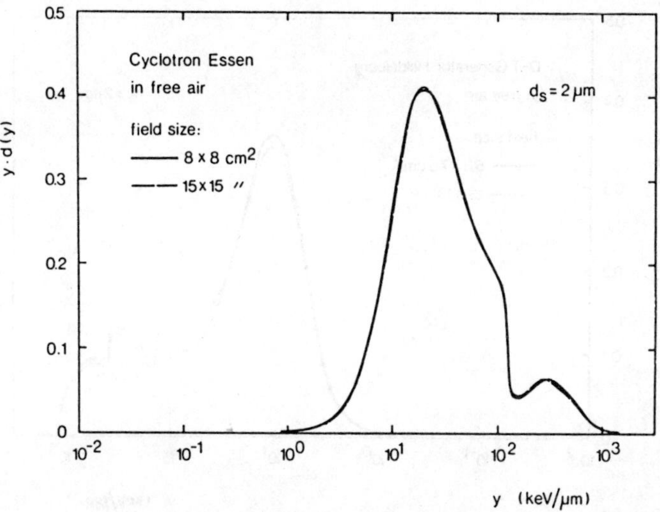

FIG.5. Comparison of two microdosimetric spectra measured with different field sizes at the cyclotron-produced neutron beam of the University Hospital, Essen (d_s = 2 μm).

In Figs 5 and 6 it is shown that there is no general answer to the question whether or not there is a dependence of field size on radiation quality. At the cyclotron facility in Essen (Fig. 5) no difference exists in the spectra for the two field sizes. In contrast, at the D-T generator in Heidelberg the increase of the field size led to changes in the spectra (Fig. 6) which correspond to a decrease in the neutron energy. It should be noted that the spectra shown in Figs 5 and 6 contain only neutron-produced events. The photon component was subtracted following a procedure described earlier [12] and the resulting spectrum of the pure neutron component was normalized again.

3.2. The influence of phantoms on microdosimetric spectra

3.2.1. Measurements inside the direct beam

All results presented so far were measured free in air, i.e. without the influence of the neutron scattering taking place in a patient or a phantom. To study this influence measurements were performed inside a water phantom ($30 \times 30 \times 30$ cm^3).

Figure 7 shows the influence of neutron scattering only in a phantom for 15.1 MeV neutrons. The measurements were carried out at the uncollimated beam in the low scattering environment of the neutron irradiation facility of GSF, Munich. As may have been expected, a dependence on depth in the phantom is

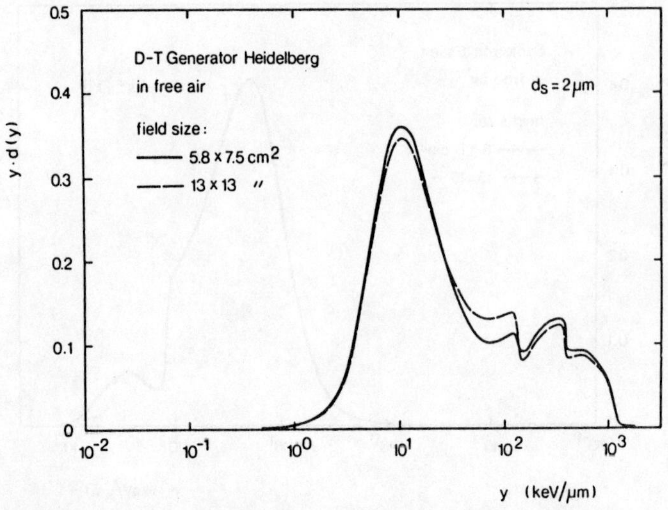

FIG.6. Comparison of two microdosimetric spectra measured with different field sizes at the D-T generator of DKFZ (d_s = 2 μm).

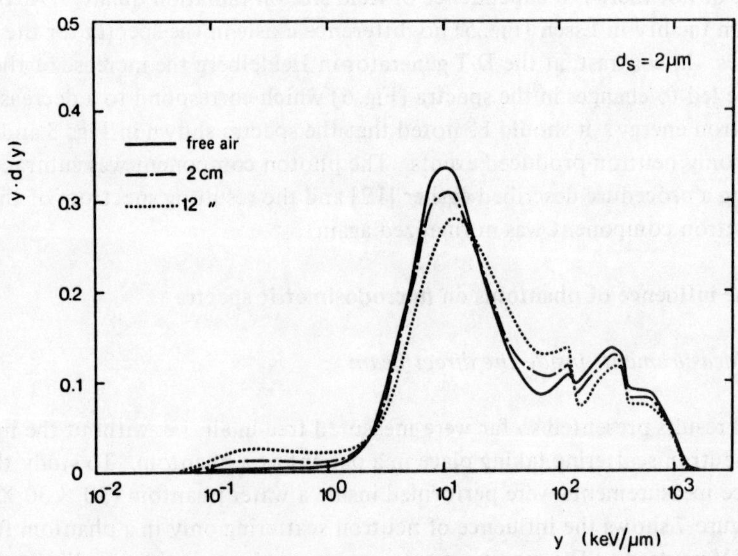

FIG.7. Microdosimetric spectra measured for 15.1 MeV non-collimated neutrons (low scatter irradiation facility of GSF) in free air and two depths in a phantom (2 cm and 12 cm) (d_s = 2 μm).

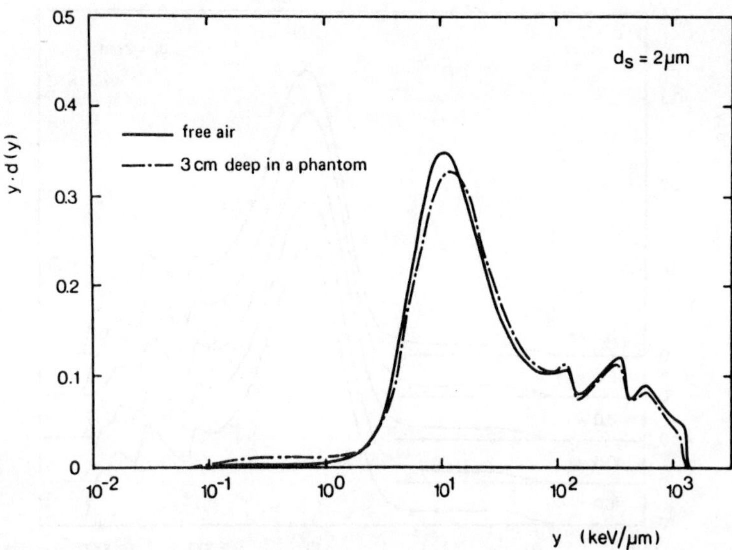

FIG.8. Microdosimetric spectra measured in the collimated beam of the D-T generator of DKFZ free in air and at 3 cm depth in a phantom (field size 5.8 × 7.5 cm²) (d_s = 2 μm).

observed which reflects an increase of the low-energy neutron component and of the photon dose fraction. Similar results have been presented by Maier et al. [14].

If similar results were to be expected at neutron therapy irradiation facilities the change in radiation quality corresponding to the change in microdosimetric spectra would have to be taken into account in treatment planning. However, at most installations such changes have not been observed [1, 2, 14]. This is shown in Figs 8 and 9. Figure 8 compares microdosimetric spectra measured free in air (solid line), and at 3 cm depth in a phantom (dashed line) for the collimated beam (field size: 5.8 × 7.5 cm²) at the D-T neutron generator in Heidelberg. The difference between the spectra is much less pronounced than that between the two corresponding spectra in Fig. 7. For depths in phantom between 3 and 16 cm and a 5 × 5 cm² field, spectra measured on the beam axis are shown in Fig. 9. The particular way of presenting the spectra was chosen because of the very small difference between them which would have made them indistinguishable in a conventional presentation. Only an increase in the photon dose fraction is recognizable.

The different dependence on penetration in a phantom for non-collimated and collimated neutron beams indicates that radiation quality inside phantoms is a complex function of neutron scattering, and thus will depend on the irradiation geometry and the materials in the environment. A possible explanation for the fact that there are no changes with depth for the collimated 14 MeV beam is that

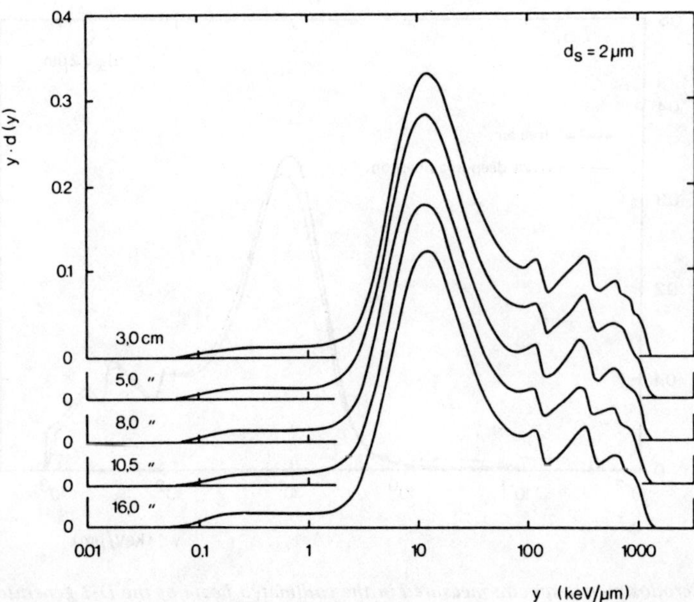

FIG.9. Microdosimetric spectra measured at 3.0, 5.0, 8.0, 10.5 and 16.0 cm phantom depths on the axis of the collimated beam of the D-T neutron generator of DKFZ (field size: 5 X 5 cm²) (d_s = 2 µm).

there is also, in absence of a phantom, a considerable low-energy neutron dose fraction due to collimator and wall scattering. More generally, changes with depth in phantom appear to be particularly pronounced for monoenergetic neutrons and are of less importance for broad neutron energy spectra. Whether or not this conclusion also holds for the very high energy neutrons to be used at recently installed therapy facilities remains to be investigated.

3.2.2. *Measurements outside the direct beam*

Regions outside the directly irradiated volume defined by the collimation of the neutron beam are also exposed to radiation. This is due to neutrons leaking through the collimator shielding (Figs 3 and 4), to neutrons scattered in the phantom, and to photons which mainly originate from neutron capture in hydrogen. To assess radiation-induced biological effects in normal tissues it is also essential to know both the total absorbed dose and the radiation quality outside the direct beam.

In Fig. 10 microdosimetric spectra are set out which have been measured at a depth of 8 cm in a phantom at lateral distances to beam axis between 0 and 15.5 cm in the collimated neutron beam (field size: 10 X 15 cm²) of the cyclotron

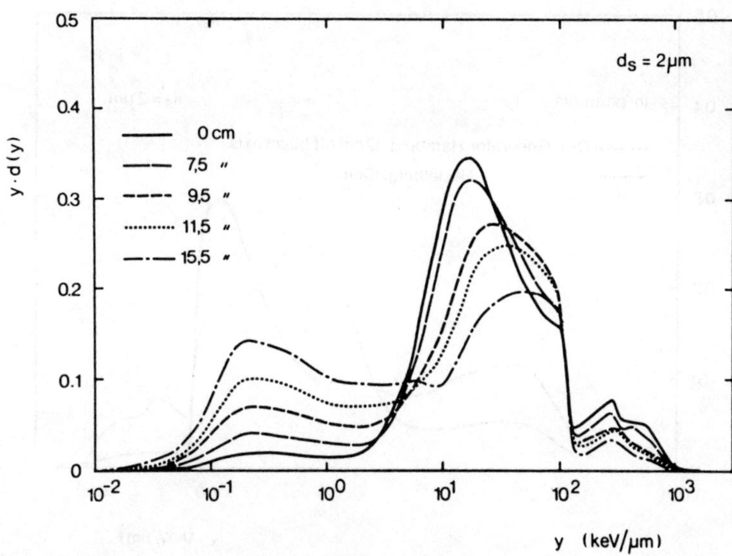

FIG.10. Microdosimetric spectra measured at a depth of 8 cm in a phantom and at lateral distances of 0, 7.5, 9.5, 11.5 and 15.5 cm to the axis of the collimated beam of the cyclotron-produced neutrons at DKFZ (field size: 10 × 15 cm²) (d_s = 2 μm).

facility in Heidelberg. As must be expected, the neutron scattering leads to significant changes with increasing distance which correspond to a substantial decrease in neutron energy and an increase of the photon dose fraction. Principally, similar results are obtained at other neutron therapy sources. The actual radiation quality in a given position, however, is influenced by the energy of the primary neutron and the field size, because neutron scattering in the directly irradiated volume depends on both parameters.

The spectra shown in Fig. 10 can be qualitatively explained by the neutron scattering in the directly irradiated volume of the phantom. There are, however, also leakage neutrons from the collimator shielding which may cause a change in radiation quality. In Fig. 11 two spectra are compared which have been measured in comparable positions outside the direct beams of the D-T neutron generators in Heidelberg (dashed line) and in Hamburg (solid line). The difference in collimator material (wood in Heidelberg and steel in Hamburg) leads to differences in radiation quality outside the direct beam, not only free-in-air (see Fig. 4) but also in a phantom [2].

3.3. Comparison of neutron therapy facilities

It is generally accepted that the comparability of clinical results from different neutron therapy facilities requires a common basis for neutron dosimetry, and a

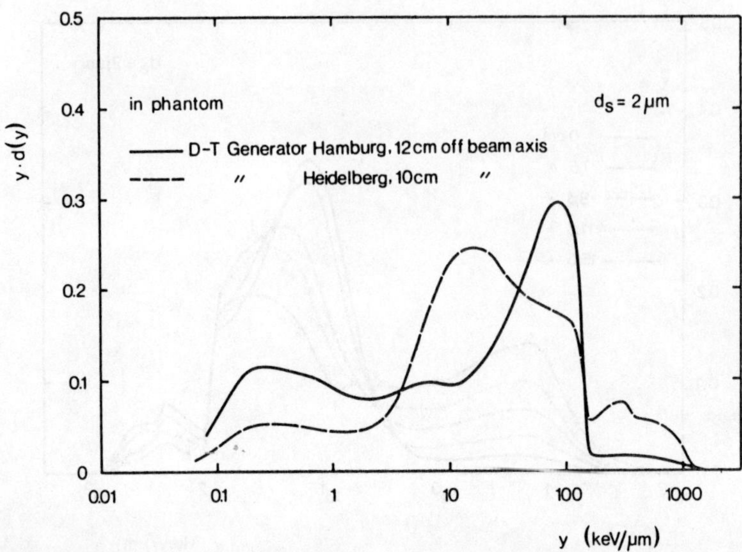

FIG.11. Microdosimetric spectra measured in comparable positions outside the direct beam and in a phantom at the D-T neutron generators of University Hospital Hamburg (solid line) and DKFZ (dashed line) (d_s = 2 μm).

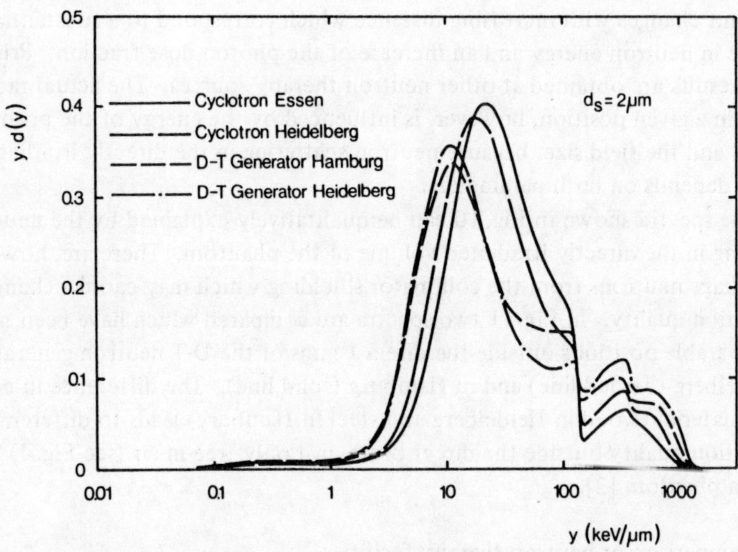

FIG.12. Microdosimetric spectra measured free in air on the axis of the collimated neutron beams of the therapy facilities at the University Hospitals in Essen and Hamburg and DKFZ, Heidelberg (d_s = 2 μm).

correspondingly considerable effort has been made towards improving the precision and accuracy of dose measurements for clinical applications. Comparatively little effort has been made to investigate the respective radiation quality. There have been, however, some radiobiological intercomparisons. In an earlier publication some results of an intercomparison of microdosimetric properties measured at four different therapy facilities were presented [2]. As an example, the comparison of the spectra measured free-in-air on the beam axis is shown in Fig. 12. The difference in mean neutron energy of the primary beams is clearly reflected in these spectra. There is, however, a recognizable difference between the two 14 MeV sources which must be explained by differences in the distance to walls and floors [2]. In general, the microdosimetric intercomparison has shown that, at a given facility, the separation of total absorbed dose into photon and neutron fractions may be an adequate radiation quality specification, but that the applicability of this procedure appears to be doubtful for comparing the clinical results obtained at different facilities.

In Table I values for y^* are compared, which have been measured in free air and at different positions inside the water phantom at the four neutron therapy facilities. As is discussed below in more detail, y^* is a microdosimetric average which takes into account the dependence of mammalian cell killing on linear energy transfer (LET) in a particular way [7]. Therefore, this average cannot be considered to be purely a physical quantity. It has been suggested that this parameter may be a useful radiation quality specification for applications in neutron therapy [1, 15]. It is remarkable that even clear differences in the measured distributions do not yield large differences in y^*. By contrast, the dose average lineal energy, $\bar{y}_D$, for the pure neutron component reflects the differences in the spectra more clearly [2].

4. DISCUSSION

The microdosimetric spectra and comparison of spectra presented in this paper have shown that measurements with low-pressure TE proportional counters are very sensitive to small changes of the primary radiation field with regard to the composition and to neutron energy. By using knowledge of the physical processes taking place in the counter wall and gas, one is able to explain the changes in the microdosimetric spectra in terms of the generally better-known parameters, relative photon dose fraction [12] and mean neutron energy [13]. The results shown illustrate that radiation quality described in terms of microdosimetric spectra is a complex function of primary neutron energy and neutron scattering in the environment. The geometry of the irradiation arrangement (collimator, field size, distance to walls and floors, phantom) therefore has a clearly detectable influence on radiation quality at any point of interest. A particular result of this

TABLE I. VALUES OF y* FOR NEUTRONS FROM THERAPY IRRADIATION FACILITIES MEASURED FREE IN AIR AND INSIDE A PHANTOM

y* has been evaluated for 2 μm simulated diameter using a value of 125 keV/μm for y_0 [7]. y_2^* in keV/μm. The field sizes were comparable (between 8 × 8 cm^2 and 10 × 10 cm^2).

Neutron therapy facility/ neutron-producing reaction		Univ. Essen d(11) + Be	DKFZ Heidelberg d(11) + D_2	Univ. Hamburg d(0.5) + T	DKFZ Heidelberg d(0.25) + T
Position in phantom					
Depth (cm)	Lateral displacement (cm)				
1.5	0	31.5	29.2	–	–
5	0	31.6	29.2	25.9	25.4
10	0	30.3	28.5	24.9	25.4
15	0	28.7	27.8	24.7	25.0
20	0	27.3	27.2	24.2	24.8
5	12	22.8	28.4	20.4	25.8
Free in air		30.9	29.1	27.3	26.4

investigation is that the collimator material not only determines the lateral dose profiles, but is also important for the radiation quality, at least outside the direct beam.

The high sensitivity of the physical detector TE proportional counter, which enables subtle changes in the physical properties of the radiation fields to be revealed, makes this instrument attractive for neutron dosimetry [12, 13]. At present it is, however, a not completely solved question as to what extent the observed changes are of significance for neutron therapy. Several aspects have to be considered to solve this problem.

The first is the question of the general suitability of the quantity of lineal energy measured for volumes with diameters of around 1 μm to serve as radiation quality parameter. As already mentioned here, this aspect is still the subject of current research [8, 16]. At present it is a widely accepted point of view that a description of the microscopic pattern of energy depositions in terms of linear energy, or energy imparted to such volumes, is not appropriate to be quantitatively linked to all radiation-induced biological effects and all radiations. Although views diverge on the suitability of alternative descriptions of the properties of energy deposition there is at least one argument in favour of the applicability of conventional microdosimetric quantities in neutron radiation. Radiobiological experiments have shown that the relative biological effectiveness (RBE) for several biological endpoints has a dependence on neutron energy which is, for a limited range of energy, similar to that of microdosimetric quantities, such as $\bar{y}_D$ [17]. A plausible argument for the suitability of microdosimetric parameters as a radiation quality specification for fast neutrons is that the range of most of the secondary charged particles is large compared to the diameters of the gas spheres. Therefore, the distance over which the energy transfer is averaged by the microdosimetric measurements is small compared to distances over which the physical properties of the charged particles change appreciably.

Although, already microdosimetric spectra represent an averaging procedure of the microscopic energy deposition pattern, they are still too complex to be used as radiation quality specifications in practical situations. A further data reduction is required which, on the other hand, further complicates the problem of the principal suitability of microdosimetric parameters. A possible one-parameter description of the microdosimetric spectra is the mean value of the dose distribution in lineal energy d(y) which is denoted $\bar{y}_D$ [7]. There is, however, an immediate objection to the general suitability of this quantity, which is that its value increases continuously with neutron energy above about 5 MeV, whereas RBE has a maximum at about 0.5 MeV and decreases with increasing neutron energy. It should be pointed out that $\bar{y}_D$ does not provide a good characterization of microdosimetric spectra. This is illustrated in Fig. 13 where two spectra with almost identical values of $\bar{y}_D$, but very different shapes, are compared. Both spectra have the same mean value of the dose distribution because $\bar{y}_D$ is strongly dependent on the frequency of events at high lineal energies.

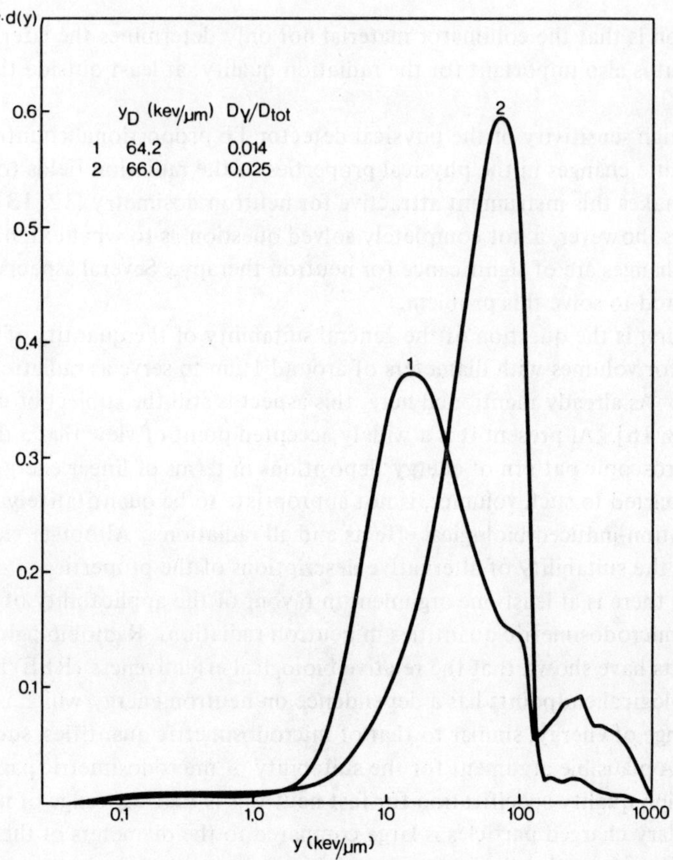

FIG.13. Microdosimetric spectra for neutrons (measured at the cyclotron facility of DKFZ) with almost identical dose average lineal energy $\overline{y}_D$ but strongly different in shape (d_s = 2 μm).

The obvious reason for the discrepancy in the neutron energy dependence of RBE and $\overline{y}_D$ is that the RBE/LET relationship for cell killing does not increase monotonously [18] but has a maximum. This has led Kellerer and Rossi [7] to introduce a quantity which they called the dose-average lineal energy corrected for saturation, y*. The definition of this quantity can be understood with the help of Fig.14, which has been taken from Caswell and Coyne [19]. y* is evaluated formally like the mean value of the dose distribution. But, instead of taking the quantity lineal energy y, a quantity, y_{SAT}, is used which is a function of y as shown in Fig.14. It has been suggested [15] that this procedure may be understood as the evaluation of the mean value of a dose distribution in lineal energy which has been weighted by the function given by y_{SAT}. This weighting function is in

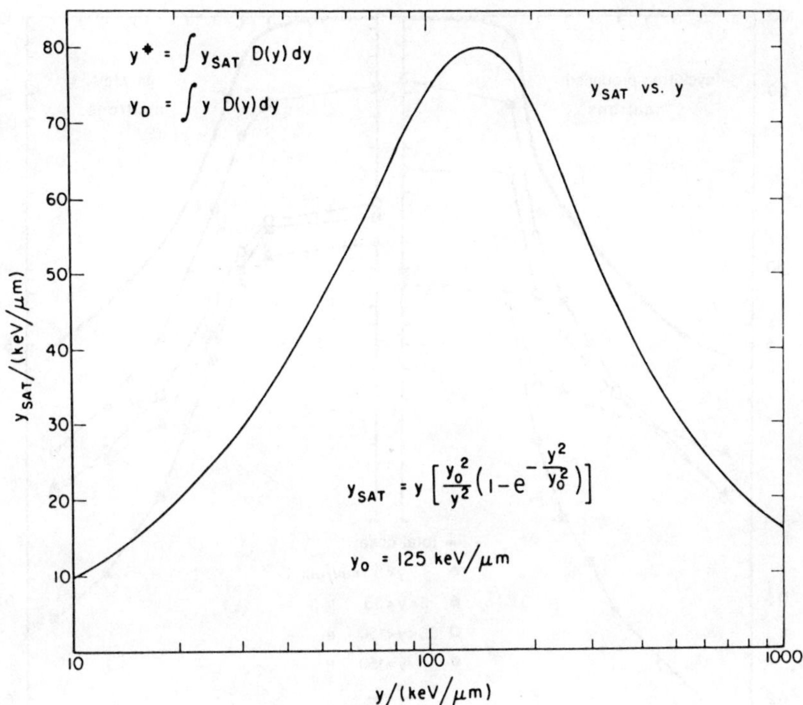

FIG.14. y_{SAT} as a function of lineal energy y. The quantities y_{SAT}, y_0 and y have been defined by Kellerer and Rossi* [7]. *(Reproduced from Ref.* [19].)

qualitative agreement with the RBE/LET relationship for T1 kidney cells found by Barendsen [18]. Expressed differently, y* is a microdosimetric average of the dose distribution weighted by a biological response function for cell killing and may therefore be expected to be applicable if cell survival is concerned. It is obvious from this consideration that y*, in contrast to $\bar{y}_D$, is not a purely physical quantity and in principle different weighting functions could be used. The potential usefulness of y* for predicting relative changes of RBE within phantoms, or for different neutron sources, has been discussed in other publications [1, 15]. For this reason values of y* for the four therapy facilities have been given in Table I.

As an alternative to the one-parameter specification of radiation quality, a separation of total absorbed dose into three or four fractions, presumably associated with different biological effectiveness, has been suggested [10, 20]. We have suggested subdividing the lineal energy spectrum into four intervals [10], and evaluating the corresponding dose fraction. An example of a result of this procedure is set out in Fig.15 where the lateral profiles at 8 cm depth in a

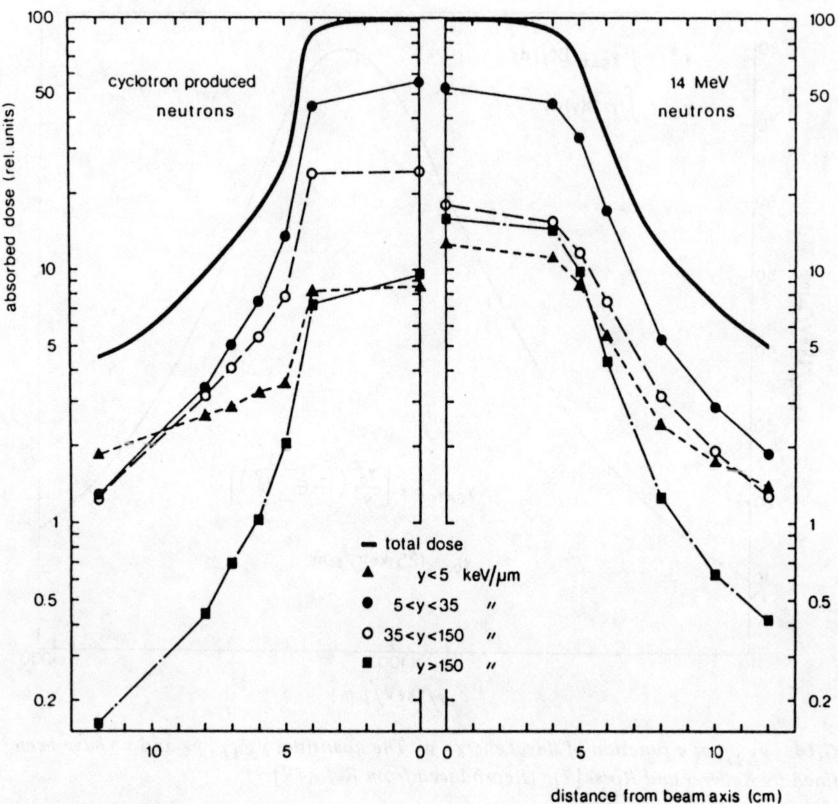

FIG.15. Lateral profiles of total absorbed dose and of dose fractions associated with different intervals of lineal energy y at 8 cm depth in a phantom for the cyclotron facility (left) and the D-T generator (right) of DKFZ. (Field size: 6 × 8 cm².)

phantom of total absorbed dose and the four dose components are shown for the two neutron irradiation units at DKFZ, Heidelberg. As a comparable but more sophisticated method it has been suggested that the dose components of gamma-rays, protons and heavy recoils be separated [20]. In practice the dose components would have to be associated with weighting factors corresponding to the differences in biological effectiveness. Such a procedure would be a simplified analogue to the evaluation of y^*.

The clinical significance of radiation quality changes as described by microdosimetric spectra also depends on the importance of other factors relevant to the therapeutical effects. These include the problem of dose fractionation and differences in the response of various types of tissue. Whether or not radiation quality changes within the patient are of relevance depends also on the level of absorbed dose. It has been argued that beyond a certain distance outside the primary beam changes of radiation quality are unimportant because of the small total absorbed dose received and because the increase in photon dose fraction and increase in neutron energy compensate each other. This argument may be invalid, however, if multiple field or dynamic treatment is applied, or if biological effects are concerned which may have a particularly high RBE at low neutron energies.

In conclusion, radiation quality is one of several problems not yet adequately solved in neutron therapy. Measurements with low-pressure TE proportional counters are expected to contribute to the solution, in particular if correlated with critical radiobiological experiments. For the limited range of application in neutron therapy, microdosimetric parameters may very well provide suitable radiation quality specifications.

ACKNOWLEDGEMENTS

The contribution of Dr. H. Schuhmacher to the microdosimetric experiments has been substantial and is gratefully acknowledged. The assistance of colleagues at the irradiation facilities at DKFZ Heidelberg, the University Hospitals of Essen and Hamburg, and GSF Munich-Neuherberg during measurements was highly appreciated. The experiments were performed with financial support from the Bundesministerium für Forschung and Technologie (BMFT) of the Federal Republic of Germany.

REFERENCES

[1] BOOZ, J., FIDORRA, J., Microdosimetric investigations on collimated fast neutron beams for radiation therapy: II. The problem of radiation quality and RBE, Phys. Med. Biol. **26** (1981) 43.

[2] MENZEL, H.G., SCHUHMACHER, H., "Comparison of microdosimetric characteristics of four fast neutron therapy facilities", Proc. 7th Symp. Microdosimetry (BOOZ, J., EBERT, H.G., HARTFIEL, H.D., Eds), EUR-7147, Harwood Acad. Publ., London (1981) 1217.

[3] INTERNATIONAL COMMISSION ON RADIATION UNITS AND MEASUREMENTS, Radiation quantities and units, Rep. No.33, ICRU Publications, Washington, DC (1980).

[4] HALL, E.J., WITHERS, H.R., GERACI, J.P., MEYN, R.E., RASEY, J., TODD, P., SHELINE, G.E., Radiobiological intercomparisons of fast neutron beams used for therapy in Japan and the United States, Int. J. Radiat. Oncol., Biol. Phys. **5** (1979) 272.

[5] BEWLEY, D.K., IAEA-AG-371/9, these Proceedings.

[6] ROSSI, H.H., Energy distribution in the absorption of radiation, Adv. Biol. Med. Phys. **11** (1967) 27.

[7] KELLERER, A.M., ROSSI, H.H., The theory of dual radiation action, Curr. Top. Radiat. Res. **8** (1972) 85.

[8] GOODHEAD, D.T., An assessment of the role of microdosimetry in radiobiology, Radiat. Res. **91** (1982) 45.

[9] ROSSI, H.H., ROSENZWEIG, W., A device for the measurement of dose as a function of specific ionization, Radiology **64** (1955) 404.

[10] MENZEL, H.G., WAKER, A.J., GRILLMAIER, R.E., BIHY, L., HARTMANN, G., "Radiation quality studies in mixed neutron gamma fields", Proc. 3rd Symp. Neutron Dosimetry in Biology and Medicine (BURGER, G., EBERT, H.G., Eds), EUR-5848, Commission of the European Communities (1978) 481.

[11] INTERNATIONAL COMMISSION ON RADIATION UNITS AND MEASUREMENTS, Average energy required to produce an ion pair, Rep. No.31, ICRU Publications, Washington, DC (1979).

[12] MENZEL, H.G., SCHUHMACHER, H., "Experimental uncertainties and applicability of the proportional counter method for the determination of photon dose fractions in neutron beams", Ion Chambers for Neutron Dosimetry (BROERSE, J.J., Ed.), EUR-6782, Commission of the European Communities (1980) 337.

[13] MENZEL, H.G., SCHUHMACHER, H., "The application of tissue-equivalent proportional counters to problems in neutron dosimetry", Proc. 4th Symp. Neutron Dosimetry in Biology and Medicine (BURGER, G., EBERT, H.G., Eds), EUR-7448, Harwood Acad. Publ., London (1981) 263.

[14] MAIER, E., KNESEWITSCH, P., BURGER, G., "Mixed field dosimetry with ionization chambers and proportional counters", Proc. 2nd Symp. Neutron Dosimetry in Biology and Medicine (BURGER, G., EBERT, H.G., Eds), EUR-5273, Commission of the European Communities (1975) 263.

[15] HARTMANN, G., MENZEL, H.G., SCHUHMACHER, H., HÖVER, K.H., GRANZOW, G., "Radiobiological intercomparison of two fast neutron therapy units using CHO-cell survival and its correlation to microdosimetry", Proc. 7th Symp. on Microdosimetry (BOOZ, J., EBERT, H.G., HARTFIEL, H.D., Eds), EUR-7147, Harwood Acad. Publ., London (1980) 1083.

[16] KELLERER, A.M., ROSSI, H.H., A generalised formulation of the dual radiation action, Radiat. Res. **75** (1978) 471.

[17] HALL, E.J., NOVAK, J.K., KELLERER, A.M., ROSSI, H.H., MARINO, S., GOODMAN, L.J., RBE as function of neutron energy: 1. Experimental observations, Radiat. Res. **64** (1975) 245.

[18] BARENDSEN, G.W., Impairment of the proliferative capacity of human cells in culture by alpha-particles with differing linear energy transfer, Int. J. Radiat. Biol. **8** (1964) 453.

[19] CASWELL, R.S., COYNE, J.J., "Energy deposition spectra for neutrons based on recent cross-section evaluation", Proc. 6th Symp. on Microdosimetry (BOOZ, J., EBERT, H.G., Eds), EUR-6064, Harwood Acad. Publ., London (1978) 1159.

[20] FIDORRA, J., BOOZ, J., Microdosimetric investigations on collimated fast-neutron beams for radiation therapy: I. Measurements of microdosimetric spectra and particle dose fractions in a water phantom for fast neutrons from 14 MeV deuterons on beryllium, Phys. Med. Biol. **26** (1981) 27.

CALIBRATION PROCEDURES OF TISSUE-EQUIVALENT IONIZATION CHAMBERS USED IN NEUTRON DOSIMETRY

B.J. MIJNHEER
Radiotherapy Department,
Antoni van Leeuwenhoek Hospital,
The Netherlands Cancer Institute,
Amsterdam, Netherlands

J.R. WILLIAMS
Department of Medical Physics and
Medical Engineering,
Western General Hospital, Edinburgh,
United Kingdom

Abstract

CALIBRATION PROCEDURES OF TISSUE-EQUIVALENT IONIZATION CHAMBERS USED IN NEUTRON DOSIMETRY.

Equations are given for the determination of absorbed dose from measurements with a tissue-equivalent (TE) ionization chamber in a neutron field. A comparison is made between these equations and similar relations applied in the absorbed dose determination of photon beams. Four methods of calibrating the TE chamber are considered, namely an exposure calibration in air in a photon beam, an absorbed dose calibration in a phantom in a photon beam, an absorbed dose calibration in a phantom in a neutron beam, and a kerma calibration in air in a neutron beam. An uncertainty analysis of the results obtained by the different calibration methods is made. It is concluded that both calibration methods in the neutron beam give the greatest overall accuracy. The major uncertainty in all techniques is due to uncertainties in the kerma data used to convert absorbed dose in the wall material to absorbed dose in tissue.

1. INTRODUCTION

Tissue-equivalent (TE) ionization chambers, flushed with a tissue-equivalent gas, are often used to determine neutron absorbed dose and neutron kerma. Standards laboratories are at present developing standard neutron beams to provide a calibration factor for these chambers under well-known conditions, but they are not yet available. Until now, the procedure for the calibration of TE ionization chambers to be used for neutron dosimetry is analogous to that used in photon dosimetry. The calibration of these TE ionization chambers should

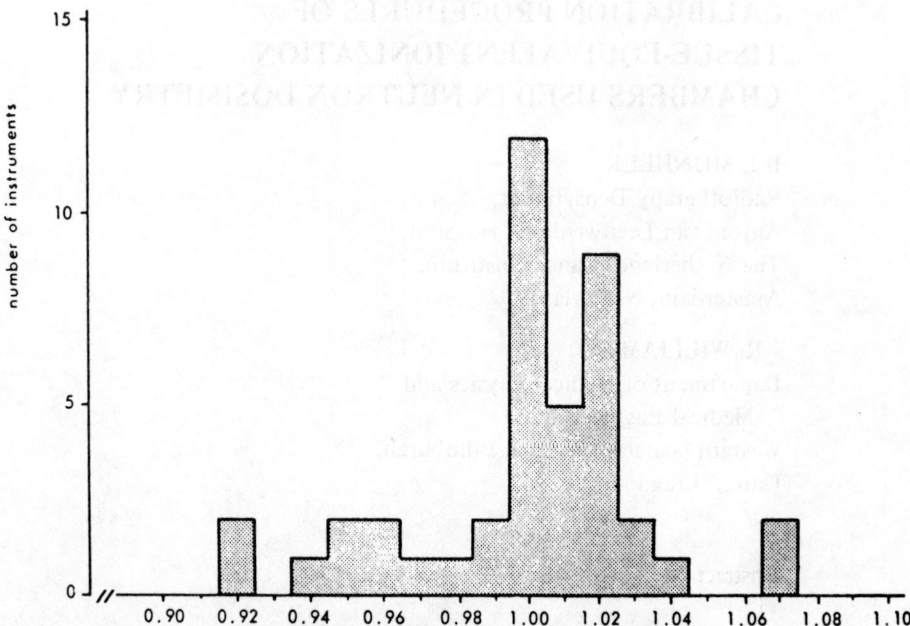

FIG.1. Ratio of calibration factors using the ENDIP ^{137}Cs source and the home gamma-ray source (from Ref. [1]).

therefore, in principle, not introduce large uncertainties. Nevertheless, during international intercomparisons, variations up to ±8% in calibration factors obtained with different photon beams have been reported (see Fig.1) [1]. As a result of these intercomparisons a standard calibration procedure has been suggested in which the TE ionization chamber is calibrated against a standard exposure chamber, which has a calibration factor traceable to a national standards laboratory. Such a procedure has been recommended in both protocols for neutron dosimetry for external beam therapy as drafted by the United States [2] and the European groups [3].

In the present paper the equations required for the derivation of $D_N + D_G$ (the total absorbed dose in tissue) are derived not only for the in air exposure calibration of the TE ionization chamber, but also for absorbed dose calibrations in a phantom irradiated by a photon or a neutron beam, and for a kerma calibration in air in a neutron beam. An uncertainty analysis of the results obtained by the different calibration methods is made. This work can be considered as an extension of an earlier publication [4] in which further details of applied correction and conversion factors and derivation of equations can be found.

2. DETERMINATION OF ABSORBED DOSE IN A NEUTRON BEAM IN A PHANTOM USING AN EXPOSURE CALIBRATION

If an exposure standard chamber is used free-in-air to calibrate the TE ionization chamber, the exposure calibration factor, N_c, of that chamber is given by:

$$N_c = \frac{X_c}{Q_c} \tag{1}$$

where X is the exposure free-in-air at the centre of the ionization chamber in the absence of the chamber, Q is the charge of one sign produced within the cavity, with subscript c referring to the calibration quality. For practical measurements with an ionization chamber the reading obtained from the chamber, R, has to be related to the charge produced within the cavity at a reference temperature and pressure by using the product of several factors, Πk_R, which are discussed in more detail elsewhere [4]:

$$Q = R \cdot \Pi k_R \tag{2}$$

To convert the reading of a chamber to absorbed dose to water, $(D_w)_\lambda$, when it is irradiated in a phantom by a photon beam of energy λ, generally a factor C_λ is employed [5]. C_λ is used in conjunction with the exposure calibration factor of the chamber at a specific calibration energy, usually ^{60}Co or 2 MV X-rays:

$$(D_w)_\lambda = R_\lambda \cdot (\Pi k_R)_\lambda \cdot N_c \cdot C_\lambda \tag{3}$$

In principle a similar expression for calculating total absorbed dose $D_N + D_G$ from the reading, R_T, of a TE ionization chamber, when used in a phantom irradiated by a neutron beam, can be used:

$$D_N + D_G = R_T \cdot (\Pi k_R)_T \cdot N_c \cdot C_N \tag{4}$$

where D_N and D_G are the neutron and gamma-ray components, respectively, of absorbed dose in a reference tissue material. With such an approach a neutron beam can be considered to be a photon beam of a quality N, having its special conversion factors included in C_N. However, questions have recently arisen over the correctness of the C_λ values given in [5]. A number of authors, e.g. [6], have re-evaluated C_λ by considering the effects of the electrons liberated in the wall of the chamber as well as those liberated in the surrounding medium. In addition, more accurate factors correcting for attenuation and scattering in the ionization chamber material in the calibration beam have recently been discussed in the literature, e.g. [7]. It is clear from these publications that these detailed calculations are necessary for the derivation of absorbed dose in photon dosimetry, and that similar considerations have to be followed in neutron dosimetry.

A general expression for calculating absorbed dose in a reference tissue t, D_t, from the reading of a TE ionization chamber when used in a phantom, can be given by:

$$(D_t)_u = \frac{R_u \cdot (\Pi k_R)_u}{m} \cdot \frac{(W_g)_u}{e} \cdot r_u \cdot d_u \cdot \left[\frac{(\mu_{en}/\rho)_t}{(\mu_{en}/\rho)_m}\right]_u \qquad (5)$$

where subscripts u and m refer to the user's radiation quality and the wall material of the TE ionization chamber, respectively; m is the mass of the cavity gas; W_g is the average energy expended per ion pair formed in the gas; e is the electronic charge; r the gas-to-wall absorbed dose conversion factor; and d the displacement correction factor and μ_{en}/ρ the ratio of the mass energy absorption coefficients. In deriving this equation it is assumed that the wall of the chamber and its buildup cap are made of the same material, which would usually be A-150 plastic. It is also assumed that they are sufficiently thick to produce charged-particle equilibrium so that the charged particles crossing the cavity will have been liberated either in the wall, the buildup cap or in the cavity itself. The choice of phantom material therefore does not affect the method of calculating the absorbed dose. Under these conditions the equation is valid both for neutrons and for photons irradiating the phantom.

A problem in applying Eq. (5) is that m is difficult to determine directly. Both in neutron and photon dosimetry at the present time the mass of gas is most commonly derived calculating the absorbed dose in the wall adjacent to the cavity, $(D'_m)_c$, which can be considered as the absorbed dose to the wall of the chamber at the mean centre of electron production, from an exposure calibration of the chamber, using the equation:

$$(D'_m)_c = X_c \cdot (\Pi k_A)_c \cdot (f_t)_c \cdot \left[\frac{(\mu_{en}/\rho)_m}{(\mu_{en}/\rho)_t}\right]_c \qquad (6)$$

where $(\Pi k_A)_c$ is the product of several correction factors, used to account for the finite size of the chamber, its buildup cap and stem when measurements are made in air, usually indicated by subscript a, and

$$(f_t)_c = \frac{(W_a)_c}{e} \cdot \left[\frac{(\mu_{en}/\rho)_t}{(\mu_{en}/\rho)_a}\right]_c \quad (7)$$

$(D'_m)_c$ can also be obtained from the charge produced within the cavity with the assumption that the chamber acts as a Bragg-Gray cavity in the calibration beam so that r can be replaced by the ratio of the mass stopping powers $(s_{m,g})_c$:

$$(D'_m)_c = \frac{Q_c}{m} \cdot \frac{(W_g)_c}{e} \cdot (s_{m,g})_c \quad (8)$$

Combining Eqs (1), (6) and (8) gives an expression for the mass of gas as a function of N_c, which can now be submitted in Eq.(5) yielding:

$$(D_t)_u = R_u \cdot (\Pi k_R)_u \cdot d_u \cdot \alpha_c$$

$$\times \frac{(W_g)_u}{(W_g)_c} \cdot \frac{r_u}{(s_{m,g})_c} \cdot \frac{[(\mu_{en}/\rho)_t/(\mu_{en}/\rho)_m]_u}{[(\mu_{en}/\rho)_t/(\mu_{en}/\rho)_m]_c} \quad (9)$$

$$\text{where } \alpha_c = N_c \cdot (\Pi k_A)_c \cdot (f_t)_c \quad (10)$$

α_c is an absorbed dose calibration factor although it is derived from an exposure calibration and the name might therefore be confusing [8].

For a photon beam $(W_g)_u$ is equal to $(W_g)_c$ and $r_u = (s_{m,g})_c$. If the absorbed dose is expressed in water, subscript w, then Eq.(9) can be rewritten as:

$$(D_w)_\lambda = R_\lambda \cdot (\Pi k_R)_\lambda \cdot N_c \cdot (\Pi k_A)_c \cdot (f_w)_c \cdot d_\lambda$$

$$\times \frac{[(s_{m,g})_\lambda}{[(s_{m,g})_c} \cdot \frac{(\mu_{en}/\rho)_w/(\mu_{en}/\rho)_m]_\lambda}{(\mu_{en}/\rho)_w/(\mu_{en}/\rho)_m]_c} \quad (11)$$

which can be shortened into Eq.(3), with C_λ defined as:

$$C_\lambda = (\Pi k_A)_c \cdot (f_w)_c \cdot d_\lambda \cdot \frac{(s_{m,g})_\lambda}{(s_{m,g})_c} \cdot \frac{[(\mu_{en}/\rho)_w/(\mu_{en}/\rho)_m]_\lambda}{[(\mu_{en}/\rho)_w/(\mu_{en}/\rho)_m]_c} \quad (12)$$

It should be noted that in photon dosimetry the walls of TE ionization chambers are usually not sufficiently thick to produce charged-particle equilibrium in the wall material during the measurement in the phantom. The two last terms of Eq.(12) should then be modified by considering also the phantom material [6].

Equation (9) has been derived for a single-component radiation beam. However, neutron beams are invariably accompanied by a photon component and in such a mixed radiation beam Eq.(9) has to be extended to include both the neutron and photon contributions to the absorbed dose. In ICRU Report 26 [9] the factors k_T and h_T are defined as being equal to the reciprocal of the product of the last three terms in Eq.(9) with the user's radiation quality being the two separate components, i.e.:

$$k_T = \frac{(W_g)_c}{(W_g)_N} \cdot \frac{(s_{m,g})_c}{r_N} \cdot \frac{\left[(\mu_{en}/\rho)_t/(\mu_{en}/\rho)_m\right]_c}{(K_t/K_m)_N} \qquad (13)$$

$$h_T = \frac{(W_g)_c}{(W_g)_G} \cdot \frac{(s_{m,g})_c}{(s_{m,g})_G} \cdot \frac{\left[(\mu_{en}/\rho)_t/(\mu_{en}/\rho)_m\right]_c}{\left[(\mu_{en}/\rho)_t/(\mu_{en}/\rho)_t\right]_G} \qquad (14)$$

where subscripts N and G refer to the neutron and photon components, respectively. It should be noted that in Eq.(13) the ratio of mass energy absorption coefficients in tissue and in the wall material has been replaced by the ratio of kerma in these materials, K_t/K_m. Using these factors, Eq.(9) can be rewritten for the mixed radiation beam:

$$R_T \cdot (\Pi k_R)_T \cdot d_T \cdot \alpha_c = k_T D_N + h_T D_G \qquad (15)$$

The reading of a TE ionization chamber in a neutron beam is sometimes interpreted as being proportional to the total absorbed dose, $D_N + D_G$. The validity of this approximation can be assessed by rewriting Eq.(15) as:

$$R_T \cdot (\Pi k_R)_T \cdot d_T \cdot \alpha_c = k_T \cdot (D_N + D_G) \cdot (1 + \delta) \qquad (16)$$

$$\text{where } \delta = \frac{D_G}{D_N + D_G} \cdot \frac{(h_T - k_T)}{k_T} \qquad (17)$$

It is usual to assume that h_T is unity and generally k_T lies in the range 0.95–1.00. For radiotherapy neutron beams D_G is generally less than 20% of the total absorbed dose within the main beam, so that δ is usually less than 0.01.

Therefore, if Eq.(16) is used with a zero value for δ the error introduced into the calculation of $D_N + D_G$ is usually less than 1%.

Equation (16) may be rewritten to give Eq.(4) where:

$$C_N = (\Pi k_A)_c \cdot (f_t)_c \cdot d_T \cdot \frac{1}{k_T} \cdot \frac{1}{1+\delta} \tag{18}$$

C_N will depend on the neutron energy and, like C_λ, on the geometry of the chamber and on the materials of which it and its buildup cap are constructed. There will in addition be a small dependence on the gamma-ray component of the neutron beam. Although in principle approximate values for C_N as for C_λ can be tabulated as a function of neutron energy, a higher accuracy can be obtained by applying the extended formula using the individual parameters contained in C_N.

3. OTHER METHODS OF CALIBRATION

Until now all equations have been based on an exposure calibration of the ionization chamber. Some standards laboratories are at the moment developing standards which will give the absorbed dose under specified conditions in a phantom irradiated with ^{60}Co gamma rays. In the future this may result in the replacement of an exposure calibration by an in-phantom absorbed dose calibration. The absorbed dose measured by such a standard will be the absorbed dose to water at the position of the centre of the measuring device when it is replaced by phantom material, $(D_w)_c$, which is the quantity normally required in photon radiotherapy. For the absorbed dose calibration factor, α_c, of the TE chamber, which is used in neutron dosimetry, the quantity required is absorbed dose to tissue at the effective centre of measurement of the TE ionization chamber. The absorbed dose measured by the standard dosimeter has therefore to be corrected to absorbed dose to tissue using the ratio of the mass energy absorption coefficients in tissue and in water, and also to be divided by a displacement factor, d_c. Then α_c can be written as:

$$\alpha_c = \frac{(D_w)_c}{R_c \cdot (\Pi k_R)_c} \cdot \left[\frac{(\mu_{en}/\rho)_t}{(\mu_{en}/\rho)_w}\right]_c \cdot \frac{1}{d_c} \tag{19}$$

where subscript w refers to water. This value of α_c, which should be identical to that derived from Eq.(10), can now be used in Eqs (15) and (16) for the derivation of D_N and D_G.

A comparison of Eqs (10) and (19) shows that the two different methods of photon calibration result in the application of different conversion factors and correction factors. To obtain the absorbed dose to tissue at the effective centre of electron production from an exposure in air or absorbed dose in water measurement, the conversion factors $(f_t)_c$ or $[(\mu_{en}/\rho)_t/(\mu_{en}/\rho)_w]_c$, respectively, have to be applied. The correction factors needed to account for the finite size of the chamber during the calibrations in air or in the phantom are $(\Pi k_A)_c$ or $1/d_c$, respectively.

If in the future, standards laboratories can provide standard neutron beams in which either the neutron and photon absorbed dose at a certain position in a phantom or the neutron or photon kerma in air, are well known, these positions can be utilized for TE ionization chamber calibrations. Absorbed dose calibration factors, α_c', can now be derived, in which the prime indicates that a neutron beam is employed for calibration purposes. The following equations will be more general and somewhat different from those given in [4]. In that paper the situation was considered that $(D_m)_N + (D_m)_G$, with m being A-150 plastic, was determined by means of an A-150 plastic TE calorimeter in an A-150 plastic TE phantom.

If a TE ionization chamber is positioned in the phantom in the neutron calibration beam, with its geometrical centre at the position where D_N and D_G are well known, then Eq.(15) can be applied for both components of the absorbed dose. If it is assumed that no large error is introduced by applying an average $(\Pi k_R)_c$ and d_c for both components, then the following equation is valid:

$$(D_N)_c + (D_G)_c \cdot \Delta_c$$

$$= (R_T)_c \cdot \left[(\Pi k_R)_T\right]_c \cdot \frac{1}{e} \cdot \frac{1}{m} \cdot (d_T)_c \cdot \left[(W_g)_N \cdot r_n \cdot (K_t/K_m)_N\right]_c \tag{20}$$

$$\text{with } \Delta_c = \frac{(W_g)_N}{(W_g)_G} \cdot \frac{r_n}{(s_{m,g})_G} \cdot \frac{(K_t/K_m)_N}{\left[(\mu_{en}/\rho)_t/(\mu_{en}/\rho)_m\right]_G} \tag{21}$$

As discussed in Section 2, an expression for the mass of gas can again be derived by rewriting Eq.(20):

$$m = \left[(W_g)_N\right]_c \cdot (r_N)_c \cdot \left[(K_t/K_m)_N\right]_c \cdot \frac{1}{e} \cdot \frac{1}{\alpha_c'} \tag{22}$$

where α_c' is the absorbed dose calibration factor given by:

$$\alpha'_c = \frac{(D_N)_c + (D_G)_c \cdot \Delta_c}{(R_T)_c \cdot \left[(\Pi k_R)_T\right]_c} \cdot \frac{1}{(d_T)_c} \tag{23}$$

In an analogous manner as discussed in Section 2, Eq. (22) can be substituted into Eq.(5) which can be used for both components of absorbed dose in the user's beam to give:

$$R_T \cdot (\Pi k_R)_T \cdot d_T \cdot \alpha'_c = k'_T \cdot D_N + h'_T \cdot D_G \tag{24}$$

in which:

$$k'_T = \frac{\left[(W_g)_N \cdot r_N \cdot (K_t/K_m)_N\right]_c}{\left[(W_g)_N \cdot r_N \cdot (K_t/K_m)_N\right]_u} \tag{25}$$

$$h'_T = \frac{\left[(W_g)_N \cdot r_N \cdot (K_t/K_m)_N\right]_c}{\left[(W_g)_G \cdot (s_{m,g})_G \cdot \{(\mu_{en}/\rho)_t/(\mu_{en}/\rho)_m\}_G\right]_u} \tag{26}$$

These definitions of k'_T and h'_T are similar to the definitions of k_T and h_T with the factor for the calibration field being appropriate to the neutron component alone. As a result, the absolute values of α'_c, k'_T and h'_T will be different from the corresponding values obtained from a photon calibration. This can easily be seen from the situation where a chamber is calibrated in a neutron beam which has the same spectrum as the user's neutron beam because then k'_T is unity. h'_T will differ from unity and be approximately equal to the reciprocal value of k_T.

If $(K_N)_c$ and $(K_G)_c$ are known accurately at a position in air in the calibration neutron beam, then α'_c can also be derived from these quantities according to:

$$\alpha'_c = \frac{\{(K_N)_c + (K_G)_c \cdot \Delta_c\}}{(R_T)_c \cdot \left[(\Pi k_R)_T\right]_c} \cdot \left[(\Pi k_A)_T\right]_c \tag{27}$$

Both equations for α'_c will yield the same value. Equation (27) is similar to the corresponding Eq.(10) for the exposure calibration factor, by rewriting $X_c \cdot (f_t)_c$ as $(K_t)_c$.

It should be noticed that in most neutron beams that might be used for calibration $D_G \cdot (D_N + D_G)^{-1}$ or $K_G \cdot (K_N + K_G)^{-1}$ would have a value of about 0.05 and Δ_c would not exceed 1.05. Ignoring Δ_c, the factor that corrects for the difference in sensitivity of the TE ionization chamber for photons and neutrons in the calibration beam, will therefore not introduce an error larger than 0.25% in the calibration factor α_c'. This also implies that for Δ_c taken equal to one, it is not necessary for the separate neutron and photon components to be known at the calibration position, but knowledge of $D_N + D_G$ or $K_N + K_G$ gives sufficiently accurate calibration factors for TE ionization chambers.

4. UNCERTAINTIES IN $D_N + D_G$ FOR THE FOUR CALIBRATION METHODS

The overall uncertainty in the determination of absorbed dose in reference tissue from a measurement with a TE ionization chamber in a neutron beam will be a composite value because of uncertainties in separate physical parameters, which are a function of neutron energy, dimensions and materials of the ionization chamber and the method of calibration. The overall uncertainty for the four different calibration methods will now be compared for three clinically employed neutron beams. It is assumed that the TE ionization chamber is well designed and not larger than about 1 cm³, thus not introducing large uncertainties in Πk_R, d and Πk_A. The overall uncertainty is determined by taking the square root of the sum of the squares of the separate uncertainties, which are estimates of one standard error.

Four different methods of calibrating a TE ionization chamber will be considered: (1) exposure calibration; (2) in-phantom calibration at a reference photon source; (3) in-phantom calibration in a reference neutron field; and (4) kerma calibration in air in a reference neutron field. The overall uncertainty in $D_N + D_G$ will be compared while the difference between k_T and h_T or k_T' and h_T' will be ignored. The expressions for $D_N + D_G$ for the four methods of calibration are summarized below:

$$(1)\quad \frac{R_T}{R_c}\cdot\frac{(\Pi k_R)_T}{(\Pi k_R)_c}\cdot d_T\cdot(\Pi k_A)_c\cdot X_c\cdot(f_t)_c$$

$$\times\frac{(W_g)_N}{(W_g)_c}\cdot\frac{r_N}{(s_{m,g})_c}\cdot\frac{(K_t/K_m)_N}{\left[(\mu_{en}/\rho)_t/(\mu_{en}/\rho)_m\right]_c}$$

$$(2)\quad \frac{R_T}{R_c}\cdot\frac{(\Pi k_R)_T}{(\Pi k_R)_c}\cdot\frac{d_T}{d_c}\cdot(D_t)_c\cdot\frac{(W_g)_N}{(W_g)_c}\cdot\frac{r_N}{(s_{m,g})_c}\cdot\frac{(K_t/K_m)_N}{\left[(\mu_{en}/\rho)_t/(\mu_{en}/\rho)_m\right]_c}$$

$$(3)\quad \frac{R_T}{R_C}\cdot\frac{(\Pi k_R)_T}{(\Pi k_R)_C}\cdot\frac{d_T}{(d_T)_C}\cdot\left[(D_N)_c + (D_G)_c\right]\cdot\frac{\left[(W_g)_N\cdot r_N\cdot(K_t/K_m)_N\right]_u}{\left[(W_g)_N\cdot r_N\cdot(K_t/K_m)_N\right]_c}$$

$$(4)\quad \frac{R_T}{R_C}\cdot\frac{(\Pi k_R)_T}{(\Pi k_R)_C}\cdot d_T$$

$$\times\left[(\Pi k_A)_T\right]_c\cdot\left[(K_N)_c + (K_G)\right]_c\cdot\frac{\left[(W_g)_N\cdot r_N\cdot(K_t/K_m)_N\right]_u}{\left[(W_g)_N\cdot r_N\cdot(K_t/K_m)_N\right]_c}$$

The estimated individual sources of error in the different physical parameters are summarized in Table I. For the photon calibrations the quoted values are discussed in [4]. The uncertainty in $(D_N)_c + (D_G)_c$ is based on a 1.5% uncertainty in a calorimetric calibration in A-150 plastic [10] combined with the quoted 2.6% uncertainty in $(K_t/K_m)_N$, assuming that the reference neutron beam would have a neutron spectrum similar to that of a d(16)+Be beam. The uncertainty in $(K_N)_c + (K_G)_c$ is based on an estimated uncertainty of 2.5% in a fluence measurement combined with a 2.0% uncertainty in the kerma factor for tissue for the reference neutron beam. The uncertainty in the kerma ratio in the user's beam relative to that in the reference neutron beam is estimated by assuming that any change would go in the same direction.

5. DISCUSSION AND CONCLUSIONS

It can be seen from Table I that a calibration of a TE ionization chamber in a neutron beam gives a smaller uncertainty than calibration in a photon beam, and that there is little difference in the overall uncertainty when the calibration is made in a phantom or in air. Table I also shows that the main contributions to the uncertainty in absorbed dose using a photon calibration come from uncertainties in kerma data, gas-to-wall absorbed dose conversion factors and W values, whereas using a neutron calibration the main uncertainty is in the kerma values. Improvement in the accuracy of neutron dosimetry with TE ionization chambers should therefore be concentrated on increasing the accuracy of the kerma data, particularly at higher neutron energies.

An additional advantage of a calibration in a neutron beam compared to a photon calibration is that any uncertainty in atomic composition of the wall materials, which is not included in Table I, will have a much smaller effect on the absorbed dose determination in the user's neutron beam. This is because

TABLE I. PERCENTAGE UNCERTAINTIES IN DETERMINING ABSORBED DOSE IN REFERENCE TISSUE FROM TE IONIZATION CHAMBER MEASUREMENTS IN THREE NEUTRON FIELDS FOR FOUR DIFFERENT CALIBRATION METHODS

Physical parameter(s)	Photon calibration		Neutron calibration	
	(1) Exposure	(2) Absorbed dose in phantom	(3) Absorbed dose in phantom	(4) Kerma in air
R_T/R_C	0.3	0.3	0.5	0.5
$(\Pi k_R)_T/(\Pi k_R)_C$	0.4	0.4	0.4	0.4
d_T	0.5	-	-	0.5
$d_T/(d_T)_C$	-	0.6	0.3	-
$(\Pi k_A)_C$	0.3	-	-	-
$[(\Pi k_A)_T]_C$	-	-	-	0.5
X_C	1.0	-	-	-
$(f_t)_C$	0.5	-	-	-
$(D_t)_C$	-	1.2	-	-
$(D_N)_C + (D_G)_C$	-	-	3.1	-
$(K_N)_C + (K_G)_C$	-	-	-	3.2
$(W_g)_N/(W_g)_C$	4.0	4.0	-	-
$[(W_g)_N]_u/[(W_g)_N]_C$	-	-	0.5	0.5
$r_N/(s_{m,g})_C$	4.3	4.3	-	-
$(r_N)_u/(r_N)_C$	-	-	0.5	0.5
$(K_t/K_m)_N$ [a]	2.6/4.5/9.7	2.6/4.5/9.7	-	-
$[(K_t/K_m)_N]_u/[(K_t/K_m)_N]_C$ [a]	-	-	0.5/2.0/7.0	0.5/2.0/7.0
$[(\mu_{en}/\rho)_t/(\mu_{en}/\rho)_m]_C$	0.1	0.1	-	-
d(16)+ Be overall[a]	6.5	6.5	3.3	3.3
d+T overall[a]	7.5	7.5	3.8	3.8
p(66)+Be overall[a]	11.4	11.4	7.7	7.7

[a] The uncertainties for the different neutron beams have been calculated by assuming that the uncertainties in all the parameters are the same except for the uncertainty in $(K_t/K_m)_N$ and in $[(K_t/K_m)_N]_u/[(K_t/K_m)_N]_C$, which are given in order of increasing neutron energy.

k'_T is less sensitive to changes in atomic composition than k_T. The differences of several per cent in responses in neutron beams, observed with individual chambers of the same design having an exposure calibration [11], will be greatly reduced.

REFERENCES

[1] BROERSE, J.J., BURGER, G., COPPOLA, M., in " A European neutron dosimetry intercomparison project (ENDIP), Results and Evaluation", Rep. EUR 6004, CEC, Luxembourg (1978) 49.

[2] AMERICAN ASSOCIATION of PHYSICISTS in MEDICINE, Protocol for neutron beam dosimetry, AAPM Rep. 7, American Institute of Physics, New York (1978).

[3] EUROPEAN CLINICAL NEUTRON DOSIMETRY GROUP, European protocol for neutron dosimetry for external beam therapy, Brit. J. Radiol. 54 (1981) 882.

[4] MIJNHEER, B.J., WILLIAMS, J.R., Phys. Med. Biol. 26 (1981) 57.

[5] INTERNATIONAL COMMISSION ON RADIATION UNITS AND MEASUREMENTS, Radiation Dosimetry: X Rays and Gamma Rays with Maximum Photon Energies Between 0.6 and 50 MeV, ICRU Rep. 14, ICRU, Washington, D.C. (1969).

[6] NAHUM, A.E., GREENING, J.R., Phys. Med. Biol. 23 (1978) 894.

[7] NATH, R., SCHULZ, R.J., Medical Physics 8 (1981) 85.

[8] NIATEL, M.-T., Private communication (1980).

[9] INTERNATIONAL COMMISSION ON RADIATION UNITS AND MEASUREMENTS, Neutron Dosimetry for Biology and Medicine, ICRU Rep. 26, ICRU, Washington, D.C. (1977).

[10] MCDONALD, J.C., MA, I.-C., MIJNHEER, B.J., ZOETELIEF, J., Medical Physics 8 (1981) 44.

[11] MIJNHEER, B.J., VAN WIJK, P.C., ZOETELIEF, J., BROERSE, J.J. in "Fourth Symposium on Neutron Dosimetry", Rep. EUR 7448, Vol. II, CEC, Luxembourg (1981) 361.

TLD-300 DETECTORS FOR SEPARATE MEASUREMENT OF TOTAL AND GAMMA ABSORBED DOSE DISTRIBUTIONS OF SINGLE, MULTIPLE, AND MOVING-FIELD NEUTRON TREATMENTS

A New Method of Clinical Dosimetry for Fast Neutron Therapy

J. RASSOW
Institut für Medizinische Strahlenphysik und Strahlenbiologie,
Universitätsklinikum Essen, Essen,
Federal Republic of Germany

Abstract

TLD-300 DETECTORS FOR SEPARATE MEASUREMENT OF TOTAL AND GAMMA ABSORBED DOSE DISTRIBUTIONS OF SINGLE, MULTIPLE, AND MOVING-FIELD NEUTRON TREATMENTS – A NEW METHOD OF CLINICAL DOSIMETRY FOR FAST NEUTRON THERAPY.

Fast neutron therapy requirements, because of the poor depth dose characteristic of present therapeutical sources, are at least as complex in treatment plans as photon therapy. The physical part of the treatment planning is very important; however, it is much more complicated than for photons or electrons owing to the need for: Separation of total and gamma absorbed dose distributions (D_T and D_G); and more stringent tissue-equivalence conditions of phantoms than in photon therapy. Therefore, methods of clinical dosimetry for the separate determination of total and gamma absorbed dose distributions in irregularly shaped (inhomogeneous) phantoms are needed. A method using TLD-300 (CaF_2:Tm) detectors is described, which is able to give an approximate solution of the above-mentioned dosimetric requirements. The two independent doses, D_T and D_G, can be calculated by an on-line computer analysis of the digitalized glow curve of TLD-300 detectors, irradiated with d(14)+Be neutrons of the cyclotron isocentric neutron therapy facility CIRCE in Essen. Results are presented for depth and lateral absorbed dose distributions (D_T and D_G) for fixed neutron beams of different field sizes compared with measurements by standard procedures (TE-TE ionization chamber, GM counter) in an A-150 phantom. The TLD-300 results for multiple and moving-field treatments (with and without wedge filters) in a patient simulating irregularly shaped (homogeneous) phantoms, are shown together with computer calculations of these dose distributions. The probable causes for some systematic deviations are discussed, which lead to open problems for further investigations owing to features of the detector material and the evaluation method, but mainly to differences in the composition of phantom materials used for the calculations (standard dose distributions) and TLD-300 measurements.

1. INTRODUCTION

Optimum radiotherapy treatment needs, as one essential basic requirement, optimum treatment planning and capability so as to realize accurately and reproducibly even complex treatment techniques and schedules. For fast neutron therapy an additional problem arises compared with photon and electron therapy. While photons and electrons as primary (and secondary) radiation quality act within a narrow and low-LET range, only weakly influenced by the particle energy spectrum, primary neutrons always act by low-LET (secondary) photons ("gammas") and high-LET heavy secondary particles produced by recoil effects and nuclear reactions ("neutrons").

The biological radiation effects in human cells are quite different for the same absorbed doses of low-LET and high-LET particles. For survival rate curves of human melanoma cells with different absorbed dose components of neutrons and photons (Fig.1), it is obvious that

The resulting survival rate can be identified by both the neutron absorbed dose D_N *and* the gamma absorbed dose D_G (or $D_T = D_N + D_G$ and D_G or D_G/D_T); and

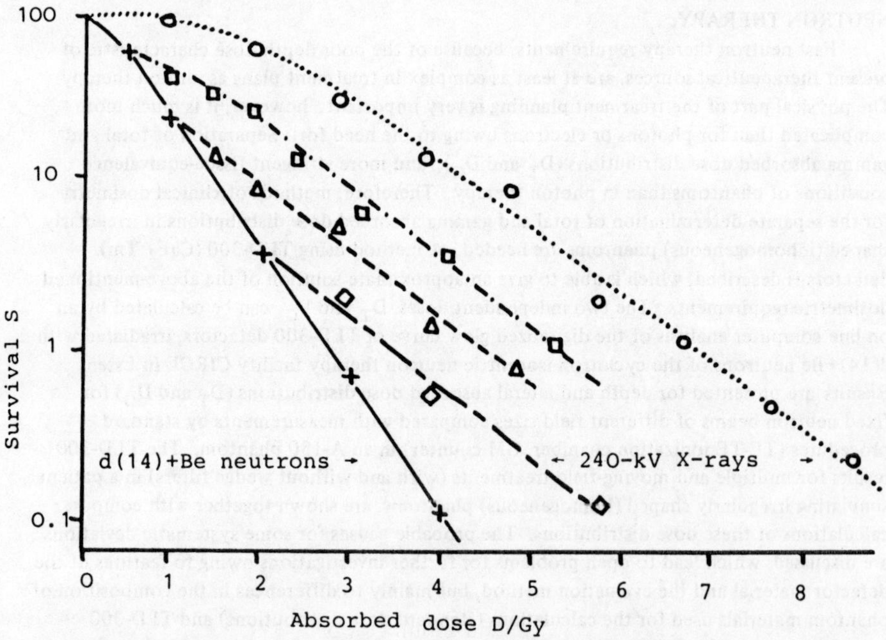

FIG.1. Survival rate for human melanoma cells MeWo for "simultaneous" (within 6 h, cycle time >30 h) neutron and X-ray irradiation "free in air". Colony forming test. After Streffer et al. [2].

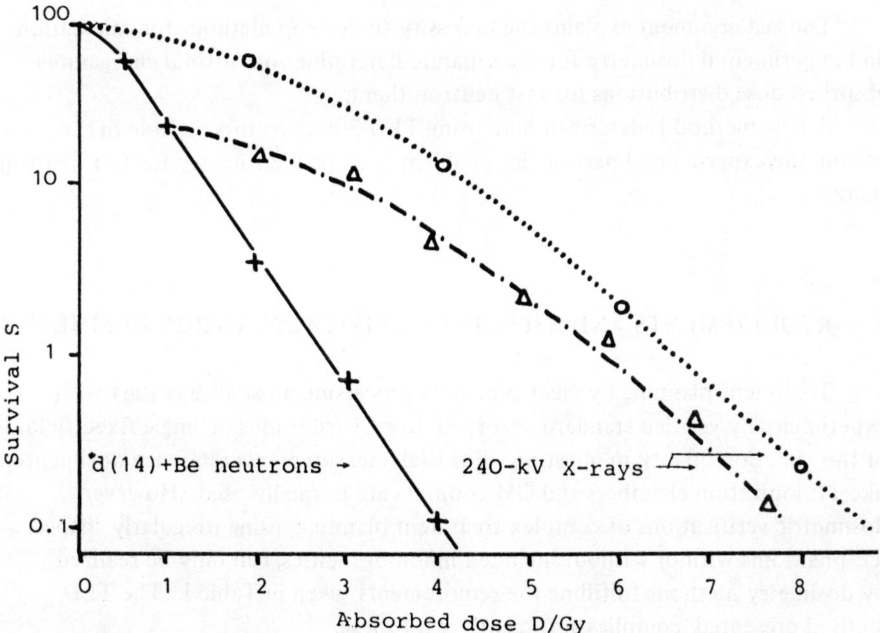

FIG.2. Survival rate for human melanoma cells Be11 (Bevey) for "simultaneous" (directly consecutive) neutron and X-ray irradiation "free in air". Colony forming test. After Streffer et al. [2].

Gamma absorbed doses D_G show recovery effects ("shoulder curves") only for high final survival rates (⩾50%) of the neutron pre-irradiation.

The result for treatment planning and the corresponding clinical dosimetry is that:

It is therapeutically essential to specify (and report) at the specification point within the target volume D_T and D_G separately (ECNEU [1]);

It is of equal therapeutical relevance for practical treatment planning and comparison of tolerance and actual therapy absorbed doses of organs at risk to know D_T and D_G separately outside the target volume as well. Especially for low absorbed dose levels outside the target volume the relative biological effectiveness of neutrons can be extremely high (Fig.1).

The superposition of neutron and photon effects can be quite different even for very similar cell types, and gamma absorbed doses D_G may show recovery effects also for low final survival rates of the neutron irradiation (Fig.2).

The last argument explains the necessity to develop methods for calculation and experimental dosimetry for the separate determination of total *and* gamma absorbed dose distributions for fast neutron therapy.

A new method is described here using TLD-300 detectors capable of solving the experimental part of this problem in clinical dosimetry for fast neutron therapy.

2. REQUIREMENTS AND ASPECTS OF CLINICAL NEUTRON DOSIMETRY

Treatment planning by electronic data processing must always start with experimentally verified standard absorbed dose distributions of single fixed fields of the radiation quality in question. For such measurements reference dosimenters like TE ionization chambers and GM counters are normally used. However, dosimetric verifications of complex treatment planning, using irregularly shaped TE phantoms with or without included inhomogeneities, can only be realized by dosimetry methods fulfilling the requirements given in Table I. The TLD method presented complies sufficiently with these.

3. COMPARISON OF CHARACTERISTIC FEATURES OF TLD METHODS FOR PHOTONS AND NEUTRONS

Table II gives a survey of those aspects which are typical and different for the clinical dosimetry for photon/electron and for neutron therapy. These results are

Particles to be measured;
Detector types for point and distribution measurements;
Requirements for phantom materials.

Table III gives a survey of characteristic features of the most common TLD methods for photon and neutron dosimetry compared with the TLD method described here. Details are discussed in the following paragraphs.

4. THE PRINCIPLE AND HANDLING OF THE NEW TLD METHOD

The new TLD method, its principle and features, together with the first results (Figs 3 to 10) are discussed in detail by Temme et al. [3].

TABLE I. REQUIREMENTS OF CLINICAL DOSIMETRY FOR MEASURING DOSE DISTRIBUTIONS OF FAST NEUTRON TREATMENT

Keyword	Requirement
Separation D_T and D_G	Separate measurement of total and gamma absorbed dose
Range: Absorbed dose	$1\ \text{mGy} < D_T < 10\ \text{Gy}$
Absorbed dose rate	$10 \frac{\mu\text{Gy}}{\text{min}} < \dot{D}_T < 1 \frac{\text{Gy}}{\text{min}}$
Detector set	Preferably only one detector with at least two measuring signals of different responses for neutrons and photons
Energy dependence	Low dependence of responses of both measuring signals on changes of energy spectrum of neutrons and photons at different phantom measuring points
Dimensions	Satisfactory small dimensions of TLD detectors for measurements: In ranges of high dose gradient, and Adjacent to surfaces of inhomogeneities
Accuracy	Satisfactory accuracy of clinical dosimetric results of dose distributions and for in-vivo point doses
Effort	Satisfactory minimal effort for measurement, calibration and evaluation using the TLD method

4.1. Handling of detectors

The handling of detectors is either the same or, for the annealing procedure, even less complicated than usual. Table IV lists some details of the handling procedure.

TABLE II. SURVEY OF CLINICAL-DOSIMETRIC ASPECTS

Primary radiation type	Photons or electrons	Neutrons	
Secondary radiation types	Electrons/photons	Neutrons/recoil particles	Photons
Dosimetric separation necessary for primary and secondary radiation types	No	Yes ("Neutrons")	("Gamma rays")
Point dose measurements Reference dosimeters	Ionization chamber (Ferrous sulphate dosimeter)	TE ionization chamber	GM counter ($C\text{-}CO_2$, Mg-Ar, film, TLD)
Field dosimeters:	Ionization chamber	TE ionization chamber	GM counter
Gas flow	No	Yes	No
Filling gas	Air	TE gas	Ne (+He, halogen)
Therapeutic dose rate possible	Yes	Yes	No
Dose distribution measurements			
Relative dosimetric methods	TLD* Semiconductor* Film	^{7}LiF (TLD-700) / ^{6}LiF (TLD-600)*	
This work		CaF_2:Tm (TLD-300)*	
Conditions on tissue equivalence of materials require congruence in:	Effective atomic number depending on predominating interaction	Percentage by weight as well for H, N as for other mainly light atoms (C and O limited compatibility)	

* Applicable for in-vivo measurements.

4.2. Evaluation of total and gamma absorbed doses D_T and D_G (or D_G/D_T) from measuring signals

In Figs 3 and 4 the glow curves for a (total) absorbed dose of 1 Gy of a photon neutron irradiation of two TLD-300 detectors are shown together with a mathematically simulated sum function composed by six single Gaussian functions for the six identified peaks. This analysis is described in Section 4.3. It is obvious that

The photon response of the detectors is much higher than the neutron response; and
The quotient of the responses of the main peaks 1 and 2 (peaks 3 and 5) is different for photons and neutrons.

These two LET depending effects result in the possibility of evaluating from the two main peaks separately the absorbed doses for neutrons D_N and photons D_G – on condition that the responses of the main peaks of the detectors remain constant for energy spectrum changes occurring between the measuring and calibration conditions.

That this pre-condition is met within certain limits can be seen in Fig.5 where there is a linear dependency between the standardized measuring signals M_1 and M_2 and the local gamma-ray content D_G/D_T (produced in different phantom depths with several field sizes). Serious deviations are found only for $D_G/D_T > 0.17$ corresponding to phantom depths $z > 20$ cm along the axis of the beam.

In Table V the symbols necessary for the equations below are defined. As "measuring signal" the mathematically determined single peak height in Section 4.3 is used.

The basic equations for the absorbed dose evaluation are given in Eqs (1), (2a) and (2b), and for the response coefficients, a_ν and b_ν, in Eqs (3a) and (3b), and for practical absorbed dose determinations in Eqs (4), (5), (6a) and (6b) (numerical values in Section 4.5).

$$D_T = D_N + D_G \tag{1}$$

$$M_1 = a_1 D_N + b_1 D_G = a_1 (D_T - D_G) + b_1 D_G \tag{2a}$$

$$M_2 = a_2 D_N + b_2 D_G = a_2 (D_T - D_G) + b_2 D_G \tag{2b}$$

$$\frac{M_1}{D_T} = a_1 \frac{D_T - D_G}{D_T} + b_1 \frac{D_G}{D_T} = a_1 + (b_1 - a_1) \frac{D_G}{D_T} \tag{3a}$$

$$\frac{M_2}{D_T} = a_2 \frac{D_T - D_G}{D_T} + b_2 \frac{D_G}{D_T} = a_2 + (b_2 - a_2) \frac{D_G}{D_T} \tag{3b}$$

TABLE III. COMPARISON OF CHARACTERISTIC FEATURES OF TLD METHODS FOR MEASUREMENT, CALIBRATION AND EVALUATION

Primary radiation type	Photons or electrons	Neutrons	
Method	One detector (conventional)	Two detectors (conventional)	One detector (this work)
Detector material: e.g.	LiF:Mg, Ti	^{6}LiF and ^{7}LiF	CaF_2:Tm
trade name	TLD-100	TLD-600 and TLD-700	TLD-300
detector dimensions: ribbon	3.2 mm × 3.2 mm × 0.9 mm	3.2 mm × 3.2 mm × 0.9 mm	3.2 mm × 3.2 mm × 0.9 mm
rod	1 mm × 1 mm × 6 mm	1 mm × 1 mm × 6 mm	1 mm × 1 mm × 6 mm
Annealing procedures			
(i) Pre-irradiation: 400°C, 1 h	Yes	Yes	Maybe no (500°C)
(ii) Pre-irradiation: 80°C, 24 h	Yes	Yes	Probably no (100°C)
(iii) Post-irradiation: 100°C, 10 min	Yes	Yes	Definitely no
Measuring procedure: Advantageous	Charge measurement within fixed (planchet-) temperature limits, digital indication	Charge measurement within fixed (planchet-) temperature limits, digital indication	Glow curve analysis by on-line computing
Less advantageous	Peak height measurement using the analogous glow curve record	Peak height measurement using the analogous glow curve record	Peak height determination using the mathematically smoothed digital glow curve information
Measuring signal	Charge (≅glow curve area)	Charge (≅glow curve area)	Single peak height or area

TABLE III. (cont.)

Separation of neutron and photon absorbed doses	Not necessary	By difference of neutron cross-section of ^{6}Li and ^{7}Li	By difference of LET dependence of measuring signal responses	
Relative methodical effort (related to TLD-100, conventional procedure):				
Measurement	1	2	<1 (no annealing)	
Calibration	1	2	1	
Evaluation	1	>2	<1 (computer)	
Measuring parameter(s)	D	D_T and D_G	D_T and	D_G
Detection limit (approximate)	0.1 mGy		0.1 mGy	0.01 mGy
Detection standard deviation: ($D_T \cong 0.01-1$ Gy)				
Reproducibility	±2%		±3%	±6%
	±3%		±5%	±10%
Corrections:				
Field size dependence	No		>200 cm^2	No
Phantom depth dependence	No		>20 cm	>20 cm

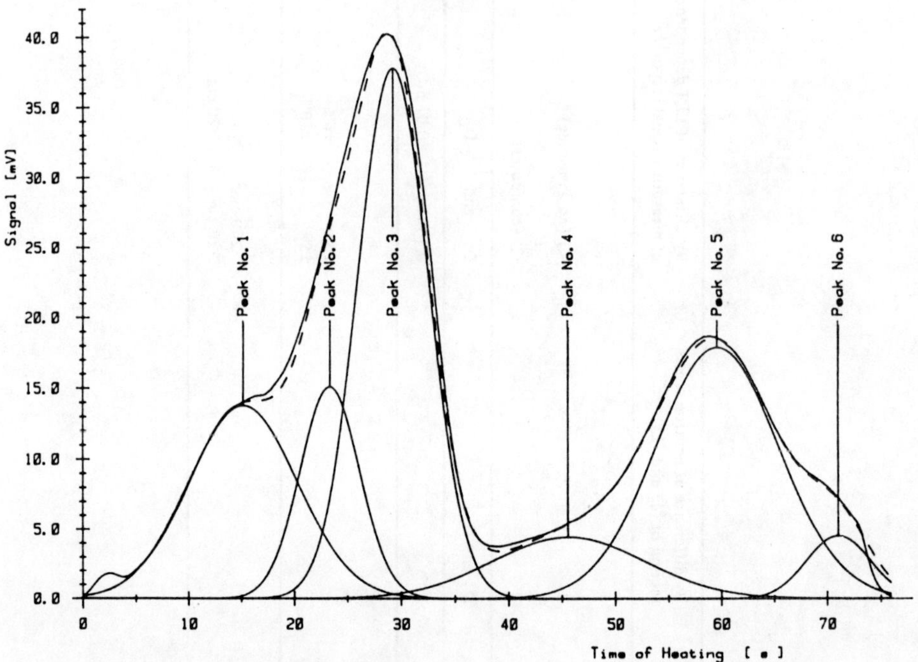

FIG.3. Analysis of a TLD-300 glow curve using Gaussian peak shape; radiation type: ^{60}Co gamma rays; phantom: polystyrene; field size: $A_c = 15$ cm × 15 cm/60 cm; focus-surface distance: $s_c = 60$ cm; phantom depth: $z_c = 5$ cm; total absorbed dose: $D_T = 1$ Gy; ——: Measured TLD signal and analysed Gaussian peaks; --------: Sum of the analysed Gaussian single peaks.

$$D_T = \frac{(b_1 - a_1)M_2 - (b_2 - a_2)M_1}{a_1 b_2 - a_2 b_1} = \frac{0.898\,M_2 - 0.697\,M_1}{0.0861} \tag{4}$$

$$D_N = \frac{b_1 M_2 - b_2 M_1}{a_1 b_2 - a_1 b_1} = \frac{0.920\,M_2 - 0.810\,M_1}{0.0861} \tag{5}$$

$$D_G = \frac{a_2 M_1 - a_1 M_2}{a_1 b_2 - a_2 b_1} = \frac{0.113\,M_1 - \;.022\,M_2}{0.0861} \tag{6a}$$

$$\frac{D_G}{D_T} = \frac{a_2 M_1 - a_1 M_2}{(b_1 - a_1)M_2 - (b_2 - a_2)M_2} = \frac{0.113\,M_1 - 0.022\,M_2}{0.898\,M_2 - 0.697\,M_1} \tag{6b}$$

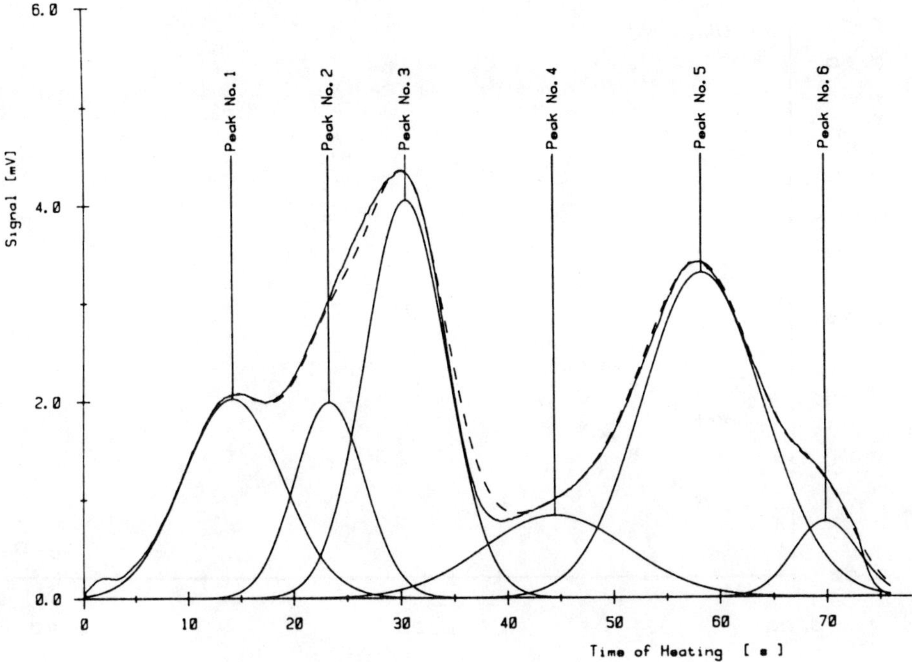

FIG.4. Analysis of a TLD-300 glow curve using Gaussian peak shape; Radiation type: d(14)+Be neutrons (CIRCE); phantom: A-150; field size: $A_0 = 10$ *cm* × *10 cm/125 cm; focus-surface distance:* $s_0 = 125$ *cm; phantom depth:* $z = 10$ *cm; total absorbed dose:* $D_T = 1$ *Gy; gamma-ray content:* $D_G/D_T = 0.107$*; ———: Measured TLD signal and analysed Gaussian peaks; - - - - - - - : Sum of the analysed Gaussian single peaks.*

4.3. Mathematical glow curve analysis for obtaining the most accurate measuring signals

To obtain the accuracy necessary for measuring the dose distributions of D_T and D_G in phantoms for neutron therapy, it is essential to optimize not only the annealing and calibration procedures but also the evaluation of the measuring signals from the glow curve of the TLD detectors in use. An adequate hardware set for this goal is the combination of a digital voltmeter and a desk-top computer (e.g. Hewlett Packard hp 3456A and hp 85F, both with IEC-bus interface HP-IB (IEEE 488) and sampling frequency of 4.67 Hz) in addition to a commercial TLD unit (e.g. Harshaw 2000). The task for this hardware set is to digitalize and store the glow curve and to make the consecutive mathematical glow curve analysis automatically. Technical details of the electronic procedure and the software configuration are described by Meissner [4].

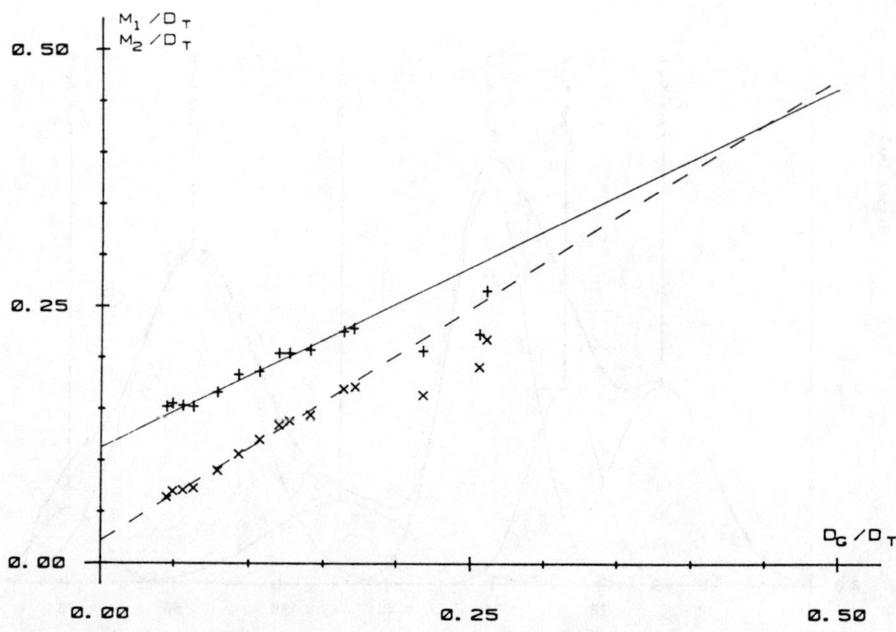

FIG.5. Standardized TLD measuring signals as function of gamma-ray content D_G/D_T; radiation type: d(14)+Be neutrons (CIRCE); phantom: A-150; detector: TLD-300 ribbon; total absorbed dose: $D_T = 1$ Gy; --------: $M_1/D_T = 0.022\ (1-D_G/D_T) + 0.920\ D_G/D_T$; ———: $M_2/D_T = 0.113\ (1-D_G/D_T) + 0.810\ D_G/D_T$.

Figure 11 gives a comparison of glow curves for TLD-300 detectors, irradiated in all cases with $D_T = 1$ Gy but different gamma-ray content D_G/D_T. It is obvious that the peaks are approximately symmetric, at least near the maxima so that they may be described by Gaussian functions of the type

$$S(t) = S_0 \exp\left[-\left(\frac{t-t_0}{B_0}\right)^2\right] \tag{7}$$

where S(t) is the measuring signal at the heating time t, and S_0 the maximum value of the digitalized samples of the Gaussian impulse at time t_0 and B_0, the width (at decay 1/e). From Eq.(7) follows the function

$$F(t) = -\ell n\left(\frac{S(t)}{S_0}\right) = \frac{1}{B_0}(t-t_0) \tag{8}$$

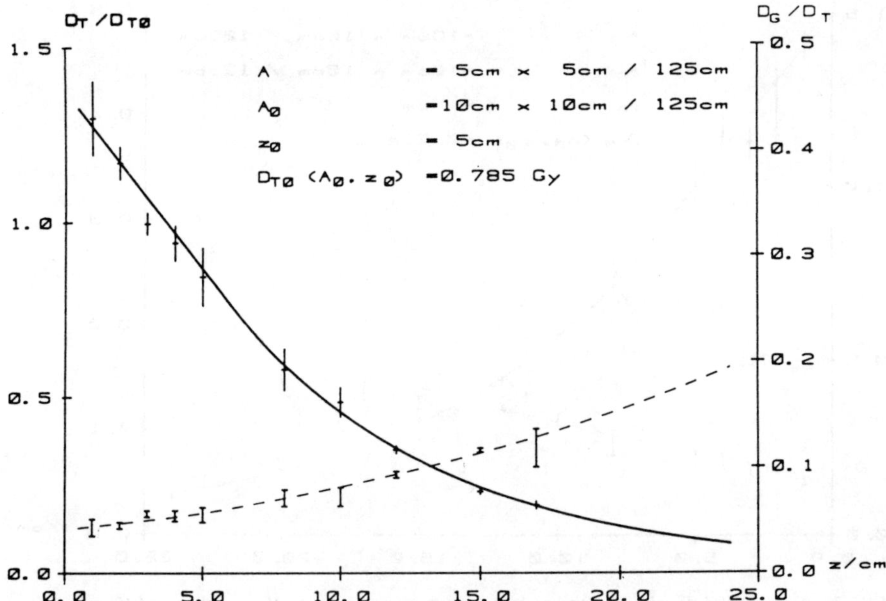

FIG.6. Normalized depth dose and relative gamma-ray content; radiation type: d(14)+Be neutrons (CIRCE); phantom: A-150; detector: TLD-300 ribbons; ++: Deviation range of indication for D_T/D_{T0}; III: Deviation range of indication for D_G/D_T; ———: TE-ionization chamber IC-18 (FWT); ---------: Geiger-Müller counter GM1 (FWT).

The right-hand side of Eq.(8) is the description of two straight lines: For $t < t_0$ the straight line decreases, and for $t > t_0$ it increases. The slopes of the two lines may be different; this would be an indicator for an overlapping effect of an adjacent Gaussian peak. For $t = t_0$ F(t) has the value 0.

As far as the assumption of a Gaussian-shaped peak profile is valid the function F(t) must follow on both sides of a peak's maximum straight lines. For example, in Fig.12 it is shown that this condition is realized for both main peaks of the glow curve.

From the abscissa position of the minimum $F(t)_{min}$ of the function F(t) the correct position of the glow peak maximum can be derived, while the deviation of $F(t)_{min}$ from the value $F(t) = 0$ supplies a correction to obtain from the maximum measured value S_0 the fitted real maximum of the glow peak S_0'.

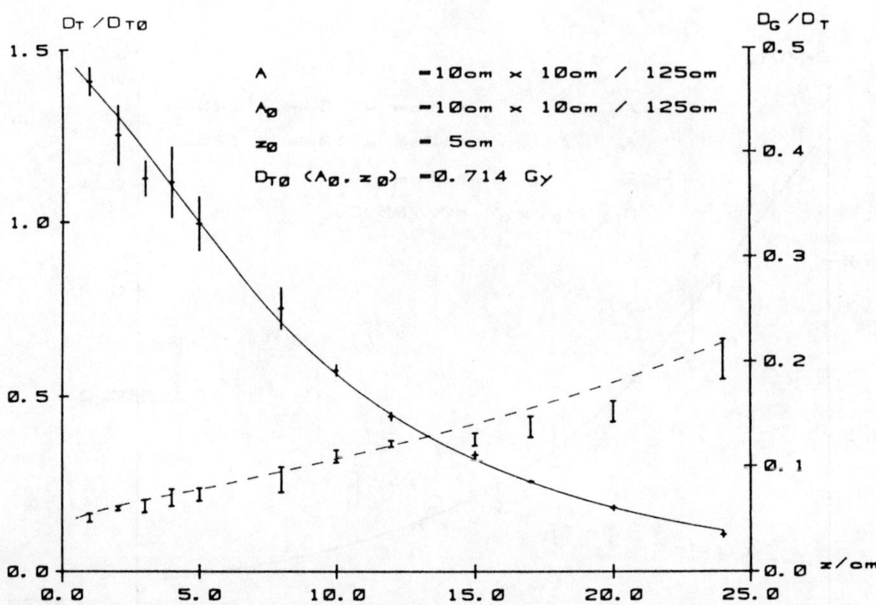

FIG.7. Normalized depth dose and relative gamma-ray content. Measuring conditions, see Fig.6.

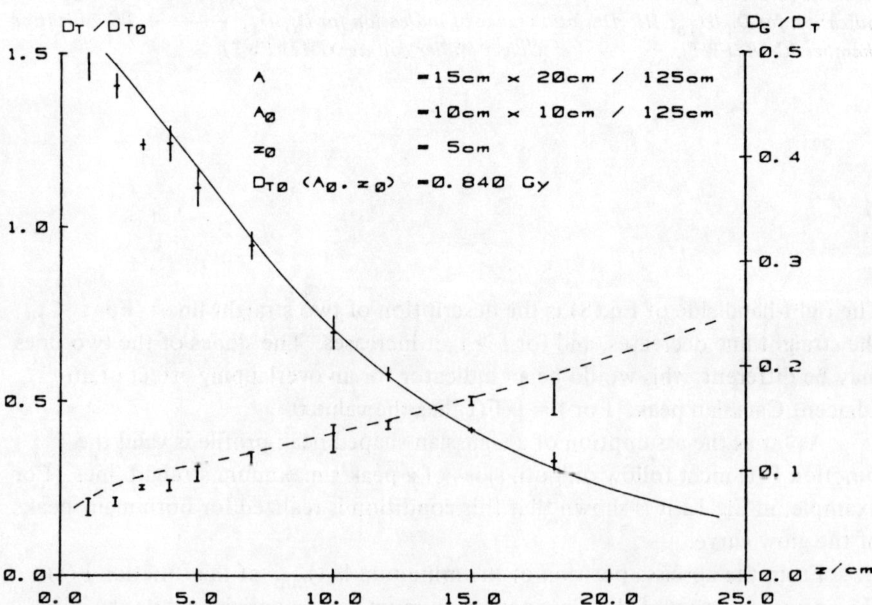

FIG.8. Normalized depth dose and relative gamma-ray content. Measuring conditions, see Fig.6.

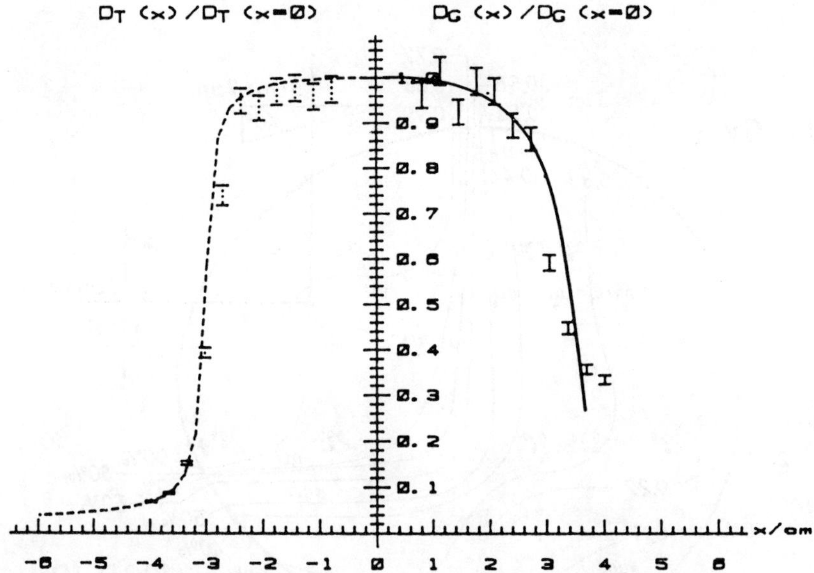

FIG.9. Normalized lateral absorbed dose distributions for $D_T(x)/D_T(x = 0)$ and $D_G(x)/D_G(x = 0)$; Radiation type: d(14)+Be neutrons (CIRCE, Essen); phantom material: A-150, 30 cm × 30 cm × 30 cm; detector: TLD-300 ribbon; ⁝⁝⁝ = Deviation range of indication of 3 detectors for $D_T(x)/D_T(x = 0)$; III = Deviation range of irradiation of 3 detectors for $D_G(x)/D_G(x = 0)$; - - - - - -: TE ionization chamber measurement (IC-18, FWT); ———: Geiger-Müller counter measurement (GM1, FWT); field size A: 5 cm × 5 cm/125 cm; phantom depth z = 0.8 cm; lateral distance x: −6 ... +6 cm.

A glow curve measurement with a temperature gradient of 4.5 K/s and a maximum heating temperature of about 350°C takes about 70 s and as a result of the sampling frequency of 4.67 Hz produces, together with the background measured values before the start of the analysable glow curve, about 400 samples. Twenty-five samples on both sides of a peak maximum are used for a least-squares fit to obtain the above-mentioned information.

The final information on all the six glow peak maxima is produced by a recursion procedure with the main steps of Table VI.

4.4. Calibration of individual photon response of measuring signals

It is an essential pre-condition for satisfactory accuracy of the TLD measuring results in neutron beams to follow a stringent procedure of photon calibration of each TLD-300 detector. It was found that for all detectors of the same annealing

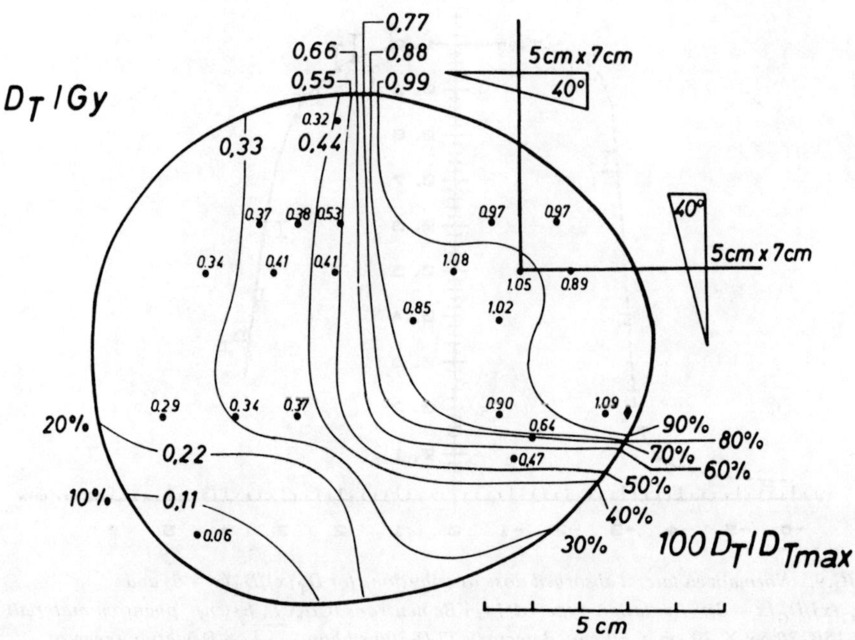

FIG.10. Calculated total absorbed isodose curves (parameter values medium-sized) compared with measured values of TLD-300 detectors for a two-wedge filter d(14)+Be neutron irradiation of a phantom out of 50% paraffin and 50% beeswax (post-operative treatment of a soft tissue fibro-sarcoma in the upper thigh (Rassow et al. [7]).

group, being handled simultaneously on the same aluminium plate carrier, small changes in sensitivity occur from one annealing cycle to the other even if the procedure is strictly reproduced. These changes, however, are the same for all detectors of the group and can thus be determined by a number (e.g. four) of "reference detectors" which are irradiated under photon reference conditions. In addition it is necessary to note deviations of individual detector features from the group-reference detector features. The corresponding "relative single detector correction factors $F_{D\nu}$" are often different for both the main peaks ($\nu = 1$ or 2) of the TLD-300 detector in question. Differences up to 15% were found. The correction factors $F_{D\nu}$ prove to be stable within ±5% for at least 50 exposition measuring annealing cycles.

The practical photon calibration procedure is schematically described by the conditions and steps in Tables VII and VIII.

TABLE IV. HANDLING PROCEDURE OF TLD DETECTORS (this work)

Keyword	Procedure
During exposition	Phantom measuring plates with milled holes for detectors
During annealing, before and after measurement and storage	Aluminium plate with milled holes for detectors
During manipulation	Suction head
Number of detectors in each annealing group	About 30–35 including 4 reference detectors irradiated under photon calibration conditions
I. Extinguishing annealing (after usage)	500°C, 1 h
II. Low-temperature peak suppression annealing (directly after annealing I without cooling down)	100°C, 2 h usual, but can be omitted applying mathematical glow curve analysis (Sect. 4.3)
III. Low-temperature peak extinguishing annealing (after irradiation and before measuring)	100°C, 10 min usual, but can definitely be omitted

4.5. Calibration of neutron-photon response coefficients

In view of Eq (4) to (6) for calculations of total absorbed dose D_T and gamma absorbed dose D_G, it is necessary to calibrate first the coefficients a_ν and b_ν ($\nu = 1$ or 2).

As already mentioned in Section 4.2, the calibration of the response coefficients must be done for as many measuring conditions as possible, varying the phantom depths and neutron field sizes and seeking the variations of the resulting standardized measuring signals M_ν on the corresponding relative gamma absorbed dose content D_G/D_T, which is measured by standard procedures with a TE ionization chamber and a GM counter.

TABLE V. SYMBOLS FOR DOSE EVALUATIONS FROM TLD MEASURING SIGNALS

Symbol	Definition	
P_1	Measuring signal, derived from main peak 1 (peak 3 in Figs 3 and 4)	
P_2	Measuring signal, derived from main peak 2 (peak 5 in Figs 3 and 4)	
M_1	$P_1\,(D_c/P_1)_{^{60}Co}$	Standardized measuring signals (measuring signal, divided by the single reference response according to Sect. 4.4)
M_2	$P_2\,(D_c/P_2)_{^{60}Co}$	
D_c	Reference absorbed dose used for photon calibration of the reference detectors according to Table IV	
T, N, G	Indices for absorbed doses in ICRU muscle tissue due to the total, the neutron, and the gamma absorbed dose, respectively	

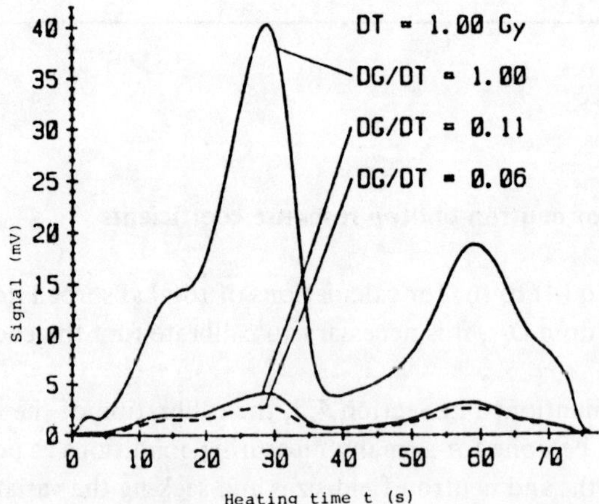

FIG.11. Glow curves of TLD-300 (CaF_2:Tm) detectors for different gamma absorbed dose content D_G/D_T of d(14)+Be neutrons (after Meissner [4]).

TABLE VI. MAIN STEPS OF RECURSION PROCEDURE OF MATHEMATICAL GLOW CURVE ANALYSIS

Step No.	Procedure
1a	Calculation of the data of the largest peak by the mathematical procedure discussed above
1b	Subtraction of the Gaussian-shaped peak given by the data of step 1a from the total glow curve
1c	Change of step 1a data so that no negative difference results in step 1b
2a to 2c	Calculation of the data of the second largest peak following the logic of steps 1a to 1c, subtracting, however, in step 2b from the differential curve resulting from step 1b
3 to 6	Calculation of the data of the next 4 peaks due to their height following the analogous sequence discussed in steps 1 and 2
7	Final test – if the sum of all 6 analysed Gaussian profiles fit satisfactorily the total glow curve. If not, the recursion is again started at the peak producing the largest deviations with subsequent recalculations of all other peak data

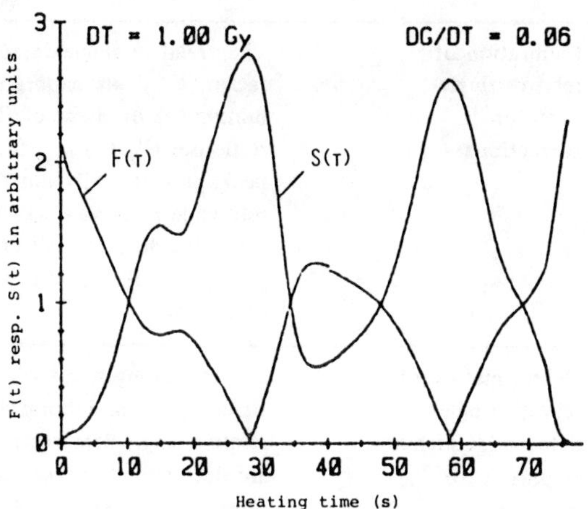

FIG.12. Time function F(t) and corresponding glow curve S(t) – see Eqs (2) and (8) (after Meissner [4]).

TABLE VII. CALIBRATION CONDITIONS FOR DETERMINATION OF THE PHOTON REFERENCE RESPONSES $(P_\nu/D_c)_{^{60}Co}$

Keyword	Specification
Radiation type	^{60}Co gamma rays
Phantom	Polystyrene cube 25 cm × 25 cm × 25 cm
Phantom measuring plate	Milled holes for ionization chamber and TLD detectors
Focus-surface distance s_c	60 cm
Phantom depths z_c	5 cm
Field size A_c at s_c	15 × 15 cm/60 cm
Calibration absorbed dose D_c	about 0.4 Gy

TABLE VIII. STEPS OF PHOTON CALIBRATION PROCEDURE

Step No.	Keyword	Procedure
1	Relative reference light correction	Deviations of the TLD measuring system sensitivity are to be corrected for by a "relative reference factor F_L" which can be determined at each measuring series by a built-in reference light test
2	Calibration of relative single-detector corrections	The "relative single-detector correction factors $F_{D\nu}$" are calibrated by determining the quotient of the individual responses $(P_\nu/D_c)_{^{60}Co}$ for both main peaks (ν = 1 or 2) and the "group-reference responses" $(P_\nu/D_c)^*_{^{60}Co}$ of the reference detectors all under conditions given in Table VI with reference light correction
3	Determination of relative single-detector reference responses for measurements	For evaluating standardized measuring signals given in Section 4.2, the necessary relative single-detector reference responses are determined on the basis of the group-reference responses of the annealing cycle in question: $(P_\nu/D_c/_{^{60}Co} = F_L \cdot F_{D_\nu} (P_\nu/D_c)^*_{^{60}Co}$

TABLE IX. CALIBRATION CONDITIONS FOR DETERMINATION OF NEUTRON-PHOTON RESPONSE COEFFICIENTS a_ν AND b_ν

Keyword	Specification
Radiation type	d(14)+Be neutrons
Phantom	A-150 plastic cube 30 cm × 30 cm × 30 cm
Phantom measuring plate	Milled holes for TLD detectors
Variation of the gamma-ray component	Different exposition phantom depths (near the beam axis)
Focus-surface distance s_0	125 cm
Field size A at s_0	5 cm × 5 cm/125 cm and 10 cm × 10 cm/125 cm
Calibration total absorbed dose D_{T0} at the exposition point	1 Gy

The calibration conditions for determining the response coefficients are listed in Table IX.

The practical determination of the response coefficients can be done by a graphical display of the measuring results in a diagram (e.g. Fig.5) M_ν/D_{T0} versus D_G/D_T taking a_ν as ordinate section, and $(b_\nu - a_\nu)$ as the slope of the linear part of the relation M_ν/D_T = factor (D_G/D_T) as it is taken as a pre-condition for Eqs (3a) and (3b).

For A-150 plastic phantom material the following response coefficients were found for the calibration conditions to Table IX:

$a_1 = 0.022$ $\qquad b_1 = 0.920$

$a_2 = 0.113$ $\qquad b_2 = 0.810$

5. APPLICATION OF TLD-300 DETECTORS FOR CLINICAL NEUTRON DOSIMETRY

One of the most important application areas of clinical dosimetry is the measurement of absorbed dose distributions in irregularly shaped tissue-equivalent phantoms with or without included inhomogeneities for all therapauetically relevant treatment techniques. The applicability of the thermoluminescent dosimetry method using TLD-300 detectors is demonstrated in the following sections.

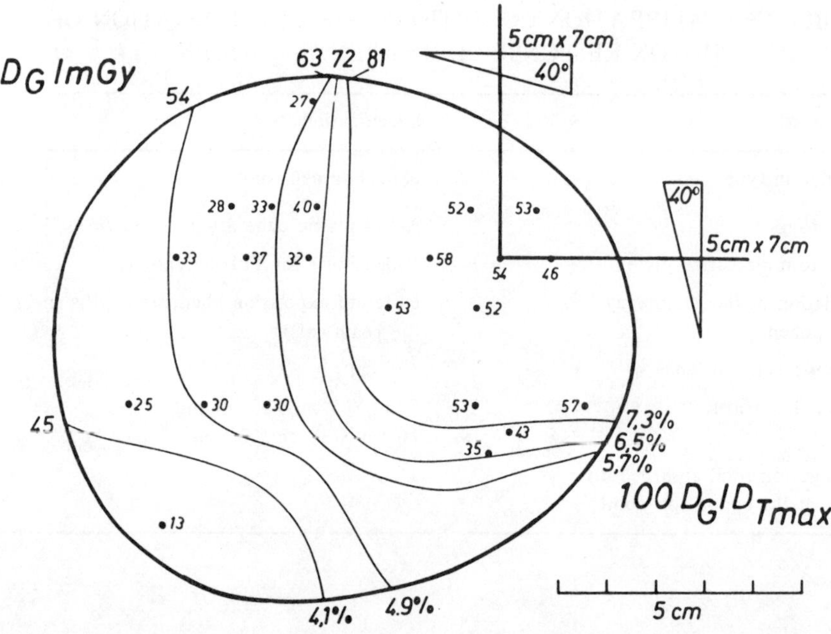

FIG.13. Calculated gamma-ray absorbed isodose curves (parameter values, medium-sized) compared with measured values simultaneously indicated by TLD-300 detectors of Fig.10.

A test is made for single fixed neutron fields with perpendicular incidence on an A-150 phantom compared with measurements with a TE ionization chamber and a GM counter (Section 5.1).and in all other cases (Sections 5.2 to 5.4) compared with calculations of dose distributions (Meissner, Rassow [5]; Baumhoer [6]). A fundamental problem arises from the fact that the calculations are made for a water phantom as a patient's body imitation, while the measurements with TLD-300 detectors could only be made in a solid phantom. This problem is still open and is explained further in Section 6.3.

5.1. Absorbed dose distributions (D_T and D_G) for d(14)+Be neutron treatments with single fixed fields

The accuracy of the TLD method can be completely experimentally tested by measurement of depth dose and lateral dose distributions in an A-150 plastic phantom of 30 cm × 30 cm × 30 cm for different field sizes of d(14)+Be neutrons using the neutron-photon response coefficients (Sections 4.5) for the TLD-300 detectors on the one hand, and a TE ionization chamber, FWT IC-18 (0.1 cm^3), and a shielded Geiger-Müller counter, FWT GM1, on the other. The

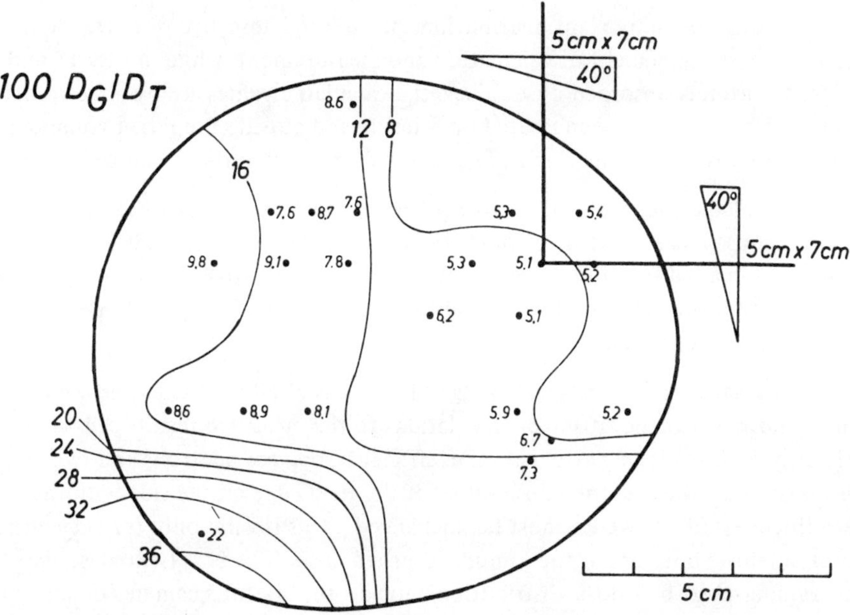

FIG.14. Calculated local gamma-ray component curves (parameter values medium-sized) compared with the quotient (in per cent) of the measured values by the TLD-300 detectors in Figs 10 and 13.

results are shown in Figs 6–9. The solid and dashed curves belong to the reference measurements while the measuring points, including the deviation range, are obtained by using always the results from three different detector readings. Deviations which seem to be systematic occur only for the largest examined field size, 15 cm × 20 cm, for phantom depth ranges z = 0 ... 6 cm (D_T) and z > 17 cm (D_G). It should be remembered that the limit of applicability for gamma absorbed dose contents is $D_G/G_T < 0.17$ (Fig.5).

5.2. Two-wedge filter fixed fields

The first case of therapeutical relevance is the post-operative treatment of a soft-tissue fibro-sarcoma in the upper thigh. The treatment planning shows that the best dose distributions with respect to the target volume can be produced by two 90° equally weighted wedge filter fields of d(14)+Be neutrons.

Figures 10, 13 and 14 show the results of a comparison of calculated and TLD-300 measured values for dose distributions of the total absorbed dose D_T (Fig.10), the gamma absorbed dose D_G (Fig.13), and the local gamma absorbed dose component D_G/D_T (Fig.14).

For the most important distributions of the total absorbed dose D_T there is a fairly good agreement of calculation and measurement, while in Figs 13 and 14, for the gamma absorbed dose values, the calculated values are by a factor of about 1.5 higher than the measured ones inside and outside the target volume in the upper right part of the thigh. The reasons for this deviations can be:

(1) The calculations are made with a program based on data of total and gamma absorbed dose measurements of single fixed d(14)+Be neutron fields for standard conditions in a half-infinite water phantom. The actual cross-section of the thigh is far away from the dimensions of a phantom with at least 5 cm phantom material outside the useful beam.

(2) Systematic examinations of wedge filter fields of all field sizes and wedge angles showed that the fitting of calculations to the measured results can be made primarily by introducing attenuation coefficients, one for neutrons and two for photons; thus, the absorbed dose values at the field edge on the side with the zero thickness of the wedge must be unchanged. This is valid only for the neutron absorbed dose, but not for the gamma absorbed dose. The best fit to describe the gamma absorbed dose distribution outside the useful beam needs an enlargement factor of about 1.5 for the gamma absorbed dose at all points, caused possibly by the additional gamma rays produced in the wedge filter and the enlargement of lower energy components in the neutron energy spectrum, producing more gamma rays in the phantom. The latter effect alone would be drastically reduced in a phantom of small dimensions.

(3) The TLD-300 ribbon detectors are irradiated from different directions and not all perpendicularly as was done within the calibration procedure. Preliminary examinations have shown that reduction of the total absorbed dose response inside a phantom at all phantom depths is about 10% for an incidence angle 90° compared with the normal incidence (0°). This effect is already taken into account in the evaluation procedure for the TLD measurements.

(4) Calculations are made under the assumption that the phantom consists of a homogeneous volume of water while the measurements are performed in a homogeneous phantom of 50% paraffin + 50% beeswax (density $\rho = 0.89\ \mathrm{g \cdot cm^{-3}}$) with one 1-mm-thick polyethylene plate ($\rho = 0.92\ \mathrm{g \cdot cm^{-3}}$) in it, containing the TLD-300 ribbon detectors. Although the hydrogen percentage by weight (paraffin: 14.9%, polyethylene: 14.37%) is higher than in water (11.19%), the density, however, is lower by a similar amount, and there remains a difference of the phantom materials for neutron interaction.

Further examination of these problems is necessary and in progress.

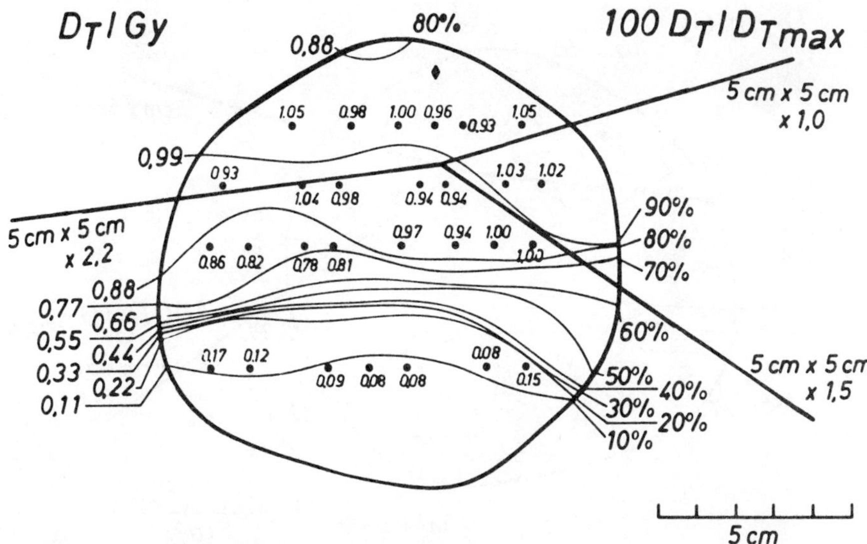

FIG.15. Calculated total absorbed isodose curves (parameter values, medium-sized) compared with measured values of TLD-300 detectors for an isocentric unequally weighted three-field d(14)+Be neutron treatment of a phantom out of 50% paraffin and 50% beeswax (glottic larynx tumour) (Rassow et al.[7]*).*

5.3. Isocentric unequally weighted three-field treatment

Very similar experimental results were obtained for an isocentric, unequally weighted, three-field treatment of glottic larynx tumour. The agreement between calculations and measurements is, for the total absorbed dose distribution (Fig.15), as good as in Fig.10, while the measured gamma absorbed dose values show a deviation from the calculated ones of a factor of about 1.25 (Figs 16 and 17); this is just half the value found in Figs 13 and 14.

Possible reasons for this feature can be:

- The treatment was made without wedge filters, so that their influences are omitted.
- The incidence angles of the beams due to the TLD-300 detectors are nowhere 90°, because four different 1 mm polyethylene plates were used with the included detectors (by contrast to the one polyethylene plate used in Figs 10, 13, 14).

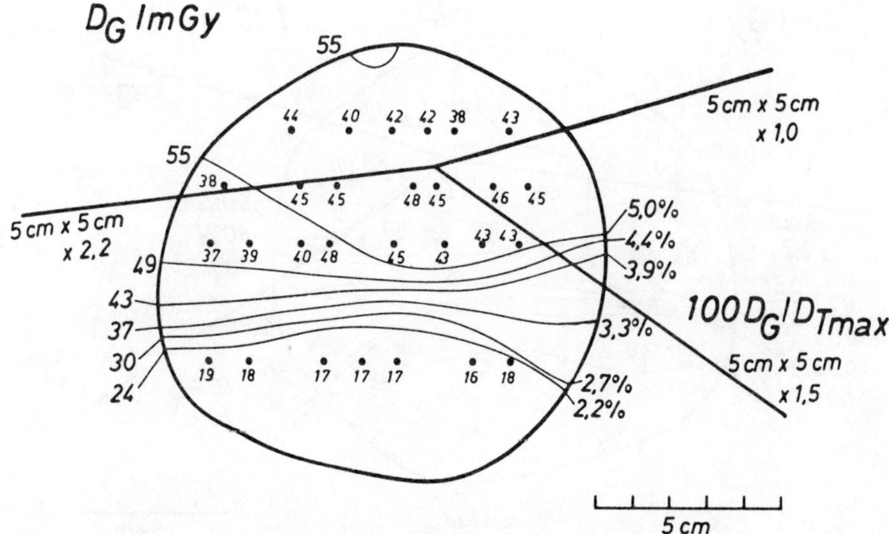

FIG.16. Calculated gamma-ray absorbed isodose curves (parameter values, medium-sized) compared with measured values simultaneously indicated by the TLD-300 detectors of Fig.15.

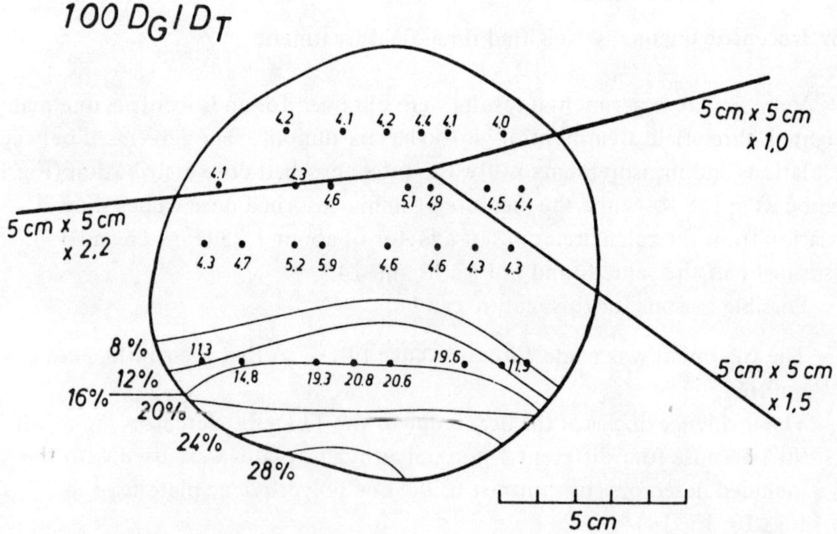

FIG.17. Calculated local gamma-ray component curves (parameter values, medium-sized) compared with the quotient (in per cent) of the measured values of the TLD-300 detectors in Figs 15 and 16.

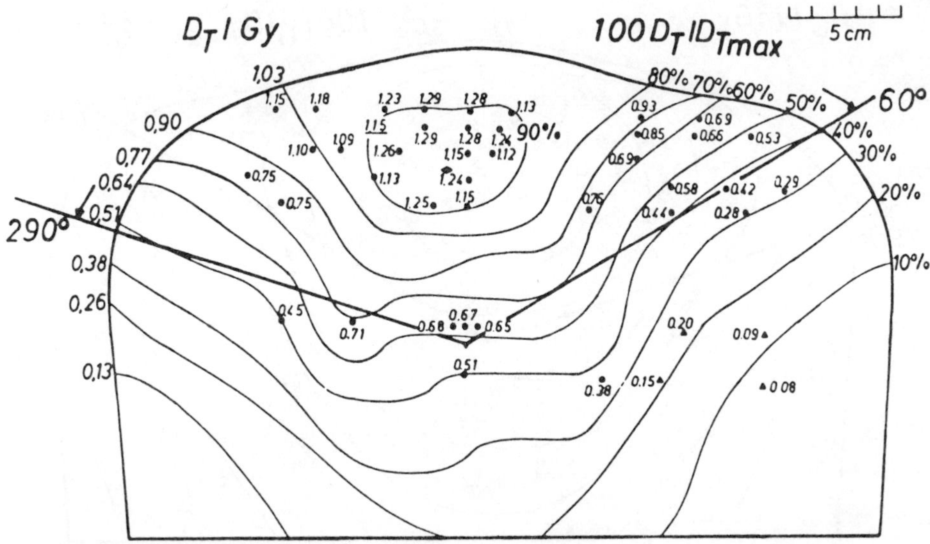

FIG.18. Calculated total absorbed isodose curves (parameter values, medium-sized) compared with measured values of TLD-300 detectors for a rotation beam d(14)+Be neutron treatment of a phantom consisting of a 25 cm × 25 cm × 17 cm block of A-150 plates and a surrounding of 50% paraffin and 50% beeswax, forming the irregularly shaped contour with a maximum thickness of 4 cm (bladder tumour) (Rassow et al. [7]*).*

5.4. Isocentric moving field

Much better agreements than in the two cases discussed in Sections 5.2 and 5.3 are also found for the total absorbed dose distribution (Fig.18) and for the gamma absorbed dose distribution (Figs 19 and 20). The treatment planning was done for a bladder carcinoma and carried out by rotation neutron therapy. The measurements were done in a phantom which consists, in its main part, of a 25 cm × 25 cm × 17 cm block of A-150 plates containing seven 1-mm-thick polyethylene sheets with milled holes for the TLD-300 detectors. These sheets are adjusted perpendicular to the drawing plane. The incidence angles of the neutron beam with respect to the TLD-300 detectors range from 0 to 70°; thus, no correction for 90° incidence was made. The irregularly shaped contour is formed by a cover of 50% paraffin and 50% beeswax with a maximum thickness of 4 cm. Most of the possible reasons, listed in Sections 5.2 and 5.3, for an explanation of the systematic deviations between calculation and measurement of the gamma absorbed dose distributions, are not applicable except for the difference of phantom materials used for calculations (water) and measurements (A-150/paraffin-beeswax).

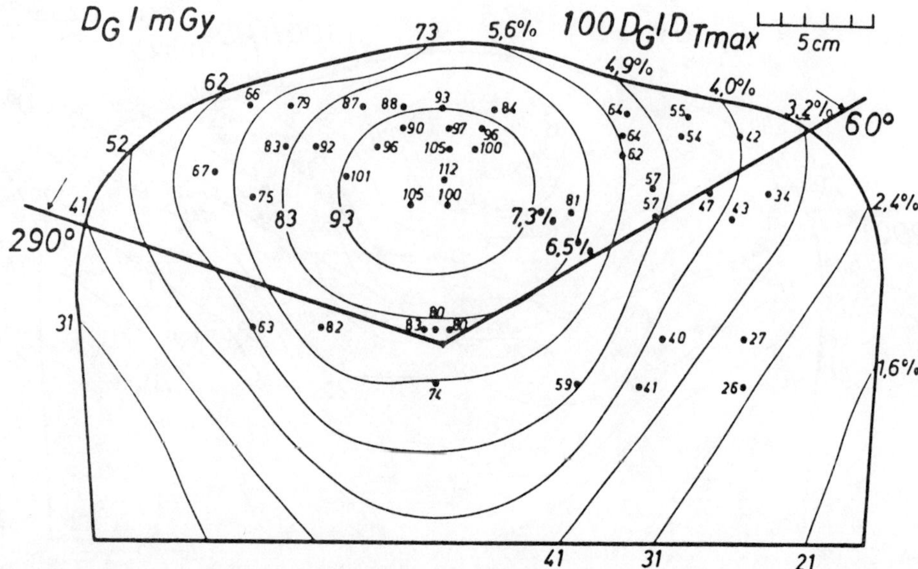

FIG.19. Calculated gamma-ray absorbed isodose curves (parameter values, medium-sized) compared with measured values simultaneously indicated by the TLD-300 detectors of Fig.18.

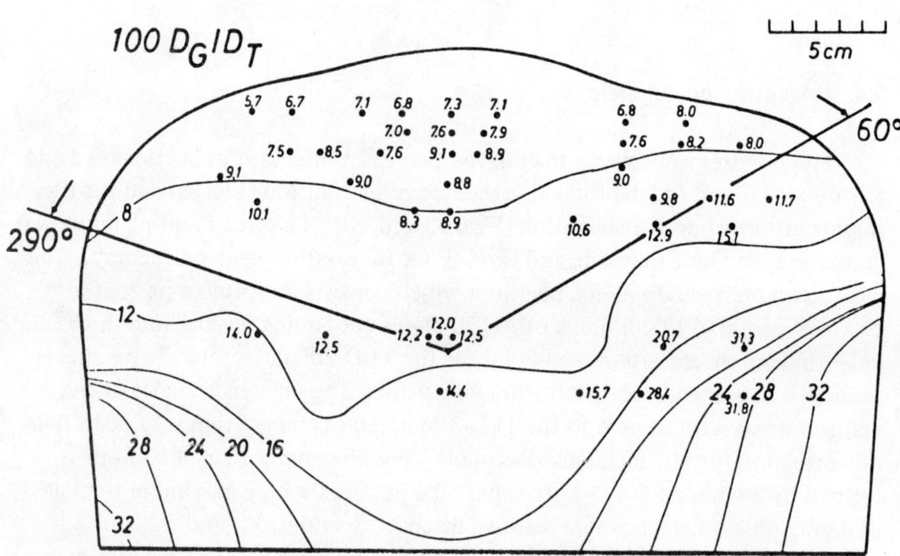

FIG.20. Calculated local gamma-ray component curves (parameter values, medium-sized) compared with the quotient (in per cent) of the measured values by the TLD-300 detectors in Figs 18 and 19.

6. PROBLEMS NEEDING FURTHER EXAMINATION

6.1. Special physical features of TLD-300 detectors

The main subjects for further investigations with respect to the physical features of TLD-300 detectors are:

(1) Dependence of the main peak response coefficients on the energy spectrum of neutrons and (less important) of photons. Different energy spectra at the measuring position are caused by
   - different primary neutron energy spectra, and/or
   - different phantom measuring points with respect to the neutron beam,
   - different field sizes of the neutron beam,
   - different phantom sizes (in the case where a variation of phantom size has an influence of the dose distribution)

(2) Dependence of the main peak response coefficients on the incident angle of the neutron beam
   - free-in-air, and
   - for different phantom depths

(3) Dependence of the main peak response coefficients on the detector's surrounding material in the thickness of the recoil proton range.

6.2. Applicability of the TLD-300 method for special clinical dosimetry problems

A few of the problems arising in the application of the TLD-300 method for therapeutically relevant conditions have already been mentioned (Sections 5.2 to 5.4). More generally they can be described as an examination of the

(1) Applicability for measuring spatial dose distributions:
   - adjacent to inhomogeneity surfaces,
   - within homogeneous phantoms – of tissue-similar materials as liquids, as rigid solids, as kneadable or mouldable plastics; and of non-soft tissue-equivalent materials as surrogates for bone, fat etc.,
   - within man-like phantoms (e.g. Alderson Rando)

(2) Applicability for in-vivo dosimetry at patient surfaces or cavities.

6.3. Reference phantom material for calculating and measuring complex neutron treatment absorbed dose distributions

The question of equivalence of phantom material to human tissue is a serious problem, especially for clinical neutron dosimetry. The differences of dependency of the nuclear reaction cross-section of all nuclides contained in the phantom

TABLE X. PRELIMINARY RESULTS: DEPTH DOSE DISTRIBUTIONS FOR A d(14)+Be NEUTRON BEAM IN DIFFERENT MATERIALS CAN BE TRANSFORMED TO ONE ANOTHER BY ALTERING THE DEPTH SCALE BY FACTORS INDEPENDENT OF PHANTOM DEPTH AND FIELD SIZE

Preliminary factors z_{H_2O}/z_M are given for transformation of the depth dose distribution in material M to that in H_2O. The element compositions and densities given are taken from White [9]; and the values for AFWL "plastinaut" are taken from Janni et al. [10]

Material	Elemental composition (% by weight)					Density	d(14)+Be neutrons kerma ratio	
	H	C	N	O	Others	(g/cm^3)	to H_2O	z_{H_2O}/z_M
H_2O	11.19			88.81		1.00	1.00	1.00
Goodman TE liquid	10.20	12.01	3.54	74.25		1.07	0.929	1.04
A-150	10.20	76.80	3.60	5.90	F:1.70; Ca:1.80	1.12	0.942	1.11
Alderson muscle	8.87	66.81	3.10	21.13	Sb:0.08	0.99	0.833	0.99
AFWL "plastinaut"	8.89	64.81	3.15	22.18	Na:0.09; Mg:0.03 P:0.27; S:0.28 Cl:0.12; K:0.17 Fe:0.005	1.10	0.836	1.00
Perspex	8.05	59.98		31.96		1.17	0.851	1.06
Polystyrene	7.74	92.26				1.05	0.751	0.94
Polyethylene	14.37	85.63				0.92–0.96	1.26	1.14
Beeswax + paraffin (1:1)						0.89	>1	1.10
ICRU muscle tissue	10.20	12.30	3.50	72.90	Na:0.08; Mg:0.02 P:0.20; S:0.50 K:0.30; Ca:0.01	1.04–1.06	0.909	

TABLE XI. OPEN PROBLEMS OF CLINICAL NEUTRON DOSIMETRY DUE TO PHANTOM MATERIALS

Keyword	Problem
Reference materials for treatment planning calculation	Which existing material should be recommended to replace patient's soft tissue in order to obtain standard dose distributions (for D_T and D_G) as a basis for complex treatment planning calculations? – Should this "reference material" be definitely a liquid (e.g. water or "Goodman liquid")? – If a solid can also be recommended, which one is compatible with the "reference liquid"?
Reference materials for dosimetric treatment planning verifications with, for example, TLD detectors	Which solid and/or kneadable, mouldable reference material should be recommended for measuring spatial dose distributions (for D_T and D_G) for complex treatments in irregularly shaped phantoms? – Is it preferable to form the main parts of a special patient simulating phantom of a solid material, while the irregularly shaped surface parts are made from a kneadable, mouldable material? Or: – Is it better to form the whole special patient simulating phantom out of a kneadable, mouldable material? Which kneadable, mouldable material should be recommended? – Which material is best for a TLD detector containing plate? – Which patient-like phantom including inhomogeneities (e.g. normal Alderson or Alderson plastinaut phantom) is preferable for dosimetric neutron treatment planning verifications?

material on the actual neutron energy spectrum, and the importance of nuclides of low atomic number as secondary particles and agents for the energy deposition, make it most desirable to have, first, a correct hydrogen content and, second, a correct nitrogen content, while (at least for neutron energies less than about 20 MeV) carbon and oxygen are more or less interchangable.

Such considerations are essential for secondary particle equilibrium, e.g. between gas filling and the wall of a TE ionized chamber for calibrating such chambers and calculating corrections to be applied to obtain correct absorbed dose determinations.

Different requirements for "tissue equivalence" are necessary for measuring whole spatial total absorbed dose (D_T) and gamma absorbed dose (D_G) distributions in phantoms.

The optimum would be reached if the identical spatial distribution of the gamma absorbed, as well as the total, dose is realized in the phantom material and the patient's body tissue. To verify complex treatment plans, e.g. by TLD measurements, it is also necessary to have phantom materials which can easily be shaped in an irregular patient equivalent contour, and which can be prepared for the exact and reproducible placement of TLD detectors.

A test of how well the above-mentioned optimum due to correct spatial dose distributions in different phantom materials is met, can be made by comparing depth dose and lateral dose distributions (for D_T and D_G). Some preliminary results are presented for d(14)+Be neutrons in Table X (Hensley and Temme [8]) showing that there exists an empirical factor, independent of field size, for a depth proportional transformation of the depth scale. This factor is, however, not easily interpretable by physical parameters like density and/or kerma ratio. As valuable as such a transformation factor is for comparing depth dose curves under standardized conditions in different phantom materials, the application of such a factor is obviously inadequate for dosimetric measurements of complex irradiations like multiple or moving field treatments of irregularly shaped phantoms.

The problems arising from these considerations that need to be solved are listed in Table XI.

REFERENCES

[1] BROERSE, J.J., MIJNHEER, B.J., WILLIAMS, J.R., European Clinical Neutron Dosimetry Group (ECNEU), European Protocol for Neutron Dosimetry for External Beam Therapy, Br. J. Radiol. **54** (1981) 882–98.

[2] STREFFER, C., van BEUNINGEN, D., BERTHOLDT, G., personal communication, 1982.

[3] TEMME, A., RASSOW, J., MEISSNER, P., "A new thermoluminescent dosimetry procedure using TLD-300 detectors for clinical dosimetry in mixed neutron gamma-ray fields", Neutron Dosimetry, Beam Dosimetry (Proc. Fourth Symp. Munich-Neuherberg, 1981) (BURGER, G., EBERT, H.G., Eds), Vol.2, Commission of the European Communities, Luxembourg, EUR 7448 (1981) 433–54.

[4] MEISSNER, P., "Einsatz eines basic-programmierbaren Tischrechners mit IEC-BUS-Voltmeter zur digitalen Glowkurvenanalyse von CaF_2:Tm bei der klinischen Dosimetrie in gemischten Neutronen-Photonen-Feldern", Medizinische Physik '81 (12. Wiss. Tag. der Dtsch. Ges. Med. Phys., Munich, 1981) (BUNDE, E., Ed.), Hüthig Verlag, Heidelberg (1982).

[5] MEISSNER, P., RASSOW, J., "Principles of a treatment planning program with separate calculation of dose distributions for neutrons and gamma-rays", Treatment Planning for External Beam Therapy with Neutrons (BURGER, G., BREIT, A., BROERSE, J.J., Eds), Urban & Schwarzenberg (1981) 162–69.

[6] BAUMHOER, W., Rechnergestützte Bestrahlungsplanung für Tumortherapie mit schnellen Neutronen an der Essener Neutronentherapieanlage CIRCE, Strahlentherapie **158** 6 (1982).

[7] RASSOW, J., TEMME, A., BAUMHOER, W., MEISSNER, P., "Dosimetrical verification of calculated dose distributions D_T and D_G for neutron therapy", Proc. World Congr. Medical Physics and Biomedical Engineering, Hamburg, 1982, Digest of the 5th World Congr. 1982, (in press).

[8] HENSLEY, F.W., TEMME, A., "Comparison of dose distributions in water, tissue-equivalent liquid, and several tissue-similar solid materials for a d(14)+Be neutron beam", ibid.

[9] WHITE, D.R., Tissue substitutes in experimental radiation physics, Med. Phys. **5** 6 (1978) 467–79.

[10] JANNI, J.F., CLARK, B.C., SCHNEIDER, M.F., BERGER, P.S., "A dose-equated manikin for space radiation research", Air Force Weapons Laboratory, Kirtland, New Mexico, Tech. Rep. No. AFWL TR-65-97 (1965) 1–22.

CLINICAL NEUTRON DOSIMETRY

J.B. SMATHERS
Department of Radiation Oncology,
University of California,
Los Angeles, California

P.R. ALMOND
Physics Department,
M.D. Anderson Hospital and Tumor Institute,
Houston, Texas

United States of America

Abstract

CLINICAL NEUTRON DOSIMETRY.

The paper reviews the present status of clinical neutron dosimetry. The approach is to consider briefly the characteristics of the neutron sources in use and the radiation fields the sources produce. The ramifications of the differences between neutron and low-LET therapy sources on treatment planning are discussed for a series of both physical and biological therapy parameters. The review concludes with a discussion of areas where new information or techniques would yield improved clinical neutron dosimetry.

I. INTRODUCTION

The radiotherapy community appears to have accepted the following suggested minimum requirements for a clinical neutron source:

- treatment time $<$ 4 minutes;
- depth dose equivalent to or better than cobalt-60 [1]

This has been translated into the quantitative requirements of:

- dose rate $>$ 20 rads/min at depth of maximum dose
- depth of maximum dose $>$ 0.5cm
- 50% of maximum dose occurs at a depth $>$11 cm.

Sources which have been or will be used clinically to achieve the above are (d,T) generatons and cyclotrons incorporating the (p,Be) and (d,Be) reactions. Without pursuing a detailed discussion of the merits of the sources or reactions, it suffices to say that average neutron energy and dose rate would appear to be insufficient criteria on which to compare neutron sources. This is because the degree to which the ion chambers obey the Bragg-Gray principle and the biological response observed in the irradiated tissues is critically dependent on the charged particle spectra which is unique to each reaction-energy combination. We shall thus begin our discussion of clinical dosimetry with a brief overview of the characteristics of the sources involved.

2. SOURCE CHARACTERISTICS

Comparative neutron spectra for the (d,T,), (d,Be), and (p,Be) reactions are illustrated in Figure 1 [2]. The (d,Be) reaction neutron spectra are for deuteron energies of 16, 30, and 50 MeV and for this reaction bracket the energies used for therapy to date. The (p,Be) reaction spectra are for a proton energy of 42 MeV and are typical of the spectra to be expected from the three clinically dedicated facilities under construction in the United States. For this reaction, the surplus of low energy neutrons can be filtered out by use of hydrogenous media, thus improving the depth dose characteristics of the reaction neutrons. The (d,T) reaction neutrons are essentially monoenergetic, 14.7 MeV.

The range of LET spectra which can result from the various sources is illustrated in Figure 2 [3]. As others will discuss this in detail, we will but call to your attention the variation in the proton/alpha ratio which occurs as the neutron spectrum extends to higher energies.

3. CLINICAL DOSE MEASUREMENTS

3.1. Basic Dosimetry

Both the AAPM and the European Neutron Dosimetry Protocols recommend the use of ionization chambers and the Bragg-Gray cavity theory for clinical neutron dose measurements [4,5].

Typically, the relationship, equation 1,

$$\text{Dose (Gray)} = \frac{W/e\ (\frac{\text{Joule}}{\text{Coul}}) \cdot Q\ (\text{coul}) \cdot S}{M(\text{kg})} \qquad \text{(Equation 1)}$$

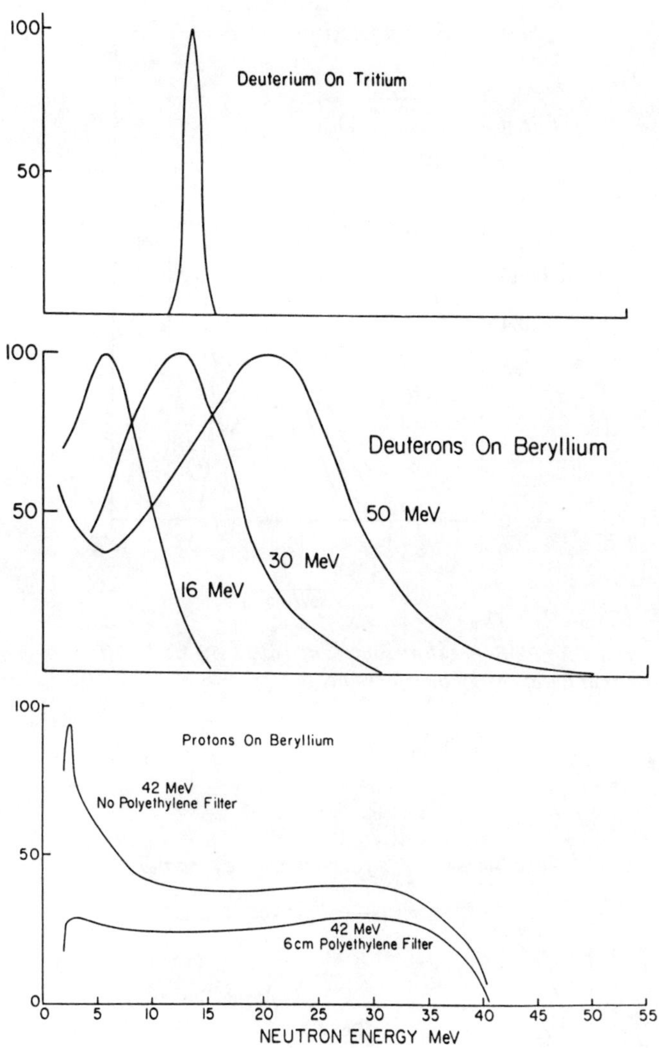

FIG.1. Neutron spectra for clinical neutron therapy sources. Peak output is normalized to 100%.

is used first to determine the mass of gas in the chamber through exposure of the chamber to a photon source with a calibration traceable to an accepted radiation standard. Knowing the dose given to the chamber and using accepted values of W/e and S for photons, the mass can be determined. Then reversing the procedure and using the determined value for M and "derived values" for W/e and S for neutrons, the neutron dose can be calculated. With the range of many of

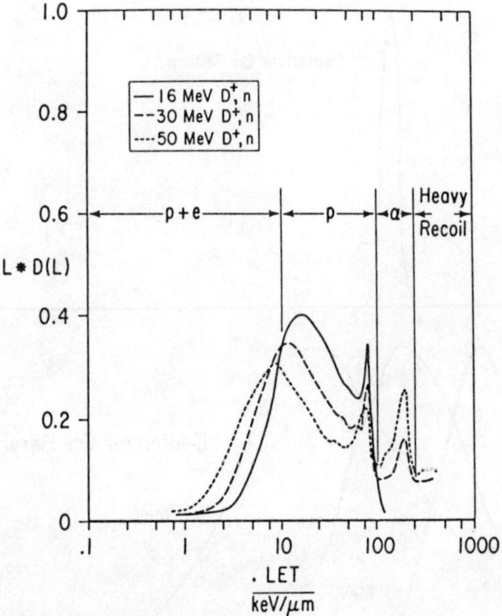

FIG.2. LET spectra for (d, Be) neutron sources of 16, 30, 50 MeV. D(L) is fraction of total dose which occurs at lineal energy transfer value, L.

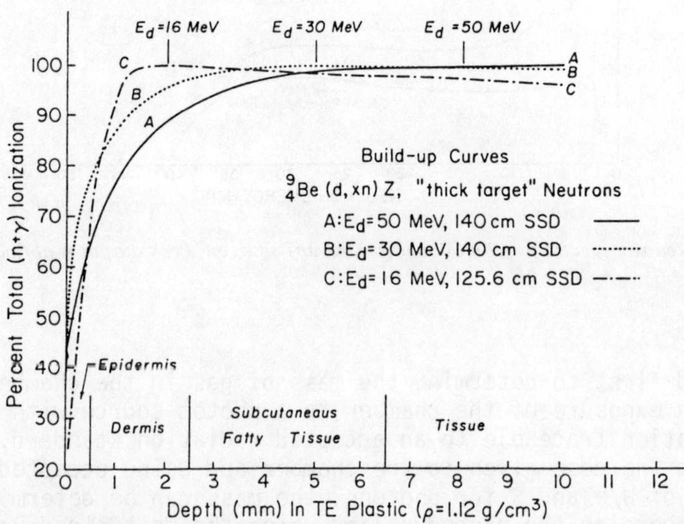

FIG.3. Total dose versus depth in first few millimetres of A-150 plastic for (d, Be) sources of 16, 30, and 50 MeV.

the charged particles created by neutron interactions violating the fundamental assumptions implicit in the Bragg-Gray theory, the close agreement of the ionization dosimetry systems with independent calorimetry measurements, within 2%, is a bit fortuitous [6].

The above not withstanding, the ionization systems are universally applied with a variety of ion chambers, commercially fabricated and individually constructed, being used. To date, one common construction characteristic has been the use of A-150 "tissue equivalent" plastic for the chamber wall and collector material [7]. No doubt this has been a contributing factor to the close agreement that has existed among the various neutron therapy facilities as to the physical magnitude of a neutron rad. Correction factors for chamber displacement, saturation, density effect, and other ion chamber corrections are discussed in the neutron dosimetry protocols and will not be considered here.

3.2. Dose Buildup Measurement

The rapidly increasing dose which occurs in the first few millimeters of tissue is determined for neutrons much as one would determine it for photon sources with the possible exception that extrapolation chambers composed of A-150 plastic are used to perform the measurements. Because of the difference in range between alphas and photons, the average LET for this region is higher than that at depths beyond maximum dose. Several efforts have been made to separate the total dose in this region into the fraction of the dose due to alphas, protons, and photons, but the matter is still not fully resolved [8, 9, 10]. The higher LET in this region may be the cause for the adverse skin reactions some have observed in neutron therapy.

The depth at which maximum dose occurs varies from a few millimeters for the 16 MeV (d,Be) spectra to about a centimeter for the 50 MeV (d,Be) and 42 MeV (p,Be) reactions, Figures 3 and 4. Using the inherent characteristic of neutron dose deposition that a large fraction of the dose is deposited by charged particles of very limited range, a thin absorber of high Z, with a low neutron cross section can be used to absorb these charged particles and as it were, restore skin sparing, Figure 5 [11, 12, 13, 14]. This concept has been used to restore skin sparing by lining the surface of the bolus material in contact with the skin, to eliminate collimator scatter, and has been discussed for intra-oral use to protect healthy tissue. Figure 6 illustrates a breast bridge incorporating this concept with the inner surface of the beam entrance side lined with lead.

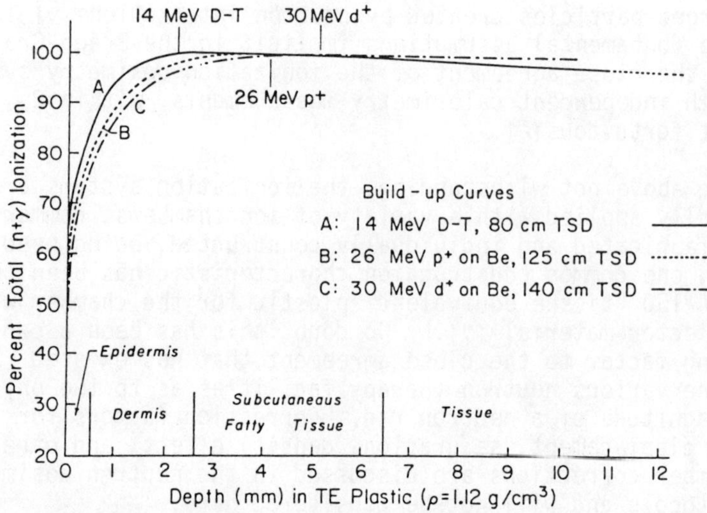

FIG.4. Total dose versus depth in first few millimetres of A-150 plastic for several neutron sources.

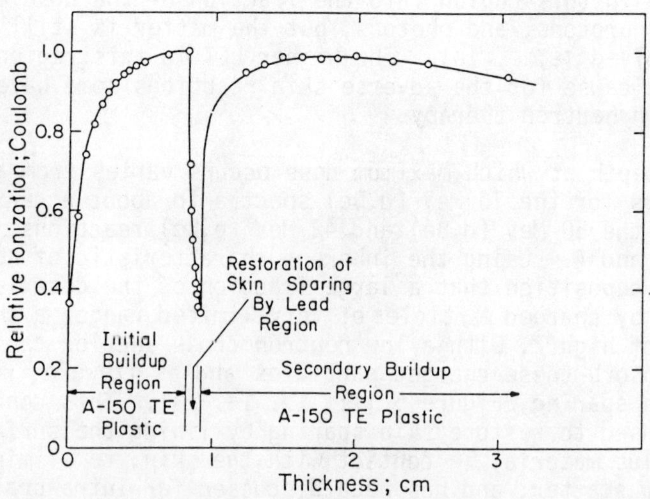

FIG.5. Restoration of "skin sparing" by use of thin lead absorbers.

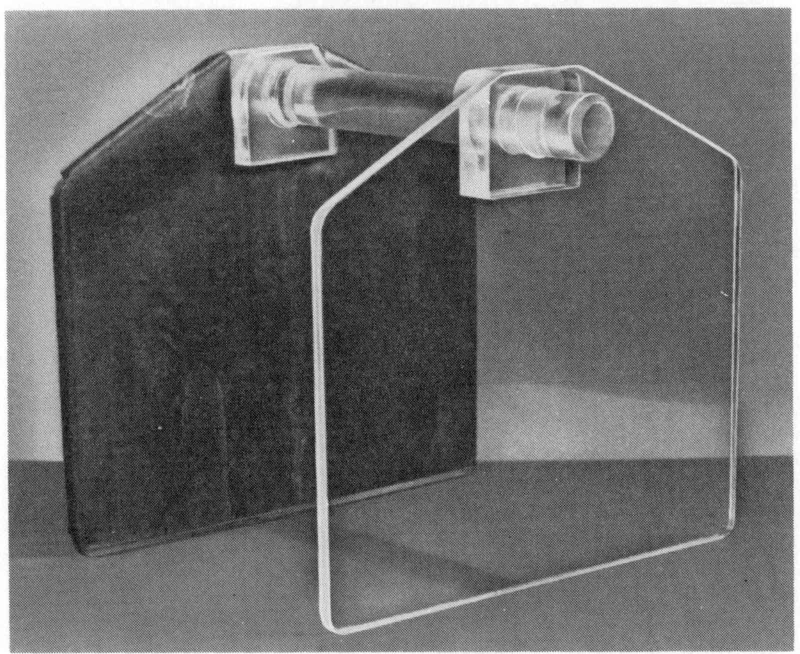

FIG. 6. Breast bridge with lead-lined entrance surface to restore "skin sparing".

3.3. Central Axis Depth Dose

A comparison of the depth dose properties of three (d,Be) neutron spectra and their approximate equivalent in photon sources is given in Figure 7. Due to a comparatively small low energy neutron population, filtering the (d,Be) spectra through hydrogenous media does not significantly improve the depth at which the 50% of maximum dose occurs. For the (p,Be) reaction this is not the situation and filtration through hydrogenous media can significantly increase the depth at which the 50% of maximum dose occurs, Figure 8. The (d,T) sources being a monoenergetic neutron source would not benefit from hydrogenous filtration, however a thin lead absorber to remove charged particles and low energy photon contamination originating in the target assembly is beneficial, Figure 9 [14].

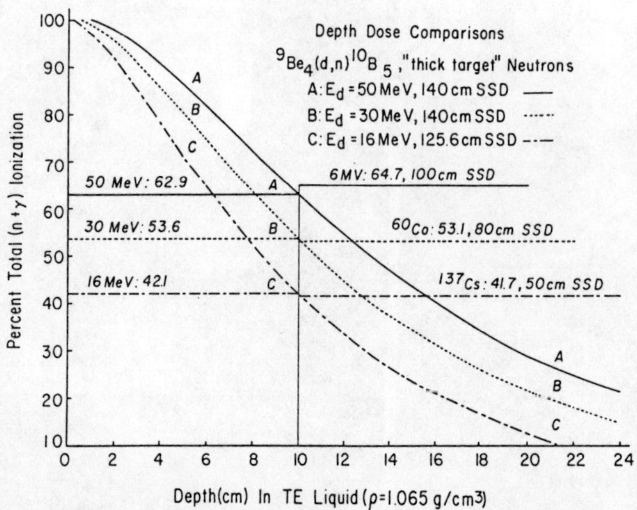

FIG. 7. Comparison of depth dose curves for (d, Be) neutron sources of 16, 30, and 50 MeV with "equivalent" photon sources indicated.

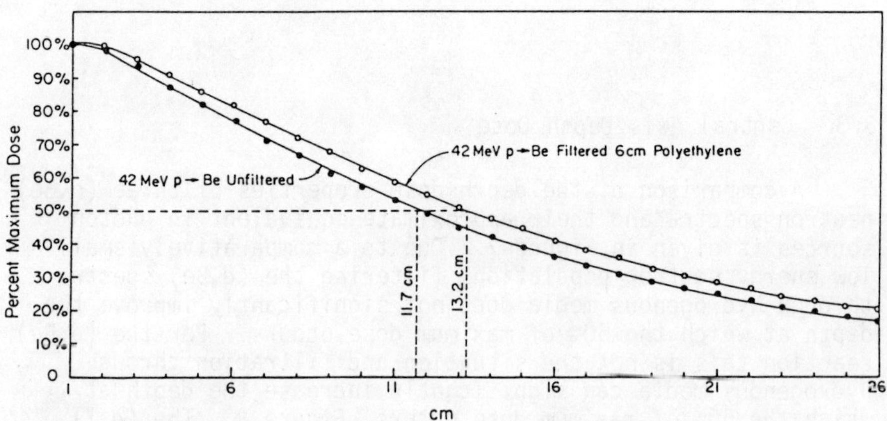

FIG. 8. Depth dose improvement obtainable by filtering the 42 MeV (p, Be) neutron spectra through 6 cm polyethylene.

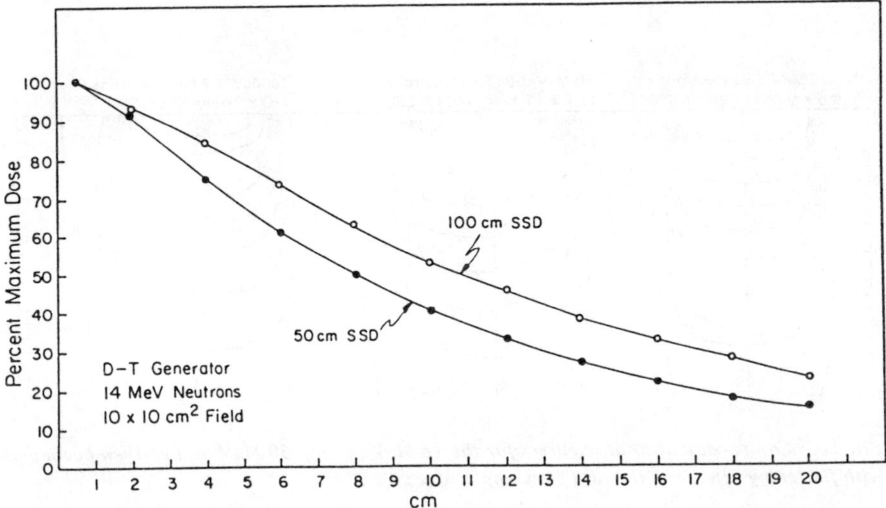

FIG.9. Depth dose for (d,T) neutron sources at TSDs of 50 and 100 cm.

3.4. Isodose Distributions

The initial reactions used for neutron therapy, (d,T) and 16 MeV (d,Be) were sufficiently isotropic in neutron emission that the resulting isodose contours yielded uniform doses within the field at depth in tissue. As the energy of the charged particles was increased to obtain higher neutron energies and increased neutron yields, the neutron emission became more forward peaked and the resulting isodose curves in phantom were more rounded and considered clinically unacceptable. Flattening filters were designed for the patient end of the fixed collimator inserts and yielded quite acceptable results.

Figure 10 compares the isodose distribution for the 16 MeV (d,Be) open field, 50 MeV (d,Be) open field and flattened field, and (d,T) open field sources. Comparison of the flattened fields, for the 50 MeV (d,Be) and 41 MeV (p,Be) sources is shown in Figure 11. The benefits of flattening the 41 MeV (p,Be) source are illustrated in Figure 12. In each case, the flattening filter, located at the distal end of the collimator, is designed to optimize the degree of field uniformity at the 70% isodose depth. For the (p,Be) source, additional polyethylene was added to increase the low energy

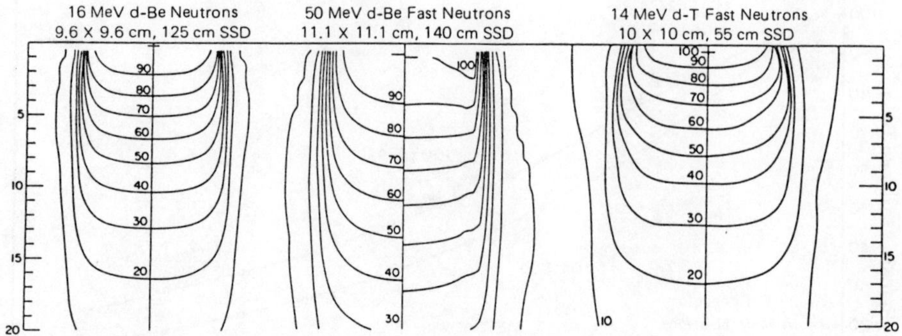

FIG.10. Comparison of isodose curves for the 16 MeV (d, Be), 50 MeV (d, Be) open beam and with flattening filter, and the (d, T) neutron sources.

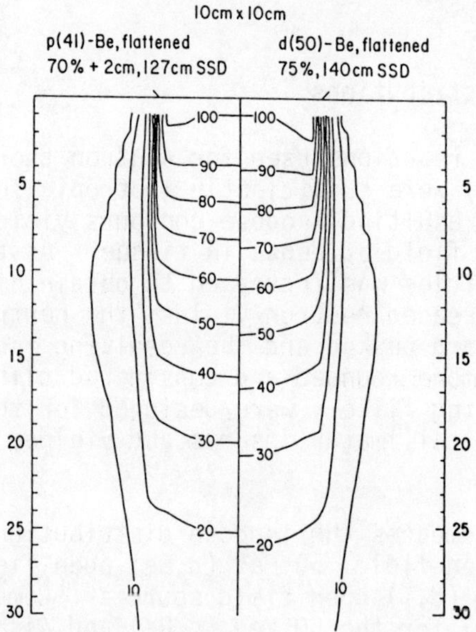

FIG.11. Comparison of neutron source isodose contours: 41 MeV (p, Be) flattened at 70% isodose depth plus 2 cm added hydrogenous beam hardening filter versus 50 MeV (d, Be) flattened at the 75% isodose depth.

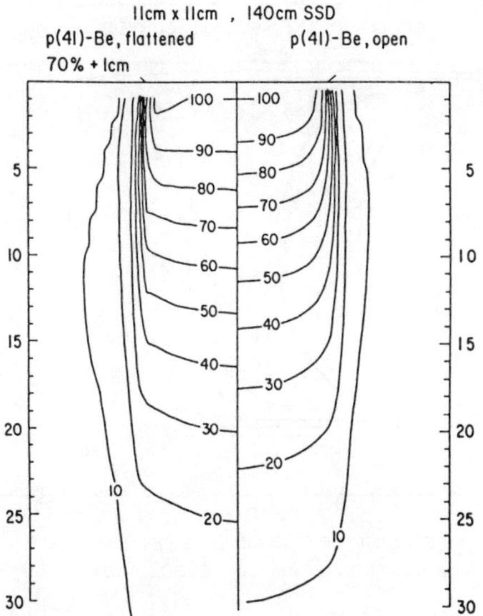

FIG.12. Comparison of 41 MeV (p, Be) neutron source isodose curves: open beam versus flattened beam at 70% isodose depth plus 1 cm added polyethylene hardening filter.

neutron filtration. The isodose curves for the most penetrating neutron beam used clinically to date are shown in Figure 13 [15]. Figure 14 compares the isodose distributions of one of the more penetrating neutron sources, 50 MeV (d,Be), with conventional photon sources. From this data, it is evident that depths of 50% of dose maximum for neutron sources can be obtained which are greater than that achieved by Cobalt-60, but it is doubtful if a neutron source will ever achieve the penetrability of 25 MV x-rays.

Wedges have been used extensively by the group at Hammersmith Hospital, and to a lesser extent by others, primarily for the treatment of head and neck tumors. Typical isodose distributions are illustrated in Figures 15 and 16. One caution, should hydrogenous materials be used in the wedge or flattening filter, the air gap between the patient and the filter must be greater than 20 cm or a high Z filter used to preclude loss of skin sparing due to charged particle emission from the filter [16]. The alternative is to use a non-hydrogenous material which does not compromise skin sparing.

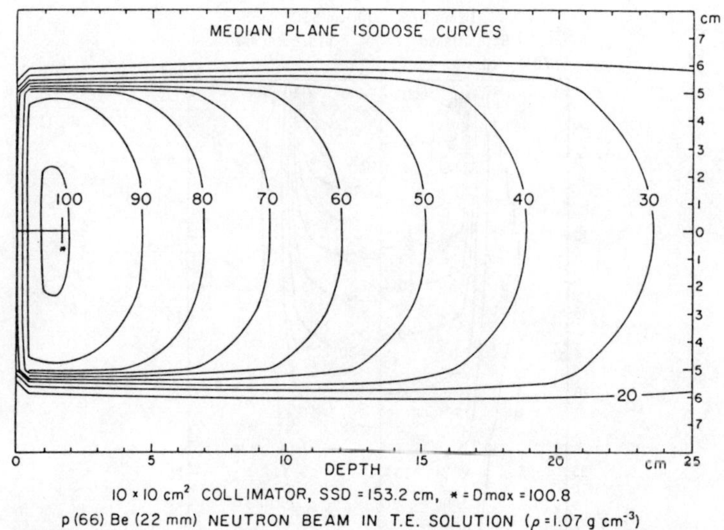

FIG.13. Isodose distribution for the 66 MeV (p, Be) neutron source.

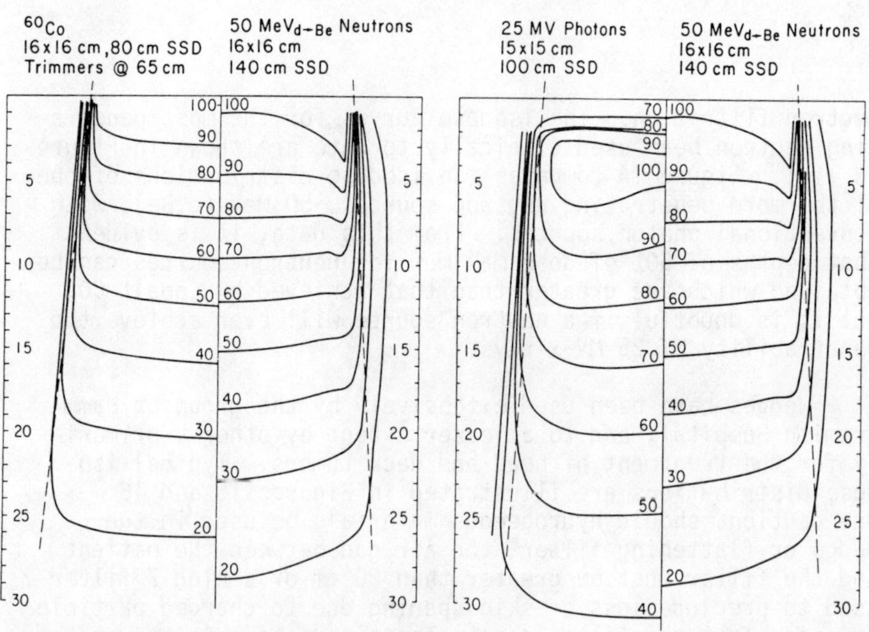

FIG.14. Comparison of isodose curves of cobalt-60 and 25 MV photon sources with the 50 MeV (d, Be) neutron source with flattening filter (75% isodose depth).

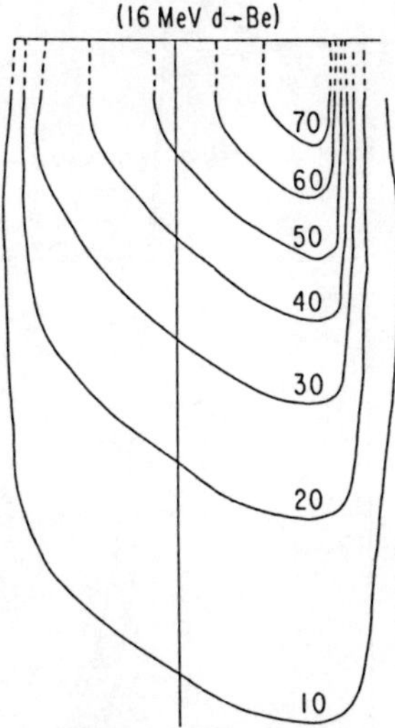

FIG.15. 16 MeV (d, Be) 45° wedge isodose distributions.

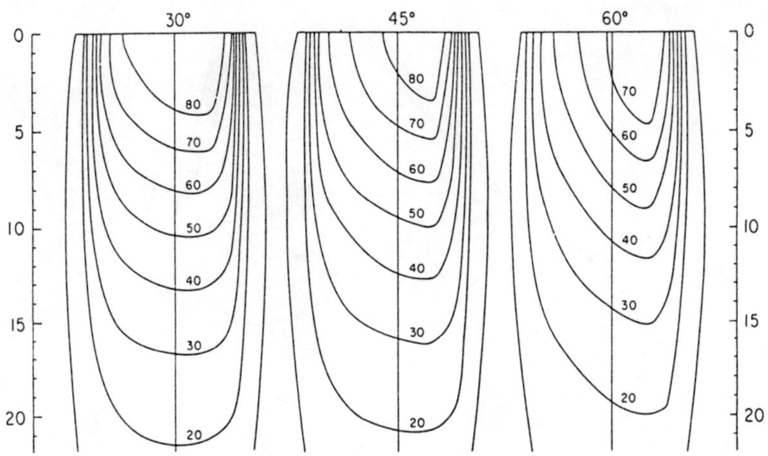

FIG.16. 50 MeV (d, Be) 30°, 45°, and 60° wedge isodose distribution.

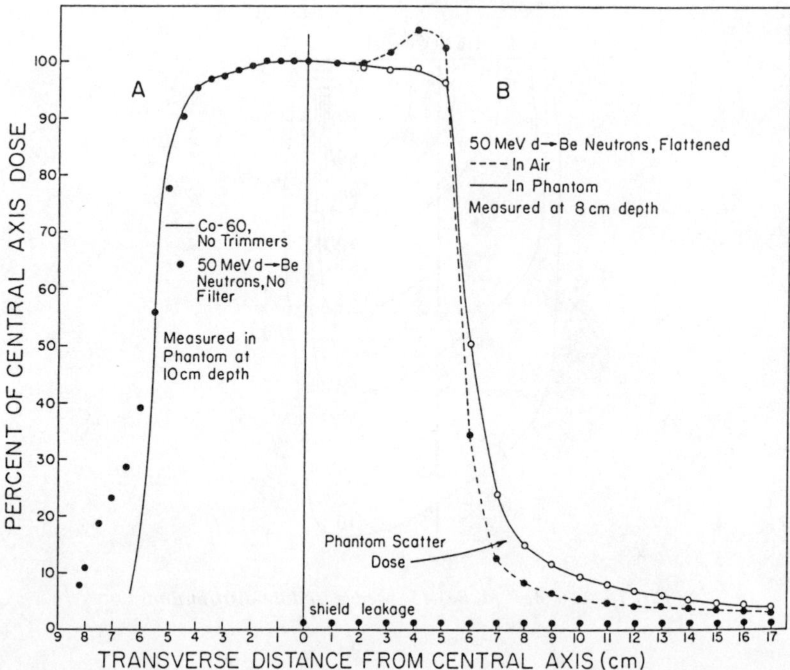

FIG.17. A: In-phantom, 10 cm depth, comparison of total dose profiles for cobalt-60 and 50 MeV (d, Be) neutron sources.
B: Comparison of in-air and 8-cm-depth total dose profiles for 50 MeV (d, Be) neutron source with phantom scatter dose magnitude indicated.

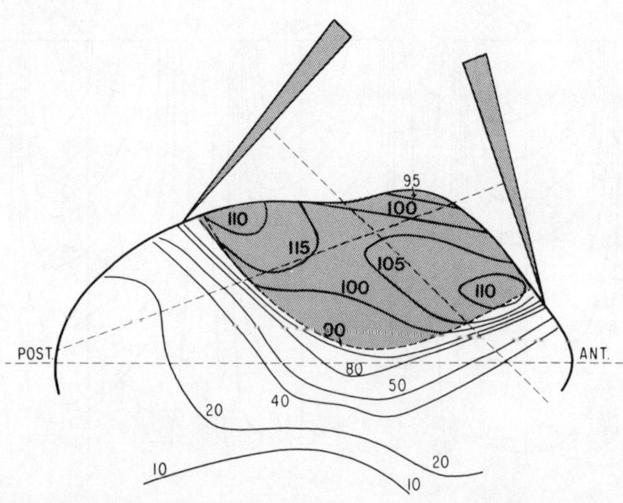

FIG.18. Illustrative treatment plan using 45° wedges and the 16 MeV (d, Be) neutron source.

The use of variable collimation should result in renewed efforts to develop an acceptable flattening filter or series of filters which can be placed immediately downstream of the target.

The in phantom scatter dose from neutrons exceeds that of cobalt and consequently source spot sizes are not a critical factor. For a 5 mm diameter spot size, the relative scatter dose compared to cobalt-60 is illustrated in Figure 17.

4. TREATMENT PLANNING

4.1. Mathematical Methods

As the clinical neutron beams are similar in characteristics to cobalt-60, it is not surprising that the mathematical approaches used to describe cobalt-60 isodose distributions have been successfully used for computerized neutron treatment planning. The concept of decrement lines has been shown to give an adequate representative of neutron distributions under a variety of conditions and table look-up with interpolation has been successfully used as well [17, 18]. The TAR concept has not been fully tested for neutrons, but there would appear to be no reason why it could not be applied as well. With the new neutron therapy systems incorporating isocentric units and the capability for rotational therapy, this may be the approach used by these facilities.

4.2. Typical Treatment Plans

Several illustrative treatment plans are included in Figures 18, 19, and 20. The wedge pair, Figure 18, is typical of the wedge plans obtainable with neutrons. The increased penumbra-scatter component of neutron fields results in larger integral volume-dose products than for photon sources. Figures 19 and 20 illustrate the problem faced by the lower energy neutron facilities as they considered the treatment of deep seated tumors.

4.3. Neutron/Gamma Ratio

Because of the variation of the clinical RBE with charged particle environment, ideally one should know the charged particle spectra vs. depth in the treatment field. As a first approximation to this, one attempts to determine the fraction of the dose contributed by neutrons and that by photons vs. location in the phantom [19, 20, 21, 22, 23].

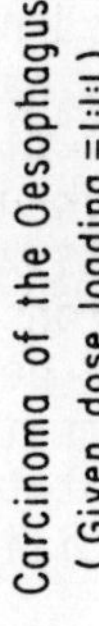

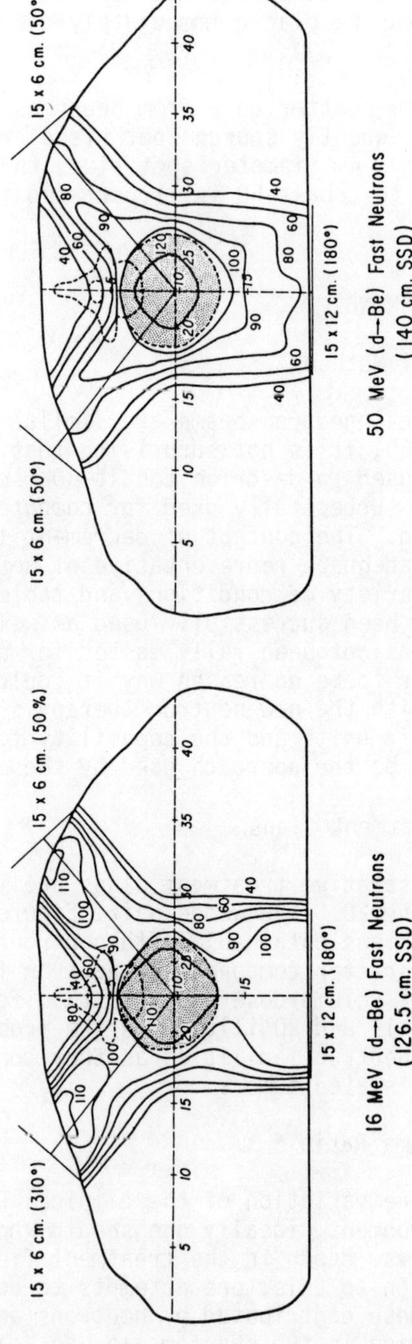

FIG.19. Carcinoma of the oesophagus treatment plan comparison for the 16 MeV (d, Be) and 50 MeV (d, Be) neutron sources. Equal loading to each field for the applied (given) dose.

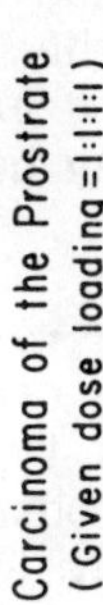

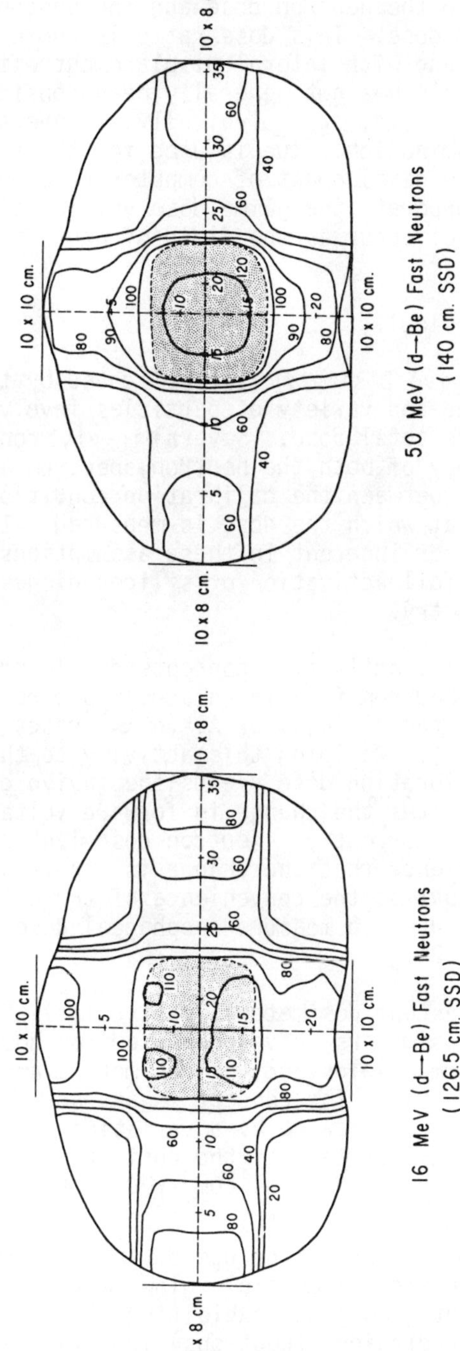

FIG.20. Carcinoma of the prostate treatment plan comparison for the 16 MeV (d, Be) and 50 MeV (d, Be) neutron sources. Equal loading to each field for the applied (given) dose.

Using this data, an average clinical REM/RAD conversion factor is applied to the neutron dose and the photon dose added to give a total REM dose. This dose ratio is known to vary with depth in tissue and with lateral displacement within the field, however this has not generally been considered in the treatment planning [22, 24]. Typically, an average photon percentage contamination value is used for the total treatment volume. The development of computer treatment planning codes which incorporate the gamma dose vs. location information will no doubt occur as the field of neutron therapy reaches maturity.

4.4. In-vivo Dosimetry

Neutron In-vivo Dosimetry is complicated by the energy range emcompassed and variety of particles involved in the deposition of the total dose. Several in-vivo concepts used assume a constancy of both the neutron spectrum and neutron/gamma dose ratio between the calibration condition and the in-vivo location at which the dose is measured. If the magnitude of the error inherent in these assumptions are acceptable, then both foil activation or silicon-diodes may be used for in-vivo dosimetry.

Using the foil activation concepts developed in reactor technology, the neutron flux in an accessible region can be determined, with the Al (n,p) or Al (n,α) reactions typically being used [25]. Ratioing this activity to that obtained from a known calibration dose yields the invivo dose. The silicon diode relates the change in forward voltage as measured at constant current vs. neutron dose [26, 27]. It has the added convenience that the change is stable and the system can be read out at the convenience of the clinical physicist. A typical use, to measure esophageal dose, is illustated in Figure 21.

Thermoluminescent dosimeters, particularly ^{6}LiF and ^{7}LiF as paired dosimeters, have been used for dosimetry in mixed neutron/gamma fields and the dosimeters are small enough for invivo use. The concept is attractive, but it's implementation is complicated by a variation of response with neutron energy for each of the dosimeters and an over response for low energy photons compared to the 1-2 MeV energy range.

Another TLD concept is the use of TLD-300, thulium doped calcium fluoride, a multiple glow peak TLD system with the different glow peaks exhibiting differing responses to high LET radiations (neutrons) [28, 29, 30].

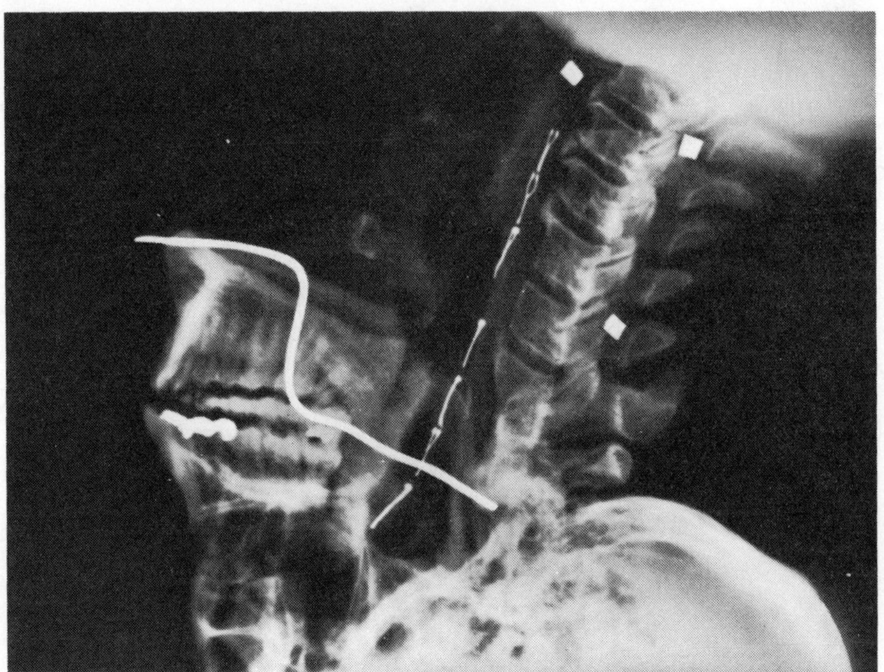

FIG.21. X-ray showing silicon diodes in place for in-vivo dosimetry of neutron dose given in treatment of oesophagus with 50 MeV (d, Be) neutron source.

4.5. Tissue Inhomogeneities

With rare exceptions clinical neutron dosimetry as practiced today does not take anatomical inhomogeneities into consideration in the treatment planning process. Clinical studies and calculations have shown that regions of high hydrogen concentration (fat) will have increased dose absorption relative to muscle with the magnitude of the increased dependent on the convolution of the neutron spectra with the hydrogen cross section [31, 32, 33]. Values for the increase in dose compared to muscle of 15 to 18% would appear to be typical.

Similarly, due to the reduced hydrogen concentration in bone (relative to muscle), the dose to bone is some 15% less than that calculated for muscle [31]. The effect this has on the charged particle spectra the bone marrow is exposed to is not known.

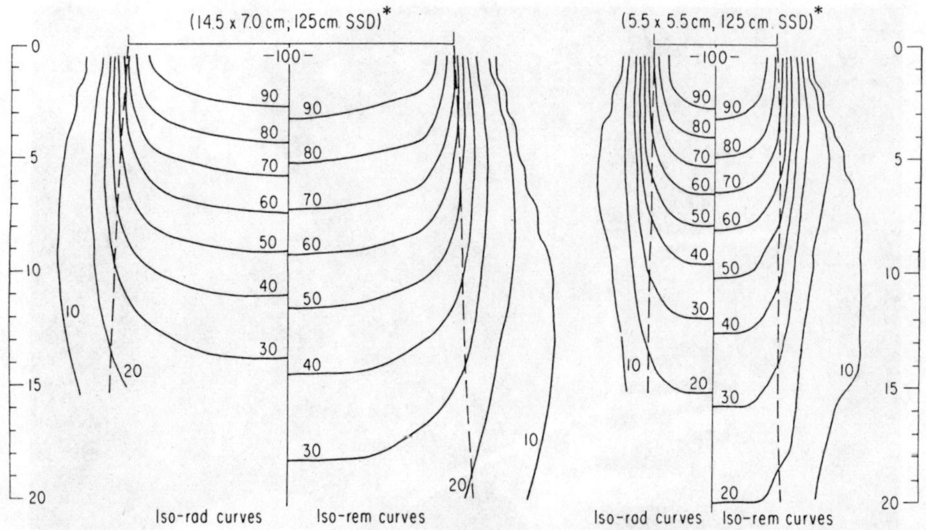

FIG.22. 16 MeV (d, Be) source iso-rem versus iso-rad distribution comparison.

Animal studies and patient invivo dosimetry for the 50 MeV (d,Be) neutron spectra have confirmed the magnitude of the increased lung dose due to lower tissue density to be about 16%. Values of similar magnitude for other neutron sources would be anticipated. In the M.D. Anderson-Texas A&M cooperative neutron trials, this was the only inhomogeneity corrected for in treatment planning.

4.6. Biologically Equivalent Dose

As neutrons are still a research treatment modality, the necessity always arises for comparing a given neutron biological dose to an equivalent photon dose. There are many proposed ways of doing this, but the one which appears to have the greatest acceptance in the United States is to multiply the neutron dose by an average RBE value and then add the photon dose component to obtain a total REM dose. If this is done for the 16 MeV (d,Be) isodose distribution, the REM vs. Rad isodose distribution is illustrated in Figure 22.

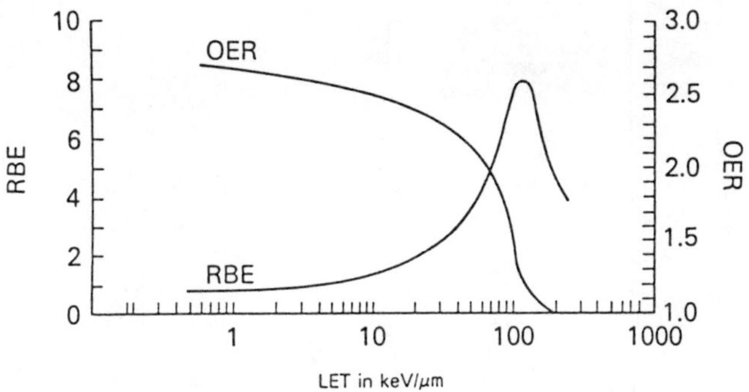

FIG.23. Oxygen enhancement ratio (OER) and relative biological effectiveness (RBE) versus LET.

5. OTHER CONSIDERATIONS

For neutrons, the radiobiological response of both tumor cells, healthy tissues, and the relative response of tumor cells as compared to healthy cells varies from that observed with low LET photon radiation. Some of the differences observed are lower OER values and higher RBE values for neutron irradiated cells, Figure 23. The fraction of the dose due to alpha particles would appear to be an important factor in the spectral averaged magnitude of these values. For neutron irradiated tissues, these parameters are reflected in part in a reduced cell cycle sensitivity and a reduced capability for sublethal repair. The above factors plus others raise the possibility that the optium fractionalization of neutron therapy may differ considerably from that used in conventional low LET therapy.

The precision with which the clinical neutron dose must be known is a factor determined primarily by the tumor control-radiation complication data for given tumors. As an example, for head and neck neoplasms treated twice weekly with neutrons only, the difference between tumor control and failure to control would appear to be 5%, Figure 24. In regions with significant inhomogeneities, this level of precision in treatment planning would be difficult to achieve and places added responsibilities on the clinical neutron physicist. The scope of the concerns the physicist must consider in neutron therapy is both broader and more difficult than those encountered in photon therapy.

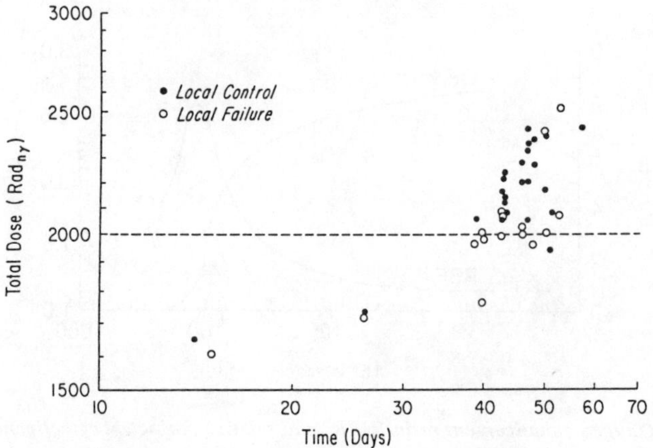

FIG.24. Total dose versus local control of head and neck neoplasms; twice weekly treatment; treatment course length in days indicated.

6. AREAS FOR FURTHER DEVELOPMENT

The following is a brief tabulation of the areas in which additional development may lead to improved clinical neutron dosimetry:

(a) determination of the charge particle spectra which exists vs. depth in the entrance dose build up region, at transcient equilibrium vs. depth in tissue, and at anatomical interfaces such as tissue-bone;

(b) development of improved neutron cross section sets throughout the energy range of 0-60 MeV with emphasis on the 20-60 MeV range;

(c) analysis of the relative benefits of using computerized axial tomography and whole body nuclear magnetic resonance scans in treatment planning;

(d) development of an internationally accepted neutron calibration standard for the neutron energy range used in therapy, i.e. $\bar{E}$ = 20 MeV or greater;

(e) improved methods of separating neutron and photon dose components and utilization of this information in clinical neutron treatment planning.

ACKNOWLEDGEMENT

This work was supported by US Public Health Service Research Contract Number CM 97315 and Research Grant 12542 from the National Cancer Institute.

REFERENCES

[1] CATTERALL, M., Proc. Int. Workshop Particle Radiation Therapy, Key Biscayne, Florida, October 1-5, 1975 (1976) 425.

[2] GRAVES, R., SMATHERS, J., ALMOND, P., GRANT, W. OTTE, V., Medical Physics 6 (1979) 123.

[3] OLIVER, G., GRANT, W., SMATHERS, J., Radiation Research 6 (1975) 366.

[4] Protocol for Neutron Beam Dosimetry, American Association of Physicists in Medicine Report #7 (1980).

[5] BROERSE, J., MIJNHEER, B., WILLIAMS, J., BRIT. J. Radiology 54 (1981) 882.

[6] MCDONALD, J., MA, I-CHANG, LIANG, J., EENMAA, J., AWSCHALOM, M., SMATHERS, J., GRAVES, R., AUGUST, L., SHAPIRO, P., Medical Physics 8 1 (1981) 39.

[7] SMATHERS, J., OTTE, V., SMITH, A., ALMOND, P., ATTIX, F., SPOKAS, J., QUAM, W., GOODMAN, L., Medical Physics 4 1 (1977) 74.

[8] BEWLEY, D., MCNALLY, N., PAGE, B., Radiation Research 58 (1974) 111.

[9] ALMOND, P., Unpublished Data

[10] WELLS, A., "Calculation of Dosimetry Parameters For Fast Neutron Radiotherapy", LA-7288-T, Los Alamos Scientific Laboratories (1978), also published in part Radiation Research 80 1 (1979) 1.

[11] SMATHERS, J., GRAVES, R., ALMOND, P., OTTE, V., GRANT, W., Medical Physics 7 1 (1980) 65.

[12] SMATHERS, J., ALMOND, P., OTTE, V., GRANT, W., Int. J. Radiation Oncology Biol. Phys. 3 (1977) 149.

[13] GREENE, D., THOMAS, R., Brit. J. Radiol. 41 (1968) 455.

[14] OTTE, V., HORTON, J., GOLDBERG, E., TRIPLER, D., SMATHERS, J., Brit. J. Radiol. 50 (1977) 449.

[15] AWSCHALOM, M., FERMILAB, Batavia, Illinois, Private Communication (1980)

[16] SMATHERS, J.B. Unpublished data (1979).

[17] SMITH, A., ALMOND, P., SMATHERS, J.B., OTTE, V., Radiology 113 (1974) 187.

[18] OTTE, V., SMATHERS, J.B., WRIGHT, R., SMITH, A., ALMOND, P., Medical Physics 3 4 (1976) 250.

[19] MIJNHEER, B., VISSER, P., WIEBERDINK, T., Monograph on Basic Physical Data For Neutron Dosimetry, EUR 5629e Commission of the European Community (1976) 145

[20] ATTIX, F., THEUS, R., ROGERS, C., Proc. Second Symposium on Neutron Dosimetry in Biology and Medicine, Neuherberg/ München September 30 to October 4 (1974) EUR 5273 d-e-f, 329.

[21] QUAM, W., JOHNSEN, S., HENDRY, G., TOM, J., HEINTZ, P., THEUS, R., Phy. Med. Biol. 23 1 (1978) 47.

[22] BEWLEY, D., Int. J. Radiat. Oncol. Biol. Phys. 3 (1977) 163.

[23] WEAVER, K., BICHSEL, H. EENMAA, J., WOOTTON, P., Medical Physics 4 5 (1977) 379.

[24] KELLERER, A., RASSOW, J., Medical Physics 7 5 (1980) 503.

[25] FIELD, S., Brit. J. Radiol. 44 (1971) 891.

[26] PRITCHARD, H., SMITH, A., ALMOND, P., SMATHERS, J.B., OTTE, V., AAPM Quarterly Bulletin 7 (1973) 92.

[27] SMITH, A., ROSEN, I., HOGSTROM, K., PRICHARD, H., Int. J. Radiat. Oncol. Biol. Phys. 2 (1977) 111.

[28] TEMME, A., RASSOW, J., MEISSNER, P., To Be Published, Proc. Neutron Dosimetry Symposium, München, June 1981.

[29] LUCAS, A., KASPER, B., Proc. 5th Inter. Conf. On Luminescent Dosimetry, February 14-17 (1977) San Paulo, Brazil.

[30] JAMES, D., GRANT, W., THOMSON, L., ALMOND, P., Medical Physics 5 4 (1978) 320.

[31] BEWLEY, D., Curr. Topic Radiat. Res. (1970) 251.

[32] HUSSEY, D., FLETCHER, G., CADERAO, J., Radiat. Res. (1975) 1106.

[33] ORNITZ, R., BRADLEY, E., MOSSMAN, K., FENDER, F., SCHELL, M., ROGERS, C., Int. J. Radiat. Oncol. Biol. Phys. In Press.

PRIMARY STANDARD LABORATORY ACTIVITIES IN NEUTRON DOSIMETRY

DOSIMETRY FOR NEUTRON THERAPY AT THE PHYSIKALISCH-TECHNISCHE BUNDESANSTALT (PTB)

G. DIETZE, H.J. BREDE, D. SCHLEGEL-BICKMANN
Physikalisch-Technische Bundesanstalt,
Braunschweig,
Federal Republic of Germany

Abstract

DOSIMETRY FOR NEUTRON THERAPY AT THE PHYSIKALISCH-TECHNISCHE BUNDESANSTALT (PTB).

A summary is given of the PTB programme in fast neutron dosimetry for therapy applications. The progress in establishing a reference standard facility for the fast neutron absorbed dose is described. The deuteron beam of the PTB cyclotron ($E_d \leqslant 14$ MeV) and the Be + d reaction are used to produce an intense neutron field. The neutron spectral fluence at the reference position has been accurately determined.

INTRODUCTION

Further development of neutron therapy requires a reliable neutron dosimetry based on accurately calibrated dosimeters which allow a comparison of clinical results from different neutron therapy centres.

One aim of the programme of the Physikalisch-Technische Bundesanstalt (PTB) is to establish a reference standard facility for the fast neutron absorbed dose which can be used for the calibration, investigation and improvement of neutron dosimeters, as well as for international intercomparisons of the neutron absorbed dose standard. This project includes a target facility for the production of an intense neutron field with well-known physical quantities (neutron spectral fluence, local fluence distribution, photon absorbed dose component) and good long-term stability and reproducibility. By using neutron kerma factors for standard tissue [1], spectral fluence information allows the tissue kerma to be determined in free air at a reference position ("fluence method"). This method is independent of absorbed dose measurements by means of ionization chambers calibrated in a well-known photon field ("photon calibration method").

Many problems in calibrating neutron dosimeters arise from the dependence of the calibration factor on the neutron field quality which, in turn, depends on the neutron spectrum of the field. Additional information is therefore required if the calibration factor of a neutron dosimeter is to be transferred to another neutron therapy facility with a different neutron field quality.

MAIN ACTIVITIES

At the PTB, studies in the field of neutron dosimetry for therapy purposes have been concentrated on four subjects described briefly as follows:

1. Collection of basic physical data

Kerma factors for carbon, oxygen, tissue and tissue-equivalent (TE) materials, and kerma ratios contribute considerably to the uncertainty of neutron dosimeter calibrations and neutron absorbed dose determinations in phantoms [2]. A detailed analysis gives an uncertainty greater than 4.3% for the kerma ratio, K_{tissue}/K_{A150}, at a neutron energy of 14 MeV [3]. To reduce these uncertainties, the accurate determination of neutron cross-sections on carbon and oxygen has been included in the PTB research programme. A large time-of-flight facility with five flight paths 12 m long has been installed [4, 5], mainly for precise measurements of elastic and inelastic neutron scattering cross-sections. Measurements on carbon for neutron energies from 6 to 14 MeV are in progress.

In addition, the cross-section of the reaction $^{12}C(n, \alpha_0)^9Be$, which strongly contributes to the carbon kerma value, has been measured for neutron energies from 7.4 to 10.0 MeV [6].

2. Investigations in monoenergetic neutron fields

The calibration of a neutron dosimeter in a neutron reference field and the transfer of the calibration to a neutron therapy facility must take into account the differences in the neutron spectra of both fields. Corrections to the calibration factor of a neutron dosimeter should be applied in this case.

Dosimeters are therefore studied in monoenergetic neutron fields with well-known properties (neutron fluence, neutron background, photon fluence etc.) in order to obtain information about effects depending on the energy of the incident neutrons.

At the low scattering target area [4], fields of "monoenergetic" neutrons up to 19 MeV have been produced and the neutron spectra of these fields have been extensively studied [7]. The $D(d,n)^3He$ and the $T(d,n)^4He$ reactions with deuterons from the Van de Graaff accelerator have been used for neutron production.

The tissue kerma rates in free air at 30 cm distance from the target, obtained with 10 μA deuteron beam current on a Ti-T target (2 mg/cm^2 Ti layer on an Ag backing with a thickness of 1 mm), are low:

Neutron energy (MeV)	Tissue kerma rate (Gy/min)	
15	5.9	10^{-4}
17	0.65	10^{-4}
19	2.3	10^{-7}

The kerma rates are calculated on the basis of the kerma factors from Caswell et al. [1].

3. Mixed-field dosimetry

In neutron fields in air or in phantoms, there are always contributions from photon radiation. The photon dose component of the total dose can be measured using the two-detector technique with a TE detector together with a neutron-insensitive detector. The k_U value of a neutron-insensitive detector, the neutron response (used as defined in IEC Publication 731 (1982); in ICRU Rep. 26 defined as "sensitivity") relative to its photon response, depends on the energy of the incident neutrons and must be known within the neutron energy range concerned. k_U values of GM counters (ZP 1100 from VALVO, MX 163/PTFE from Alrad Inst.) and of an Mg(Ar) chamber (type MG-2 from Exradin) have therefore been measured at various neutron energies [8, 9]. The dead-time behaviour of GM counters, which should be considered in intense neutron-photon fields, has been further studied [10].

4. Neutron reference field from Be + d

A facility for the production of a collimated intense neutron field has been established. Deuterons from the PTB cyclotron ($E_d \leqslant 14$ MeV) are used to bombard a thick Be target. The Be + d reaction has been chosen for several reasons:

(1) A high neutron beam intensity with a tissue kerma rate in air up to 0.2 Gy/min at a distance of 80 cm from the beryllium target for a deuteron beam with an energy up to 14 MeV can easily be obtained.

(2) For these dose rates only a simple target construction with a small water-cooling system is necessary.

(3) The neutron field is stable and reproducible over a long period if the deuteron beam parameters are fixed.

(4) The deuteron beam charge can in addition be used as a neutron fluence monitor if the deuteron energy is constant.

(5) The broad energy spectrum of the neutrons reduces the influence of sharp resonances in neutron cross-sections on the calibration conditions.

(6) The neutron energies that mainly contribute to the total absorbed dose are below about 10 MeV, with a mean neutron energy between 5 and 6 MeV. For these energies, the uncertainties of the kerma factors of carbon and oxygen are much smaller than for energies above 10 MeV. The kerma factors therefore contribute less to the total uncertainty of a dosimeter calibration [2] in this field than, for example, in a 14 MeV neutron field.

The deuteron beam of the cyclotron is guided through the beam line system with two 90° magnets and a switching magnet on to the Be target situated in the large experimental hall. The Be target is surrounded by a movable collimator allowing measurements either under low scattering conditions or in a collimated neutron beam of various field sizes.

The length of the collimator in the beam direction is 60 cm. The reference position for calibrations has been chosen on the beam axis at 80 cm distance from the target. Measurements at distances greater than 60 cm are possible. The distance from the target to the wall of the experimental hall is about 11 m in this particular direction. In the following sections, some details are given which concern the determination of the neutron spectral fluence and the standard tissue kerma measured in air at the reference position in the neutron beam.

THE SPECTRAL NEUTRON FLUENCE FROM THE REACTION Be + d

The spectral source strength of neutrons from a Be + d reaction at 0° to the incident deuteron beam direction has been measured for deuterons with an energy of 11.7 MeV, using a 2-mm-thick Be target and the time-of-flight facility (Fig.1). The flight path was 12 m long. A large collimator [11] shielded the neutron detector against background radiation. The deuteron beam from the cyclotron had a pulse duration of 1 ns and the beam pulse repetition rate was strongly reduced to 44.1 kHz. In the energy range between 10 keV and 20 MeV, the neutron spectrum could therefore be unambiguously identified by analysing the time-of-flight spectrum.

For neutron detection, an NE 213 scintillation detector (5.07 cm in height and 5.06 cm in diameter) with a conventional n-γ pulse-shape discrimination technique, was used. A lower discriminator threshold allows only neutrons of energies above 300 keV to be detected. The detector was absolutely calibrated for several neutron energies by means of a proton recoil telescope.

The detector efficiency ϵ_d, defined as the quotient of the number of detected events in the scintillator divided by the number of neutrons incident on the scintillator, depends on the neutron energy and on the pulse-height threshold for the detection of an event. The energy dependence of the

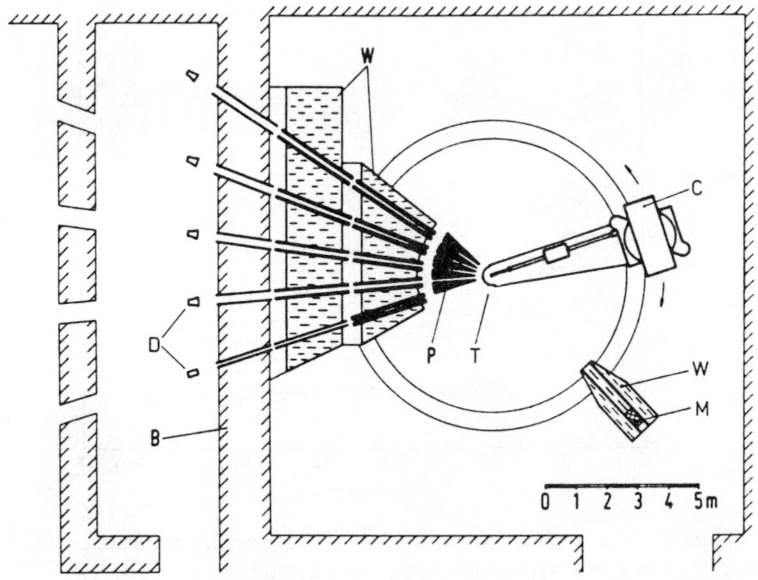

FIG.1. Schematic diagram of the time-of-flight facility.
C: cyclotron, T: Be target, W: water tanks, B: concrete wall, M: monitor, P: polyethylene collimator system, D: NE 213 scintillation detectors.

detector efficiency, which must be known for the analysis of a time-of-flight spectrum, was calculated by using the computer code NEFF4 [12] for neutron energies from 100 keV to 20 MeV and five different pulse-height thresholds (Fig.2). The threshold values were accurately determined by measurements of Compton electron spectra from various photon sources [13]. The agreement between the experimental data and the efficiency calculations was better than 5% for all neutron energies below 20 MeV.

The neutron time-of-flight spectra from ^{9}Be + d at zero degrees to the incident beam direction were measured with five different detector thresholds. Each of these spectra was analysed independently using the different efficiency functions. Good agreement was obtained for the neutron spectral fluence at the detector position. The resulting spectra at energies between 0.3 and 16 MeV, corrected for the neutron absorption in air are given in Fig.3. A comparison with spectrum measurements of Lone [14] at 12.0 MeV shows a strong peak at E_n = 0.8 MeV in both spectra because of the inelastic scattering ^{9}Be(d,d')^{9}Be* → n + ^{8}Be → 2α. The higher total yield measured by Lone corresponds mainly to the higher incident deuteron energy. For neutron energies below 1.5 MeV there is a difference in the absolute neutron yield and in the shape of the spectrum.

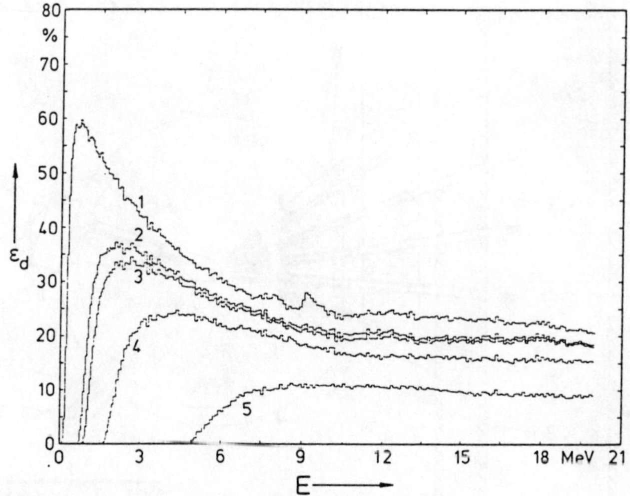

FIG.2. Calculated neutron detector efficiency of an NE 213 scintillation detector (5.07 cm in height, 5.06 cm in diameter) for various bias values, L, given in electron energies:
1. L = 0.025 MeV;
2. L = 0.153 MeV;
3. L = 0.204 MeV;
4. L = 0.482 MeV;
5. L = 2.124 MeV.

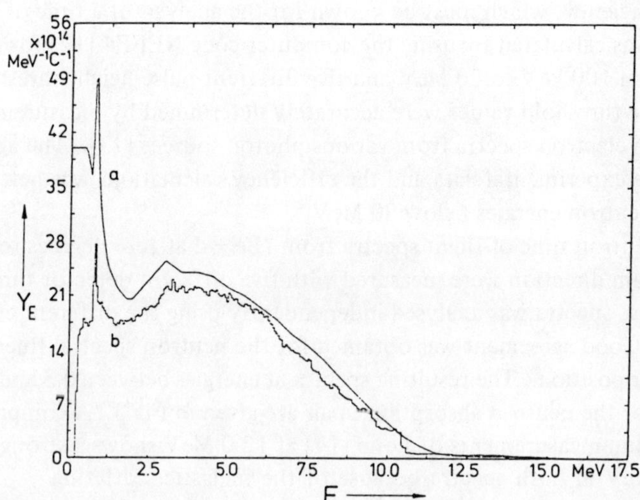

FIG.3. Spectral neutron yield $Y_E(E)$ from Be + d (11.7 MeV) at 0 degrees to the incident deuteron beam direction.
(a) Data from Lone et al. [14] at 12.0 MeV. (b) This work.

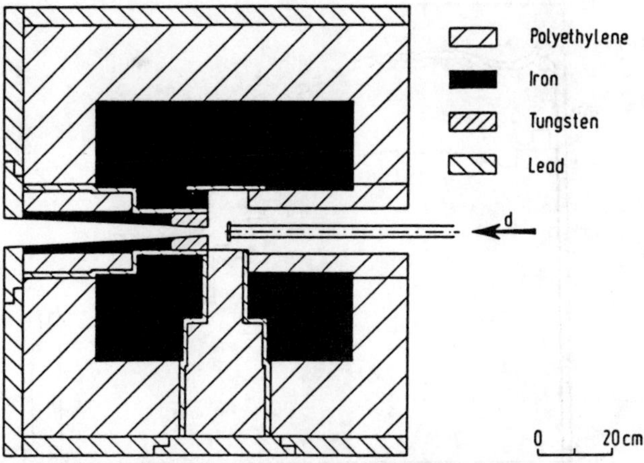

FIG.4. Diagram of the collimator at the calibration facility.

THE COLLIMATOR

The Be target at the calibration facility is surrounded by a movable collimator. The field size at the reference position can be changed from 5 cm × 5 cm up to 20 cm × 20 cm. A field size of 10 cm × 10 cm is usually used. The collimator contains iron, polyethylene with B_4C, lead and a liner of tungsten and iron (Fig.4). The influence of the collimator on the spectral fluence at the reference position was determined by means of Monte Carlo calculations with the code COLLI [15]. The measured undisturbed spectral neutron yield from ^{9}Be + d was taken as the source spectrum and angular distribution of the emitted neutrons was considered to be isotropic for these calculations.

In Fig.5 the results are given for the reference position. Obviously the collimator has little influence on the shape of the neutron energy spectrum, whereas the neutron fluence is increased by about 14% (Table I).

TISSUE KERMA DETERMINATION

The standard tissue kerma in air at the reference position was determined by two methods – the "fluence method" and the "photon calibration method".

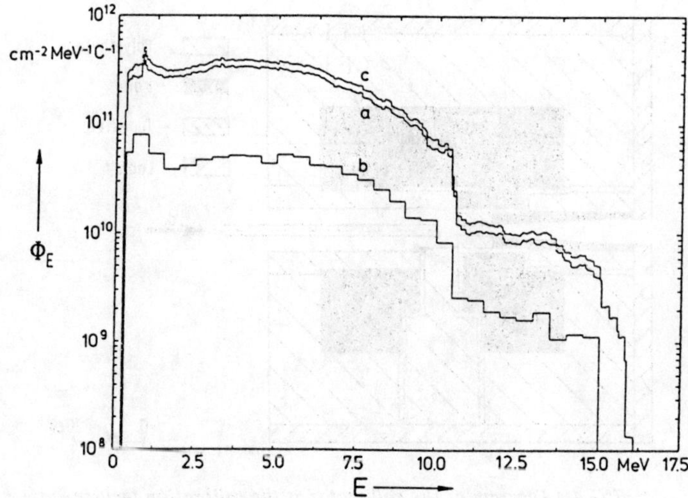

FIG.5. Spectral neutron fluence $\phi_E(E)$ per incident beam charge from Be + d (11.7 MeV) at a distance of 80 cm from the target on the central neutron beam axis.
(a) unscattered neutrons; (b) collimator scattered neutrons; (c) all neutrons (a + b).

With the fluence method the tissue kerma per incident beam charge K is calculated by

$$K = \int_0^{E_{max}} \phi_E(E)\, K_F(E)\, dE$$

$\phi_E(E)$ is the neutron spectral fluence per incident beam charge and $K_F(E) = K/\phi$ are the kerma factors of ICRU muscle tissue which were taken from Caswell et al. [1]. An uncertainty arises from the lack of information about the neutron spectrum below 300 keV neutron energy. Rough assumptions have been made on the shape of the spectrum to test the influence of this fraction to the total kerma value (Fig.6). The spectral fluence below 300 keV neutron energy may be (1) increasing, (2) constant, (3) decreasing or (4) zero. The results of a calculation (see Table II) show that the tissue kerma value varies by less than 1%, if these different assumptions on the neutron spectrum below 300 keV are considered.

The second method applied is a measurement of the tissue kerma by ionization chambers which have been calibrated in a photon reference field.

TABLE I. FLUENCE ϕ AND TISSUE KERMA K PER INCIDENT BEAM CHARGE IN AIR AT THE REFERENCE POSITION FOR E_d = 11.7 MeV
$\bar{E}$ is the fluence averaged neutron energy

Fluence per incident beam charge	ϕ ($cm^{-2} \cdot C^{-1}$)	ϕ/ϕ_{total} (%)	$\bar{E}$ (MeV)	K ($Gy \cdot C^{-1}$)	K/K_{total} (%)
"direct"	2.519×10^{12}	85.8	4.55	102.9	86.0
"scattered"	0.417	14.2	4.49	16.8	14.0
"total"	2.936	100	4.54	119.7	100.0

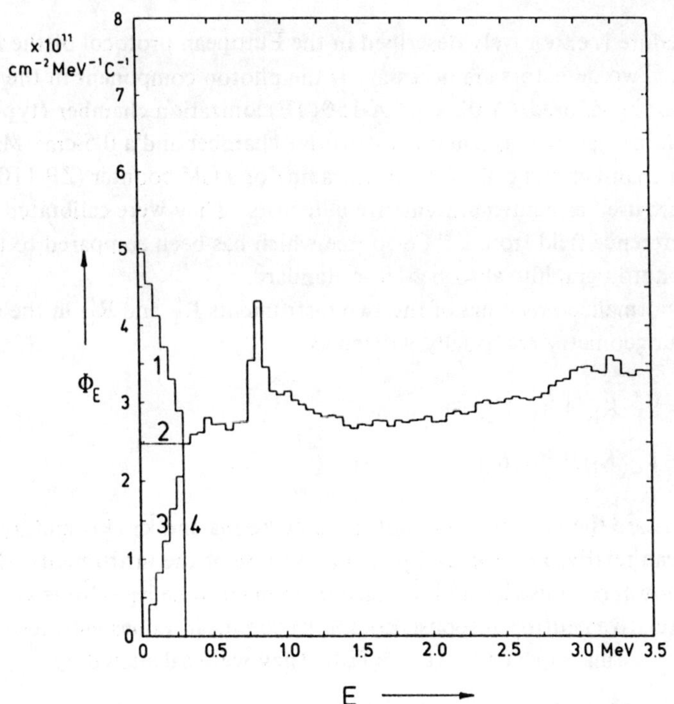

FIG. 6. Spectral neutron fluence $\phi_E(E)$ per incident beam charge from Be + d (11.7 MeV) at a distance of 80 cm from the target with various assumptions on the spectral fluence below 300 keV.
1. "high"; 2. "constant"; 3. "low"; 4. "zero".

TABLE II. FLUENCE ϕ AND TISSUE KERMA K PER INCIDENT BEAM CHARGE FOR VARIOUS ASSUMPTIONS ON $\phi(E)$ FOR $E < 300$ keV

$\bar{E}$ is the fluence averaged mean neutron energy. K_c is the kerma given by the constant distribution for $E < 300$ keV

ϕ $E < 300$ keV	ϕ $(cm^{-2} \cdot C^{-1})$	$\bar{E}$ (MeV)	K $(Gy \cdot C^{-1})$	$(K-K_c)/K_c$ (%)
Zero	2.512×10^{12}	4.56	102.83	−0.6
Low	2.543	4.51	103.15	−0.3
Constant	2.587	4.44	103.42	0
High	2.630	4.37	103.70	+0.3

The procedure is extensively described in the European protocol of the ECNEU group [2]. Two detectors are necessary if the photon component in the neutron field cannot be ignored. A 0.5-cm³ A-150(TE) ionization chamber (type T-2 from Exradin) was used as a neutron-sensitive chamber and a 0.5-cm³ Mg(Ar) ionization chamber (type MG2 from Exradin) or a GM counter (ZP 1100 from Valvo) were used as neutron-intensitive detectors. They were calibrated in a photon reference field from a ^{60}Co-source which has been compared to the national photon graphite absorbed dose standard.

The normalized readings of the two instruments R'_T and R'_U in the case of a free-air geometry are usually written as

$$R'_T = \bar{k}_T \cdot K_N + \bar{h}_T \cdot K_G$$

$$R'_U = \bar{k}_U \cdot K_N + \bar{h}_U \cdot K_G$$

K_N and K_G are the neutron and photon tissue kerma and $\bar{k}_T$, $\bar{k}_U$ and $\bar{h}_T$, $\bar{h}_U$ are the mean relative neutron and photon response of the instruments. The relative photon response $\bar{h}_T$ and $\bar{h}_U$ can be assumed to be $\bar{h}_T = \bar{h}_U = 1$.

The relative neutron response $\bar{k}_U$ and $\bar{k}_T$ are mean values with respect to the spectral kerma $K_E(E) = K_F(E) \cdot \phi_E(E)$. They were calculated by

$$\bar{k}_U = 1/K \cdot \int_0^{E_{max}} k_U(E) \cdot K_F(E) \cdot \phi_E(E)\, dE$$

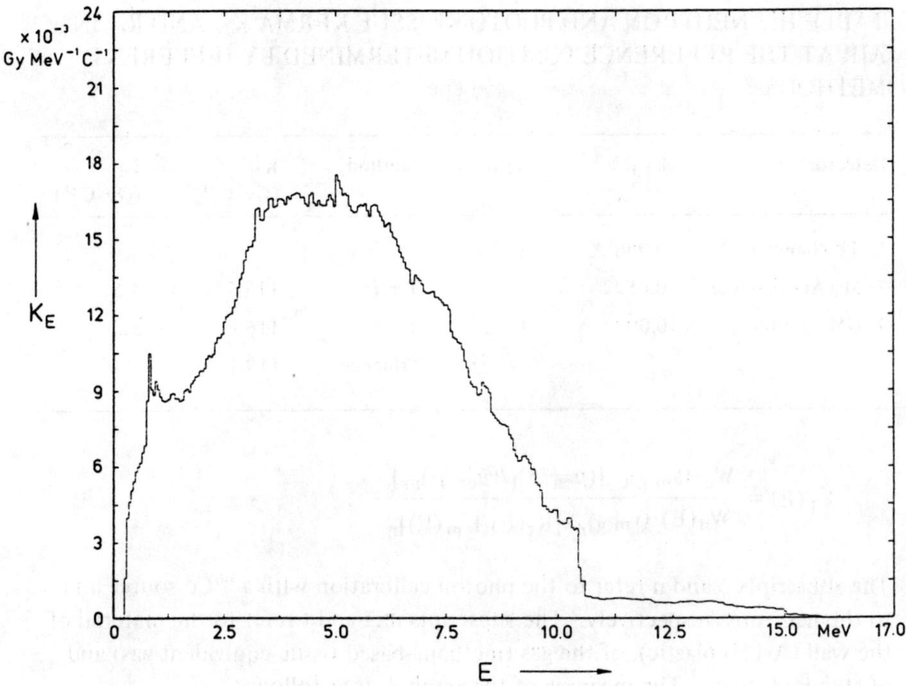

FIG. 7. Spectral tissue kerma $K_E(E)$ per incident beam charge Q in air at the reference position from Be + d (11.7 MeV).

with

$$K = \int_0^{E_{max}} K_F(E) \cdot \phi_E(E)\, dE$$

Figure 7 shows the spectral kerma per incident beam charge obtained with the neutron fluence data at the reference position. For the GM counter, $k_U(E)$ values were taken from Guldbakke [8]. The measured data were connected by straight lines to obtain a continuous function. The calculation gives a mean value $\overline{k}_U = 0.0097$. For the Mg(Ar) chamber, the $k_U(E)$ data from Waterman et al. [16] were used. A continuous function was obtained by the same method as mentioned above. The calculated mean value is $\overline{k}_U = 0.0522$.

The formula for the relative neutron response $k_T(E)$ of a TE ionization chamber is given by the expression [17]

TABLE III. NEUTRON AND PHOTON TISSUE KERMA K_N AND K_G IN AIR AT THE REFERENCE POSITION DETERMINED BY DIFFERENT METHODS

Detector	$\bar{k}_{T,U}$	$\bar{h}_{T,U}$	Method	K_N ($Gy \cdot C^{-1}$)	K_G ($Gy \cdot C^{-1}$)
1. TE chamber	0.966	1			
2. Mg(Ar) chamber	0.0522	1	1 + 2	115.5	4.5
3. GM counter	0.0097	1	1 + 3	116.8	3.2
	–	–	"fluence"	119.7	–

$$k_T(E) = \frac{W_c \cdot (s_{m,g})_c \cdot [(\mu_{en}/\rho)_t/(\mu_{en}/\rho)_m]_c}{W_n(E) \cdot (r_{m,g})_n \cdot [K_t(E)/K_m(E)]_n}$$

The subscripts c and n refer to the photon calibration with a ^{60}Co source and to the neutrons, respectively. The subscripts m,g and t refer to the material of the wall (A-150 plastic), of the gas (methane-based tissue-equivalent gas) and of standard tissue. The meaning of the symbols is as follows:

W – average energy required to produce an ion pair in the gas;

$s_{m,g}$ – mass stopping power of the wall relative to the gas for particles produced in the chamber wall;

$r_{m,g}$ – gas-to-wall absorbed dose conversion factor for ionization chambers [18];

μ_{en}/ρ – mass energy absorbtion coefficient.

Recommended values from the ECNEU report [2] have been taken for some quantities:

$W_c/e = 29.3\ J \cdot C^{-1}$

$(s_{m,g})_c = 1.00$

$(\mu_{en}/\rho)_t/(\mu_{en}/\rho)_{mc} = 1.001$

$(r_{m,g})_n = 0.99$

$(r_{m,g})_n$ is almost constant for the neutron energies considered. Only $W_n(E)$ and K_t/K_m have been energy averaged. The calculation gives a mean value $\bar{k}_T = 0.9656$.

The ionization chamber measurements have been carried out with deuterons of 11.7 MeV and a beam current of 5 μA corresponding to a kerma rate of about 3.6×10^{-2} Gy·min^{-1} at the reference position. The ionization chamber reading has been corrected for temperature and pressure, for lack of saturation and for wall attenuation effects. The neutron and photon tissue kerma per incident beam charge were determined twice, using the GM counter and the Mg(Ar) chamber as the neutron-insensitive detector.

The results are given in Table III. The values for the low photon kerma component K_G differ by about 35%. This may be due to the uncertainty of k_U for the Mg(Ar) chamber used. For the neutron kerma, the value obtained by the "fluence method" is slightly higher than both of the other values, but the difference is less than 2.5%.

Further measurements at the neutron reference field are in progress and, in addition, a detailed analysis of the uncertainties is necessary before this facility can be used for neutron therapy dosimeter calibrations and other investigations.

REFERENCES

[1] CASWELL, R.S., COYNE, J.J., RANDOLPH, M.L., Radiat. Res. **83** (1980) 217.

[2] BROERSE, J.J., MIJNHEER, B.J., WILLIAMS, J.R., Br. J. Radiol. **54** (1981) 882.

[3] JAHR, R., BREDE, H.J., Phys. Med. Biol. **25** (1980) 923.

[4] BREDE, H.J., et al., Nucl. Instrum. Methods **169** (1980) 359.

[5] BÖTTGER, R., et al., in Proc. Int. Conf. Nuclear Data for Science and Technology, Antwerp, 1982. To be published.

[6] DIETZE, G., BREDE, H.J., KLEIN, H., SCHÖLERMANN, H., in Proc. Fourth Symp. Neutron Dosimetry, Vol.1, EUR-7448, Luxembourg (1981) 373.

[7] DIETZE, G., GULDBAKKE, S., MENZEL, H.G., SCHUHMACHER, H., BÜHLER, G., in Proc. Eighth Symp. Microdosimetry, Jülich, 1982. To be published.

[8] GULDBAKKE, S., JAHR, R., LESIECKI, H., SCHÖLERMANN, H., Hlth. Phys. **39** (1980) 963.

[9] JAHR, R., DIETZE, G., GULDBAKKE, S., LESIECKI, H., SCHLEGEL-BICKMANN, D., in Proc. Fourth Symp. Neutron Dosimetry, Vol.1, EUR-7448, Luxembourg (1981) 452.

[10] GULDBAKKE, S., KLEIN, H., in Proc. Fourth Symp. Neutron Dosimetry, Vol.2, EUR-7448, Luxembourg (1981) 385.

[11] SCHLEGEL-BICKMANN, D., DIETZE, G., SCHÖLERMANN, H., Nucl. Instrum. Methods **169** (1980) 517.

[12] DIETZE, G., KLEIN, H., PTB-Bericht ND-22 (1982).

[13] DIETZE, G., KLEIN, H., Nucl. Instrum. Methods **193** (1982) 549.

[14] LONE, M.A., FERGUSON, A.J., ROBERTSON, B.D., Nucl. Instrum. Methods **189** (1981) 515.

[15] SCHLEGEL-BICKMANN, D., PTB-Bericht ND-14 (1979).

[16] WATERMAN, F.M., KUCHNIR, F.T., SKAGGS, L.S., KOUZES, R.T., MOORE, W.H., Phys. Med. Biol. **24** (1979) 721.

[17] MIJNHEER, B.J., WILLIAMS, J.R., Phys. Med. Biol. **26** (1981) 57.

[18] BICHSEL, H., RUBACH, A., in Proc. Third Symp. Neutron Dosimetry in Biology and Medicine, EUR-5848, Luxembourg (1978) 549.

IAEA-AG-371/8

THE PROGRAMME OF THE UNITED STATES NATIONAL BUREAU OF STANDARDS IN DOSIMETRY STANDARDS FOR NEUTRON RADIATION THERAPY*

L.J. GOODMAN, J.J. COYNE, R.S. CASEWELL
Centre for Radiation Research,
National Bureau of Standards,
Washington D.C.,
United States of America

Abstract

THE PROGRAMME OF THE UNITED STATES NATIONAL BUREAU OF STANDARDS IN DOSIMETRY STANDARDS FOR NEUTRON RADIATION THERAPY.

This report discusses two aspects of the neutron dosimetry programme at the United States National Bureau of Standards (NBS), namely the plans and progress towards establishing dosimetry standards for neutron radiation therapy, and an investigation of the neutron and gamma-ray tissue kerma rates from a ^{252}Cf source. Neutron radiation therapy is being clinically tested at a number of centres in the world. To maximize the chances of success of this radiation therapy modality, good physical dosimetry is needed. To facilitate exchange of therapy experience between institutions, the United States dosimetry standards base must be accurate and consistent with the international standards system. The purpose of the NBS programme is to improve the accuracy and consistency of measurements of absorbed dose for neutron radiation therapy by providing national dosimetry standards and improved data on neutron interactions with tissue and tissue-equivalent materials. A longer-term goal is to develop a calibration facility at NBS where neutron dosimeters can be calibrated and their energy dependence studied. As an initial step in establishing the neutron dosimetry standards programme, it was decided to make neutron and gamma-ray measurements of a ^{252}Cf source at NBS, whose neutron emission rate had been measured and was known with an uncertainty of 1%. Neutrons emitted by spontaneous fission in ^{252}Cf sources are used for interstitial and intracavitary applications to neutron therapy, and for calibrating various dosimeters used for radiation protection. Besides being convenient sources of fission neutrons, such sources also emit copious amounts of gamma rays. Both radiation components must be accurately assessed for proper application.

1. INTRODUCTION

This report discusses two aspects of the neutron dosimetry programme at the United States National Bureau of Standards (NBS), namely the plans and progress towards establishing dosimetry standards for neutron radiation therapy,

* This investigation was supported by PHS Grant Number CA26313-03 awarded by the National Cancer Institute, DHHS.

and an investigation of the neutron and gamma-ray tissue kerma rates from a ^{252}Cf source.

Neutron radiation therapy is being clinically tested at a number of centres in the world. To maximize the chances of success of this radiation therapy modality, good physical dosimetry is needed. To facilitate exchange of therapy experience between institutions, the United States dosimetry standards base must be accurate and consistent with the international standards system. The purpose of the NBS programme is to improve the accuracy and consistency of measurements of absorbed dose for neutron radiation therapy by providing national dosimetry standards and improved data on neutron interactions with tissue and tissue-equivalent materials. A longer-term goal is to develop a calibration facility at NBS where neutron dosimeters can be calibrated and their energy dependence studied.

As an initial step in establishing the neutron dosimetry standards programme, it was decided to make neutron and gamma-ray measurements at NBS of a ^{252}Cf source whose neutron emission rate had been measured and was known with an uncertainty of 1%. Neutrons emitted by spontaneous fission in ^{252}Cf sources are used for interstitial and intracavitary applications to neutron therapy, and for calibrating various dosimeters used for radiation protection. Besides being convenient sources of fission neutrons, such sources also emit copious amounts of gamma rays. Both components of radiation must be accurately assessed for proper application.

2. DOSIMETRY STANDARDS FOR NEUTRON THERAPY

If the clinical trials of neutrons as a radiation therapy modality are to have maximum chance of success, and if results are to be comparable from hospital to hospital, it is important that neutron absorbed dose measurements be on the same neutron measurements scale. Neutron dosimetry intercomparisons have shown discrepancies of from 5 to 20% among laboratories experienced in neutron dose measurement [1–3]. Therefore, recommendations have been made for national standards laboratories to develop standards for neutron dosimetry. Thus, Report 26 of the International Commission on Radiation Units and Measurements (ICRU) [4] states: *"It is recommended that standards laboratories develop and establish standard instruments and/or radiation fields which will be directly applicable to absorbed dose calibrations for neutron energies applied in the life sciences."* ICRU Report 27 [1] recommended that: *"National standards laboratories should be encouraged to offer neutron dosimetry standards in the near future."*

An accuracy of 5% in dose to the tumour, which is generally considered desirable [5, 6], requires an accuracy of physical dosimetry to about 3%, and national standards should be somewhat better. The intercomparisons mentioned above clearly indicate that the desired accuracy is not being obtained. The United States neutron therapy installations have chosen to adopt nearly identical ionization chamber dosimetry techniques in order to be on the same scale. This scale, however, is somewhat arbitrary since many factors needed to determine neutron absorbed dose absolutely are not well known and depend on the neutron spectrum which will be different at various installations. A very important uncertain factor is W, the mean energy required to produce an ion pair in the cavity gas.

To maximize the chances of success of this radiation therapy modality, good physical dosimetry is needed; and to facilitate exchange of therapy experience between institutions, the US dosimetry standards base must be accurate and consistent with the international standards system. The goals of the NBS programme are to develop a national standard for neutron absorbed dose measurement, together with appropriate transfer instrumentation to carry the calibration to hospitals and clinics doing radiotherapy, and to develop a monoenergetic neutron calibration facility at NBS for neutron dosimeter calibration and energy dependence studies.

Choosing a standard instrument for neutron dosimetry is complicated by the problem of making measurements in a mixed neutron and gamma-ray field because all neutron radiation fields are invariably accompanied by gamma rays. Neutron instrumentation may be used which discriminates against gamma rays, or pairs of instruments may be used, one sensitive to neutrons and gamma rays and the other ideally sensitive to gamma rays only, the fast neutron dose being obtained by subtraction. Practical gamma-ray dosimeters have some small sensitivity to neutrons so that separation of the two components of the radiation field requires the solution of a pair of simultaneous equations. The instrument sensitive to both neutrons and gamma rays is usually a tissue-equivalent (TE) ionization chamber. The neutron-insensitive gamma-ray dosimeter may be photographic film, a Geiger-Müller (GM) counter, a non-hydrogenous ionization chamber, or a non-hydrogenous proportional counter.

A national standard instrument, while embodying the most accurate calibration information available, also needs to be convenient for use with secondary standard instruments sent to NBS for calibration. For this reason ionization chamber instruments are more suitable as a standard than, for example, calorimeters which are too insensitive for use with many of the available radiation fields. Therefore, we propose that the national standard be an extremely well-calibrated TE ionization chamber.

To measure the gamma-ray component of the neutron field, several neutron-insensitive gamma-ray dosimeters are being tested, including Mg-Ar ionization chambers, a miniature GM counter, and a graphite porportional counter.

Towards this end the following instruments have been acquired[1]:

(a) Three 1 cm^3 spherical TE ionization chambers (Far West Technology Model IC-17);

(b) Three 0.5 cm^3 thimble TE ionization chambers (Exradin Model T2);

(c) One 1 cm^3 spherical magnesium ionization chamber (Far West Technology Model IC-17M);

(d) One 0.5 cm^3 thimble magnesium ionization chamber (Exradin Model MG2);

(e) One miniature Geiger-Müller counter with photon energy compensation shield and ^{6}Li thermal neutron shield (Far West Technology Model GM-1).

A 13-mm-diameter spherical graphite proportional counter (Far West Technology Model LET-$\frac{1}{2}$) is on order.

These instruments are undergoing tests to study their sensitivity, stability, reliability, and general suitability for making measurements in mixed neutron and gamma-ray fields. The neutron-insensitive gamma-ray dosimeters, namely the Mg-Ar ionization chambers, the GM counter and the graphite proportional counter, are being tested and intercompared to select one instrument for the measurement of the gamma-ray absorbed dose in the presence of neutrons.

The TE ionization chambers are undergoing redundant calibration by three independent methods outlined as follows:

(1) The TE ionization chambers and the neutron-insensitive gamma-ray dosimeters have been calibrated in NBS standard gamma-ray beams, from ^{60}Co and ^{137}Cs. The neutron absorbed dose calibration is obtained using the best available information on W, stopping power, and kerma factors. The major uncertainty in such a calibration is contributed by W since the mean energy required to produce an ion pair is not well established for the TE gases used in the TE chambers [7]. Furthermore, it is necessary to make a kerma factor correction because the TE materials available do not approximate closely enough the tissue composition desired [8, 9]. Improved corrections to these problems are being investigated theoretically.

(2) Except for the question of sensitivity, the calorimeter is attractive as a primary standard measuring neutron absorbed dose. Since the calorimeter has no inherent gamma-ray discrimination, a pair of calorimeters is required, one of TE plastic and one of graphite. The calorimeter pair will be compared to the TE ionization chamber plus gamma-ray dosimeter using intense neutron sources with different neutron energy spectra. Information on the thermal defects for the calorimeters is taken from the literature.

---

[1] In the interests of accuracy and clarity in describing various items of instrumentation, mention is made of commercial sources. This in no way implies endorsement of such products by the US Government, or by the IAEA.

(3) It is very important, in order to understand the response of neutron dosimeters, to be able to irradiate them with monoenergetic neutrons of various energies and study the energy dependence. Using known fluences of neutrons of known energy and kerma-to-fluence factors, one can predict the reading of the dosimeter and compare it to the observed reading. Fluences will be measured using a pulse ionization double-fission chamber. The NBS 3-MV Van de Graaff accelerator monoenergetic neutron facility can be used to provide neutrons of many energies from below 0.1 MeV to 20 MeV. Monoenergetic neutrons are not available at every energy, but at a sufficient number of energies for one to be confident of dosimeter response versus energy. An intense source of 14 MeV neutrons in a well-shielded laboratory 5.5 m × 5.5 m × 4 m has been constructed using a new beam line from the Van de Graaff.

By virtue of these three largely independent methods one can achieve a very high degree of certainty that the calibration of the working standard is within known limits of uncertainty. The present practice at US clinical centres corresponds to method (1), but has problems of uncertainty in the mean energy per ion pair, W, and in relative stopping powers of wall and gas in cavity chambers, as well as uncertainty owing to kerma factors at high neutron energies. Agreement between the three independent methods, and intercomparisons with the Bureau International des Poids et Mesures (BIPM), and with other national standards laboratories, will ensure the accuracy and consistency of the US dosimetry standards for neutron therapy. An intercomparison under the sponsorship of Section III, Neutron Measurements, of the BIPM Consultative Committee for Measurement Standards for Ionizing Radiations (CCEMRI), is planned for 1983 to take place at the National Physical Laboratory, United Kingdom.

The experimental standards programme is supported by a theoretical neutron dosimetry programme aimed at improving knowledge about physical data and correction factors necessary for interpreting the experimental measurements. The programme of theoretical calculations consists of five parts:

(1) Calculations of the initial energy spectra of the secondary particles generated by the neutron interactions in the tissue or tissue-equivalent materials are made using nuclear cross-section information. The general technique for doing this has been developed [10] and initial spectra have been calculated at many energies up to 20 MeV.

(2) A second important quantity is the slowing-down spectrum, which is an equilibrium or average spectrum of the secondary particles as they slow down from the initial energies to rest [10]. The slowing-down spectra are obtained from the initial spectra and stopping-power data for the secondary particles. Both these spectra are needed for microdosimetry calculations and also are used directly as input to biological models of neutron effects.

(3) A table of kerma factors (kerma per unit fluence) up to 30 MeV was prepared and included in ICRU Report 26 [4]. These data have been updated in a more recent paper [8]. A compilation of kerma factors up to 30 MeV has been prepared for 44 compounds and mixtures and is in press [9]. Data compilations such as ENDF/B-4 do not yet provide all the data needed to evaluate kerma factors. Therefore, certain models and assumptions have to be put into the calculations. In addition, there is very little cross-section information above 20 MeV, except for hydrogen. To obtain cross-section information for the kerma calculations in this energy region one is forced to use a combination of limited total cross-section information, nuclear model calculations, extrapolation from lower energies, and guesswork. We shall continue to build and maintain a capability for theoretical calculation and interpolation of the needed cross-sections, consistent with available experimental data, as discussed in para. (5) below.

(4) The most important numerical factor entering into ionization chamber neutron dosimetry is undoubtedly the mean energy required per ion pair, W. Calculations of W for the common methane-based TE chamber gas have been made using experimental values of W and energy deposition calculations for neutron energies up to 20 MeV [7].

(5) In connection with all the calculations mentioned above, improved theoretical nuclear cross-section data, especially above 20 MeV, are necessary. Calculational ability is essential to ensure that the dosimetric calculations will not be held up by lack of input data. Therefore, the final element of this programme is to develop and maintain a theoretical nuclear cross-section data calculational competence. Studies in neutron dosimetry calculations versus nuclear data have been reported [11].

It is hoped that refined experimental measurements, supplemented by careful theoretical dosimetry calculations, will yield neutron dosimetry standards which will enhance the efficacy of neutron radiotherapy.

3. DOSIMETRY OF A ^{252}Cf SOURCE

In June 1980 Drs Johan J. Broerse and Johan Zoetelief of the Radiobiological Institute TNO, the Netherlands, visited NBS. To derive maximum benefit from their visit, it had been previously agreed that they would bring their neutron dosimetry instruments to NBS in order to carry out joint comparison measurements of a large (~2 mg) ^{252}Cf source known as NS-100. This source was especially prepared for NBS by the Oak Ridge National Laboratory (ORNL) and is relatively lightly encapsulated in about 3.2 mm aluminium and 0.5 mm stainless steel. Californium sources of this strength are usually obtained from the Savannah

River Laboratory (SRL) and are doubly encapsulated in ~2.5 mm stainless steel or zircaloy and platinum-rhodium alloy. Sources of this latter type were used during the International Neutron Dosimetry Intercomparison (INDI) [1] and also during the European Neutron Dosimetry Intercomparison Project (ENDIP) [2]. During these intercomparisons, irradiations with the ^{252}Cf source were carried out with additional material surrounding the source in order to facilitate remote transport.

The dosimeters provided by the TNO group consisted of five TE ionization chambers of different sizes and designs, and a photon-energy-compensated miniature GM counter. The NBS instruments were composed of three spherical TE ionization chambers of identical design, a Mg-Ar ionization chamber, and a GM counter similar to the TNO instrument. All TE ionization chambers used a steady flow of methane-based TE gas from the same gas cylinder, and all the instruments were calibrated using the same NBS standard ^{137}Cs gamma-ray beam.

Measurements of the NS-100 source were performed in a room large enough to make room backscattered neutrons negligible, i.e. less than 0.5% of the TE chamber response. All measurements were made with the dosimeters at a nominal distance of 30 cm from the source. The results of these measurements may be briefly summarized as follows: The neutron tissue kerma rate in free air determined by the TNO group had an average value for the five instruments that differed by less than 3% from the value determined by the average response of the three instruments of the NBS group. The gamma-ray tissue kerma rates in free air measured by the two GM counters agreed to within 1.4%. The gamma-ray tissue kerma rate in free air based on the response of the Mg-Ar chamber was about 15% higher than indicated by the GM counters. The significance of this higher value was not immediately appreciated.

The neutron emission rate of the NS-100 ^{252}Cf source had been previously measured at NBS using the manganese sulphate bath method and was known with an uncertainty of 1%. From this emission rate, the known neutron spectrum, a weighted kerma factor computed for the spectrum, and a decay correction, the neutron tissue kerma rate in free air was calculated. The neutron tissue kerma rate determined with the TE ionization chamber and GM counter was about 40% higher than the kerma rate calculated from the neutron emission rate! No reason for this large and unexpected discrepancy was evident.

Based on an available gamma-ray spectrum for a ^{252}Cf source [12], for which no uncertainty was given and which was probably for a source in the heavier SRL encapsulation, Charles Eisenhauer of NBS calculated that the gamma-ray tissue kerma rate in free air should be 0.56 of the neutron kerma rate. Using these data, the gamma-ray tissue kerma rate was calculated from the neutron emission rate of NS-100 and the value agreed to within less than 0.5% with the value determined from the GM counter measurements. However, because of the high value of the measured neutron kerma rate, the ratio of gamma-ray to neutron kerma rates was only 0.40. Thus, the good agreement for the gamma-ray data served to confuse the neutron data even more.

TABLE I. SUMMARY OF DATA FOR NS-100 ^{252}Cf SOURCE

Method	Relative kerma		Kerma ratio
	Neutron	Gamma ray	$\frac{\text{gamma ray}}{\text{neutron}}$
Calculation from neutron emission rate determined by manganese sulphate bath and fission counter	100	100	0.56
Measurement with TE ionization chambers and GM counter	140	100	0.40
Measurement with TE ionization chambers and 3 mm wall magnesium ionization chamber	125	115	0.48
Measurement with TE proportional counter	107	178	0.9
Measurement with TE ionization chamber and 1 mm wall magnesium ionization chamber	104	164	0.88
Measurement with TE ionization chamber and 1 mm wall magnesium ionization chamber with 2.2 mm steel around source	103	98	0.53

Further measurements with NS-100 were made ten months later. At this time only NBS instrumentation was used, namely one of the three spherical TE ionization chambers used previously, a new thimble TE ionization chamber, and the GM counter. The spherical chamber had a 5-mm-thick TE wall and the thimble chamber had a 1-mm-thick wall. During the same measurement session, Emmert D. McGarry of NBS re-measured the neutron emission rate of the source using a pulse ionization fission counter. The outcome of these measurements confirmed the previous results, that is the neutron kerma rate agreed with the previous value within 1.3%, and the gamma-ray kerma rate was the same as the earlier value. Similarly, the neutron emission rate determined with the fission counter was the same as the earlier value. In addition, the inverse-square attenuation measured with one of the TE chambers over a range of distances checked to within 1.5% showing that there was no anomalous background contribution. Thus, we were still left with the 40% discrepancy in neutron kerma rate.

Earlier, new measurements had been made in collaboration with Joseph C. McDonald and Hans Menzel using a small spherical TE proportional counter. Evaluation of the ionization event spectrum yielded a total tissue kerma rate which agreed with the earlier TE ionization chamber value within 5%. The neutron kerma rate indicated by the proportional counter and the neutron emission rate agreed to within 7%. The ratio of gamma ray to neutron kerma rate calculated from the event spectrum was 0.9. Based on these results it was now possible to propose a tentative hypothesis to explain the earlier discrepancy, namely that the photon spectrum of NS-100 contains a significant fraction of low-energy radiation not measured by the GM counter. Indeed, examination of the manufacturer's specifications for the GM counter indicated a fairly uniform response for photons of energies down to 70 keV and then a sharp drop in sensitivity so that at 50 keV its response was less than the 1% of that at higher energies.

To test this hypothesis, a series of measurements was made in April 1982 employing additional steel sleeves around the NS-100 source to simulate thicker encapsulation which would act to filter out low-energy radiation. A thimble TE chamber with a 1 mm wall thickness was used in combination with a Mg-Ar chamber also with a 1 mm wall. It is important to note that the magnesium chamber used during the first set of measurements had a 3 mm wall and was therefore less sensitive to low-energy radiation.

The results of these latest measurements were very satisfying. With no added material around the source, the ionization chamber measurements gave a neutron kerma rate which agreed with the neutron-emission-rate-calculated value within 4%, while the gamma-ray to neutron kerma ratio was 0.88. With the addition of as little as 2 mm steel around the source this ratio dropped to 0.53, the neutron kerma rate changed by only 1%, and the gamma-ray kerma rate was now only 2% different from that calculated from the neutron emission rate. Table I summarizes the NS-100 data.

The reason that the earlier measurements had indicated good agreement between the gamma-ray kerma rates measured with the GM counter and calculated from the neutron emission rate is that the energy compensation shield around the GM tube acts as a filter for low-energy radiation which was also excluded from the calculated photon source spectrum. The drop in gamma-ray kerma rate with the addition of 2 mm steel confirms the hypothesis that the NS-100 source in its light encapsulation emits significant low-energy radiation, presumably photons.

It is planned to use thin foils to find the attenuation coefficient of the soft radiation in order to determine the type and effective energy of these radiations.

This problem with a ^{252}Cf source has served to illustrate the caution which must be exercised in the use of the GM dosimeter, and the difficulties which may be encountered when measuring sources in unconventional encapsulation. The virtues of a TE proportional counter for mixed radiation dosimetry were also clearly demonstrated.

ACKNOWLEDGEMENTS

The authors thank the following colleagues for their help with the ^{252}Cf problem: Johan J. Broerse, Charles Eisenhauer, Emmert D. McGarry, Hans Menzel, Joseph C. McDonald, and Johan Zoetelief.

REFERENCES

[1] INTERNATIONAL COMMISSION ON RADIATION UNITS AND MEASUREMENTS, An International Neutron Dosimetry Intercomparison, ICRU Rep. 27, Washington, D.C. (1978).

[2] BROERSE, J.J., BURGER, G., COPPOLA, M., Radiation Protection. A European Neutron Dosimetry Intercomparison Project (ENDIP): Results and Evaluation, EUR 6004, Commission of the European Communities (1978).

[3] SMITH, A.R., ALMOND, P.R., SMATHERS, J.B., OTTE, V.A., ATTIX, F.H., THEUS, R.B., WOOTTON, P., BICHSEL, H., EENMAA, J., WILLIAMS, D., BEWLEY, D.K., PARNELL, C.J., Dosimetry intercomparisons between fast-neutron radiotherapy facilities, Med. Phys. **2** (1975) 195.

[4] INTERNATIONAL COMMISSION ON RADIATION UNITS AND MEASUREMENTS, Neutron Dosimetry for Biology and Medicine, ICRU Rep. 26, Washington, D.C. (1977).

[5] HERRING, D.F., COMPTON, D.M.J., "The degree of precision required in the radiation dose delivered in cancer radiotherapy," Computers in Radiotherapy (GLICKSMAN, A.S., COHEN, M., CUNNINGHAM, J.R., Eds), The British Institute of Radiology (1971).

[6] SHUKOVSKY, L.J., FLETCHER, G.H., Time-dose and tumor volume relationships in squamous cell carcinoma of the tonsillar fossa, Radiology **107** (1973) 621.

[7] GOODMAN, L.J., COYNE, J.J., W_n and neutron kerma for methane-based tissue-equivalent gas, Radiat. Res. **82** (1980) 13.

[8] CASWELL, R.S., COYNE, J.J., RANDOLPH, M.L., Kerma factors for neutron energies below 30 MeV, Radiat. Res. **83** (1980) 217.

[9] CASWELL, R.S., COYNE, J.J., RANDOLPH, M.L., Kerma factors of elements and compounds for neutron energies below 30 MeV, Int. J. Appl. Radiat. Isot. (1982).

[10] CASWELL, R.S., COYNE, J.J., Interaction of neutrons and secondary charged particles with tissue: secondary particle spectra, Radiat. Res. **52** (1972) 448.

[11] CASWELL, R.S., COYNE, J.J., "Energy deposition spectra for neutrons based on recent cross section evaluations," Proc. 6th CEC Symp. on Microdosimetry (BOOZ, J., EBERT, H.G., Eds), Vol. 2, Harwood Academic Publishers (1978).

[12] STODDARD, D.H., Radiation Properties of Californium-252, USAEC Rep. DP-986, Savannah River Laboratory, Aiken, South Carolina (1965).

IAEA-AG-371/7

ACTIVITY OF THE BUREAU INTERNATIONAL DES POIDS ET MESURES (BIPM) IN NEUTRON DOSIMETRY

V.D. HUYNH
Bureau International des Poids et Mesures,
Sèvres, France

Abstract

ACTIVITY OF THE BUREAU INTERNATIONAL DES POIDS ET MESURES (BIPM) IN NEUTRON DOSIMETRY.

The Bureau International des Poids et Mesures (BIPM) is a co-ordinating centre for national laboratories working in metrology. In the area of neutron dosimetry, BIPM was asked to study the performance of tissue-equivalent ionization chambers of various designs in order to select a set of instruments which could serve as reference and transfer instruments. The other activities of the neutron measurement group at BIPM are also described.

INTRODUCTION

The Bureau International des Poids et Mesures (BIPM) is a co-ordinating centre for national laboratories working in the field of metrology. One of its principal tasks is to organize periodically international comparisons with a view to achieving uniformity on a world-wide scale and a long-term basis.

In neutron dosimetry, the Comité Consultatif pour les Etalons de Mesure des Rayonnements Ionisants (CCEMRI), which is a committee for advising on the work to be carried out at BIPM on radiation measurements, proposed that the coming neutron dosimetry intercomparisons be performed in two stages. First, the National Physical Laboratory (NPL, Teddington, United Kingdom) will organize a "pilot laboratory" intercomparison of neutron dosimeters, presumably tissue-equivalent (TE) ionization chambers, using its intense (10^{11} s^{-1}) 14 MeV neutron dosimetry facility. This intercomparison would take place during 1983. A second comparison, to be co-ordinated by BIPM, is planned to start in 1984. It will be carried out by circulating instruments to the participants. The BIPM was asked to study the performance of various TE ionization chamber designs in order to select a set of chambers which could serve as reference and transfer instruments. The sensitivity to neutrons measured in air (charge/kerma) is to be reported by the participating laboratories. Each laboratory should make its own determination of the gamma-ray component in its neutron field. The ^{60}Co γ-calibration factors for the transfer instruments will be provided by BIPM.

NEUTRON MEASUREMENTS AT BIPM

The BIPM neutron measurements can be summarized as follows:

Measurement of the emission rate of neutron sources with the manganese bath method;
Measurement of neutron fluence rate (2.5 MeV and 14.7 MeV) with the associated particle method;
Study of TE ionization chambers as reference and transfer instruments for international comparison of kerma measurements;
Calibration of the BIPM (d + T) neutron field in terms of tissue kerma in free air.

It should be pointed out that the kerma calibration in the BIPM (d + T) neutron field can be obtained, on the one hand, by the TE ionization chamber measurements, associated with a calibrated Geiger-Müller counter to separate the γ-ray component and, on the other hand, by the absolute neutron fluence measurements by applying an appropriate evaluated fluence-to-kerma conversion factor.

To provide the γ-calibration factors of the TE ionization chambers, a ^{60}Co γ-ray beam calibrated in terms of exposure with a very high long-term stability is available at BIPM, in the X- and γ-ray measurement group. The relative standard deviation over a period of several years has been observed to be about 10^{-4} for exposure measurements.

ABSOLUTE NEUTRON FLUENCE MEASUREMENTS

The neutron fluence is measured by the method of associated particle counting [1]. For (d+T) neutrons, the number of α particles is measured by a silicon surface barrier detector, the effective solid angle of which is defined by a 4-mm-dia. collimator close to the detector and by its distance from the target (1 m). The detector is placed at an angle of 150° with respect to the incident deuteron beam. A 150 kV SAMES type electrostatic generator is used to produce the (d+T) neutrons with a deuteron beam of 140 keV on a thick Ti-T target (0.65 mg$\cdot$cm^{-2} of tritiated titanium on 0.5 mm copper backing). A total neutron emission rate of 1.6×10^9 s^{-1} is produced for a deuteron beam current of 20 μA with a fresh target. The corresponding coincidence angle of neutrons is 26.8° and the calculated average neutron energy is (14.65 ± 0.05) MeV. The overall uncertainty of the fluence determination is estimated to be 1.4% (standard deviation).

MEASUREMENT OF THE NEUTRON SENSITIVITY OF A GEIGER-MÜLLER COUNTER

The sensitivity of the BIPM Geiger-Müller counter, type ZP 1311, to 14.61 MeV neutrons and to ^{60}Co photons has been measured.

The sensitivity (count rate/fluence rate) of the counter to neutrons, ϵ_n (GM), was measured using coincidences between the neutrons and their associated α-particles. The principle of the method is given in Ref. [2]. In the case where the counter axis was parallel to the neutron cone axis, the value of ϵ_n (GM) obtained was

$$\epsilon_n(\mathrm{GM}) = (4.65 \pm 0.46)\ 10^{-4}\ \mathrm{cm}^2$$

By using a fluence-to-kerma (in tissue) conversion factor of 6.63×10^{-11} Gy·cm^2 for 14.61 MeV neutrons [3], the response of the counter was

$$(\mathrm{GM})_n = (7.01 \pm 0.70)\ 10^6\ \mathrm{Gy}^{-1}$$

The sensitivity (count rate/exposure rate) of the counter to photons was calibrated in the ^{60}Co field of BIPM. By using a conversion factor of 9.66×10^{-3} Gy/R = 37.44 Gy·kg·C^{-1} to transform the calibration in terms of exposure to a calibration in terms of kerma in tissue (calculated by BIPM), the response of the counter was

$$(\mathrm{GM})_c = (461.70 \pm 1.85)\ 10^6\ \mathrm{Gy}^{-1}$$

and the ratio, $(\mathrm{GM})_n/(\mathrm{GM})_c = k_U$, was

$$k_U = 0.0152 \pm 0.0015$$

All uncertainties are given in standard deviations.

TE IONIZATION CHAMBER MEASUREMENTS

At present, three TE ionization chambers are under investigation at BIPM. Two of them are of the T2 type, manufactured by Exradin, and the third is of the TNO type, 1 cm^3, constructed and offered by the Radiobiological Institute, Rijswijk. These chambers, flushed with TE gas, are to be tested for suitability at BIPM in the (d+T) neutron field as well as in the calibrated ^{60}Co γ-ray field, in order to select a set of chambers which could serve as reference and transfer instruments. The goal is for BIPM to provide equipment capable of retaining a calibration over a long period.

(a) Measurements in (d+T) neutron field

At a distance of 30 cm from the target and with a TE gas flow rate of 15 $cm^3 \cdot min^{-1}$, the currents obtained by the two Exradin T2 chambers, with an additional 1 mm thick cap, are respectively 1.377×10^{-13} A (chamber No. 199) and 1.388×10^{-13} A (chamber No. 191) for a fluence rate of 14.65 MeV neutrons of 10^5 $cm^{-2} \cdot s^{-1}$. The relative standard deviation over a period of two months has been observed to be about 1%. For the TNO chamber, with an additional 0.75 mm thick cap, the corresponding measured current is 2.131×10^{-13} A. The average leakage current of the three chambers is about 3×10^{-16} A.

The preliminary measurements have been carried out concerning the saturation characteristics, effect of change in polarity, effect of gas flow rate on instrument response and effects of attenuation and scattering of wall thickness.

For the monitoring purposes, in addition to the associated α-particle counting which is the principal reference, we use also a De Pangher long counter placed at 0° and at a distance of 2.21 m, a Geiger-Müller counter (ZP 1311 type) placed at 0° and at a distance of 80 cm (position chosen to avoid obstructing the usual beam for long counter), and a stilbene scintillator, with its neutron-gamma discrimination system, placed at 67.7° and at a distance of 2.0 m.

The ionization currents are measured using the Townsend method with an automatic device [4]. For a given gas flow rate, the pressure in the cavity volume of the chamber is determined by the mean value of two pressures measured in the gas flow circuit at the points situated respectively before (inlet) and after (outlet) the chamber.

(b) Measurements in ^{60}Co γ-ray field

The two Exradin T2 chambers are at present under investigation in the ^{60}Co γ-ray beam of BIPM. The chambers are studied under flushing by TE gas, or by air from a bottle (mixture of gases in the same proportions as in dried atmospheric air). They are also studied open to atmospheric air without flushing. Some preliminary results can be mentioned here for chamber No. 191.

– An increase in the chamber response of about 3% is observed when the TE gas flow rate increases from 10 $cm^3 \cdot min^{-1}$ to 50 $cm^3 \cdot min^{-1}$, and an increase of 0.1% per 10 $cm^3 \cdot min^{-1}$ in the range of 50 to 90 $cm^3 \cdot min^{-1}$.

– After the functioning of the chamber filled with air, a time interval of about 90 minutes is required for pre-flushing with TE gas (flow rate of 15 $cm^3 \cdot min^{-1}$) before the measurements can start, in order to obtain a constant response of the chamber. Inversely, after the functioning of the chamber filled with TE gas, a long period of four hours is necessary for pre-flushing with air (flow rate of 15 $cm^3 \cdot min^{-1}$), in order to obtain a constant response of the chamber flushed with air. In this case, the current does not depend on the gas flow rate.

– The response of the chamber open to atmospheric air, corrected to dry air, is in agreement with the response obtained when flushing with the air from the bottle. Nevertheless, this response is 0.2% lower than the measurement made in 1981 and the similar decrease for chamber No. 199 from 1981 to 1982 is even higher. These differences are not yet explained.

– The ratio of the response of chamber No. 191 filled with TE gas (flow rate of 90 $cm^3 \cdot min^{-1}$) to that of the chamber filled with dry air is 1.165.

REFERENCES

[1] HUYNH, V.D., Metrologia **16** 1 (1980) 31.

[2] FOWLER, J.L., COOKSON, J.A., HUSSAIN, M., SCHWARTZ, R.B., SWINHOE, M.T., WISE, C., UTTLEY, C.A., Nucl. Instrum. Methods **175** 2 (1980) 449.

[3] INTERNATIONAL COMMISSION ON RADIATION UNITS AND MEASUREMENTS, ICRU Rep. 26, Neutron Dosimetry for Biology and Medicine (1977).

[4] HUYNH, V.D., Comité International des Poids et Mesures, Procès-Vervaux **48** (1980) 76.

FAST NEUTRON DOSIMETRY AT THE NATIONAL PHYSICAL LABORATORY (NPL)

V.E. LEWIS, D.J. THOMAS
Division of Radiation Science and Acoustics,
National Physical Laboratory,
Teddington, Middlesex,
United Kingdom

Abstract

FAST NEUTRON DOSIMETRY AT THE NATIONAL PHYSICAL LABORATORY (NPL).

The aim of the programme of the National Physical Laboratory is to establish a reference standard facility for fast neutron absorbed dose, to develop appropriate secondary and transfer devices for measurements in other neutron fields, and to measure the ancillary physical data necessary for accurate neutron dosimetry. Progress in setting up the reference facility is described. The facility is based on a dedicated 150 kV accelerator used for the production of a collimated beam of d + T neutrons. Other, lower intensity neutron fields produced in low-scatter environments are also available. These are useful for the development, investigation and calibration of the instruments used to standardize the fields of the collimated beam.

1. INTRODUCTION

In 1973 the British Committee on Radiation Units and Measurements [1] agreed that a national reference standard for the measurement of neutron absorbed dose should be established at the National Physical Laboratory (NPL). The NPL Neutron Physics section then had two accelerators, both of which were used to produce low-intensity fields in a large low-scatter environment. The original plan included the use of one of them, the 3.5 MV Van de Graaff, for the production of an intense field with neutrons having energies up to 6 MeV (but with a mean energy of around 2 MeV) by the action of 3 MeV deuterons on a beryllium target. This idea was later abandoned for practical reasons; also the spectrum was not very appropriate.

It had also been decided to set up an intense d + T neutron field. This was appropriate at the time since two hospitals in the United Kingdom had d + T therapy machines, NRPB had a facility, and an expansion was expected in this area. It is of course a very suitable reaction. However, it was necessary to acquire another accelerator as the existing 150 kV SAMES machine did not produce a high enough intensity nor was it in a convenient position. Work started on this new generator – the Dosimetry Machine – and the first neutrons were produced in 1976. Work was later stopped owing to a shortage of man-power. At that point it was decided that the output, although reasonable for that type of accelerator, was not adequate and that a duoplasmatron ion source was required. To develop this a collaboration was entered into with the Radiation Centre of Birmingham University.

In the meantime the existing neutron fields have been used to develop and calibrate the necessary instrumentation. These are well standardized, virtually monoenergetic, and have very low photon contamination. The d + T neutron fields produced using the SAMES accelerator have been particularly useful. Secondary standard techniques used in neutron fluence and photon dosimetry were extended to fast neutron dosimetry.

2. LOW-INTENSITY d + T NEUTRON FIELD

An analysed beam of 150 keV deuterons from the SAMES incident on a tritiated titanium target foil produces d + T neutrons with energies from about 13.6 MeV at 150° to 14.7 MeV at 0° to the deuteron beam axis. The total output (into 4π) is limited to about 2×10^9 s^{-1} by the need to restrict radiation levels outside the building. The fluence rate is typically 1.5×10^6 $cm^{-2} \cdot s^{-1}$ at 30 cm yielding a dose rate (in tissue) of about 40 $mGy \cdot h^{-1}$. The fluences produced are routinely measured with an accuracy of better than 2% using associated alpha particle monitoring; higher accuracies are readily obtained if required. With the appropriate choice of target backing it is possible to obtain a low-energy neutron component of less than 1%. Under these conditions the most accurate method of deriving kerma is that of applying the appropriate fluence-to-kerma conversion factor to the measured fluence, as discussed by Thomas and Lewis [2].

The photon dose component is measured using energy-compensated Geiger-Müller counters whose relative neutron dose sensitivities (k_U values) were determined absolutely by Lewis and Young [3]. A recent joint project with PTB, TNO and AvL, suggested by NPL and financed through CENDOS (Mijnheer et al. [4]), has shown very satisfactory agreement between the k_U values determined by different techniques for the same GM counter.[1] (This also applied to the k_U values for lower energy neutrons determined using the technique of Lewis and Hunt [5].)

---

[1] PTB = Physikalisch-Technische Bundesanstalt (Fed. Rep. Germany).
TNO = Institutes of the Organisation for Health Research TNO (Rijswijk, Netherlands).
AvL = Antoni van Leeuwenhoekhuis (Netherlands).
CENDOS = Co-operative European Research Project on Collection and Evaluation of Neutron Dosimetry Data.

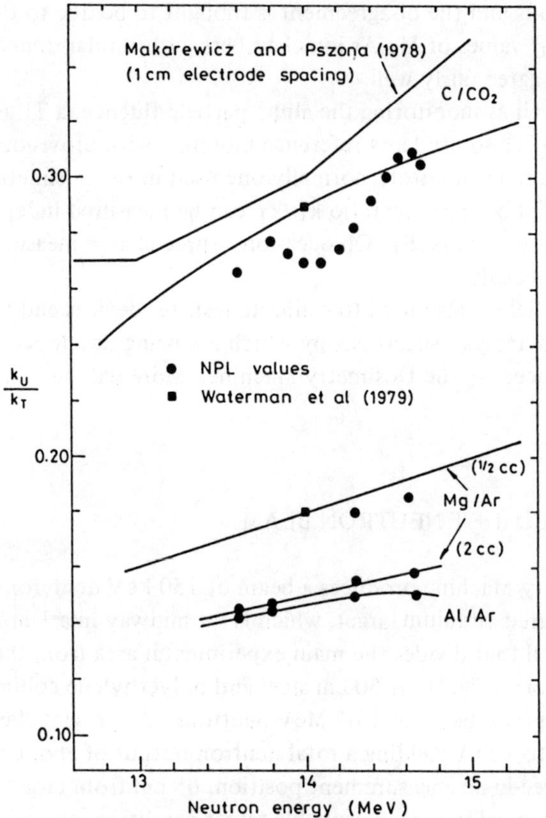

FIG.1. Variation of k_U values with energy.

Knowing the neutron and photon dose components it is relatively straightforward to measure the neutron sensitivities of dosimeters of interest over this energy range. The accuracy, even for that of a tissue-equivalent (TE) ion chamber (k_T value), is higher than can be derived via a photon calibration and calculations using ancillary physical data. In fact the k_T values measured for TE ion chambers agreed very well with those calculated. This cannot be said of the k_U values of the "neutron insensitive" devices, partly because few of these have been calculated. Measurements have been made for a number of devices including various TLD phosphors [6], TSEE materials, and NPL-designed 2-cm^3 graphite/carbon dioxide, magnesium/argon, and aluminium/argon ion chambers. The ion chamber results are compared in Fig.1 with the calculated values for C/CO_2 [7] and some measured values for C/CO_2 and Mg/Ar chambers [8]. (The solid lines through the latter are interpolations.) The Mg/Ar combination is very inhomogeneous and the disagreement is thought to be due to the difference in volume. The k_U values of Mg/Ar ion chambers with similar dimensions (0.5 cm^3 volume) agree fairly well.

Usually, as well as monitoring the alpha particle fluence, a TE ion chamber in a fixed position is also used as a reference monitor, with provision for a second TE ion chamber monitor; normally one used in the dosimetry of other fields. This is useful because the ratio k_U/k_T can be measured independently of D_N as discussed by Lewis [9]. On occasions, three charge measuring systems are in use simultaneously.

The SAMES field is also used to calibrate transfer devices and the techniques used for activation analysis spectroscopy which are being developed and extended to the fields produced by the Dosimetry Machine. More will be said about this later.

3. COLLIMATED d + T NEUTRON BEAM

The Dosimetry Machine produces a beam of 150 keV deuterons incident on a rotating tritiated titanium target, which is set midway in a 1-m-thick concrete shield wall that divides the main experimental area from the neutron dosimetry laboratory (Fig.2). A 50 cm steel and polyethylene collimator provides an appropriate beam of 14.7 MeV neutrons. At present the deuteron current is limited to 1 mA yielding a total neutron output of about 5×10^{10} s^{-1}. At the standard free-in-air measurement position, 65 cm from target, the dose rate (in tissue) is typically, for a reasonable target condition, about 200 $mGy \cdot h^{-1}$. The limitation is due to the conventional rf ion source. A duoplasmatron ion source designed for currents of 5 mA has been built, but so far it has not been possible to install it.

Two transmission-type ion chambers mounted in the collimated beam and feeding the electrometers E1 and E2 are used as reference monitors. A GM counter is mounted outside the collimator and a ^{238}U fission counter is being installed inside the wall cavity close to the target for additional monitoring. The same electronics are used for the monitoring system as are used for the SAMES field. There are two charge measuring systems for the ion chambers coupled to a third used for ion chambers being calibrated and also to scalers (S1, S2, S3, ...) for the GM counter, time, integrated beam current, the alpha particle/fission counter, and other devices.

Located in the dosimetry laboratory is a 1.0 TBq ^{137}Cs irradiator/calibrator used for photon calibration of dosimeters. The dose rate at one metre is about 72 $mGy \cdot h^{-1}$ (mid-1982) with a calibration directly traceable to national standards held elsewhere in the division. Standard currents of 3.5 pA and 42 pA are also available in this room. They are derived from radium sources in two

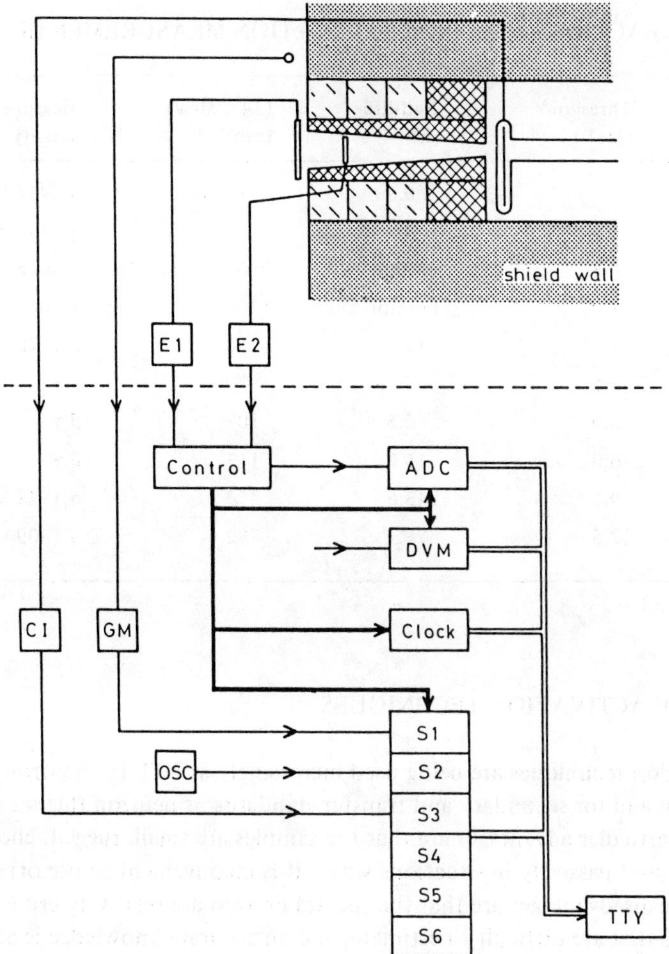

FIG.2. NPL neutron dosimetry facility.

large re-entrant ionization chambers and used for testing and calibrating charge measuring systems.

The neutron fields, free-in-air and at different depths in a water-filled phantom (30 cm cube with lucite walls) are measured primarily with three TE ion chambers and several GM counters. Other devices such as non-hydrogenous ion chambers and TLD are also used, but of course with a lower accuracy as the neutrons are not monoenergetic and there is considerable variation of their k_U values with energy. Spectral measurements using activation techniques are being made in order to obtain the effective neutron sensitivities. At the same time the accuracy of activation measurements alone for the determination of absorbed dose is being investigated.

TABLE I. REACTIONS USED FOR ACTIVATION MEASUREMENTS

Reaction	Threshold (MeV)	Half-life (h)	(14.7 MeV) (mb)	Measured activity
$^{115}In(n,\gamma)$	–	0.9	Very low	γ,(415 keV)
$^{115}In(n,n')$	0.4	4.9	60	γ,(335 keV)
$^{237}Np(n,f)$	0.3	Prompt	2400	Fission
$^{238}U(n,f)$	1.0	Prompt	1200	Fission
$^{64}Zn(n,p)$	1.8	12.7	150	β^+,(511 keV)
$^{27}Al(n,p)$	3.5	0.15	70	β,γ
$^{56}Fe(n,p)$	5.0	2.5	108	β,γ
$^{27}Al(n,\alpha)$	6.0	15.0	113	β,γ
$^{93}Nb(n,2n)$	9.5	243.6	450	γ,(943 keV)
$^{90}Zr(n,2n)$	12.5	78.4	780	γ,(909 keV)

4. USE OF ACTIVATION TECHNIQUES

Activation techniques are being used increasingly at NPL for neutron spectroscopy and for secondary and transfer standards of neutron fluence and dose. The particular advantages are that the samples are small, rugged, cheap, and may be used passively in situations where it is inconvenient to use other devices. The disadvantages are that the spectra derived are relatively crude with uncertainties that are difficult to estimate, and an accurate knowledge is also required of excitation functions, detector efficiencies, decay scheme data, etc. However, some simplification and improvement in accuracy are achieved by comparing the activities induced in the fields under investigation with those induced in the NPL standard fields. The reactions used at NPL are summarized in Table I. The fission reactions are obtained using pulse fission counters. The activities with simple β-γ decay schemes are measured absolutely using a $4\pi\beta$-γ counter combination. For the remaining reactions gammas with appropriate energies (in parentheses) are monitored with Ge(Li) detectors.

The simplest application is the comparison of fluences of similar fairly monoenergetic fields. For d + T neutrons iron or aluminium activation is widely used, the choice depending on the fluence level or the requirement for a longer-lived activity. We have used these at other establishments in the UK, bringing the active samples back to NPL for analysis. Unfortunately the rapid variation of excitation functions from 14 to 15 MeV can introduce undesirable

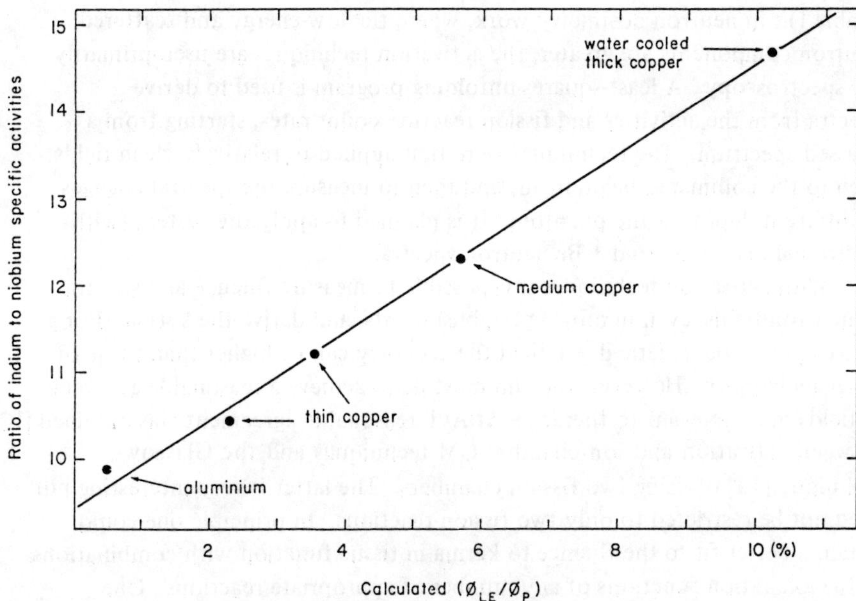

FIG.3. Variation of indium activity with types of target. The niobium activity is a measure of the primary neutron fluence as the threshold is above the highest energy of the secondary neutrons. The variation of the ratio of specific activities is proportional to the fraction of low-energy neutrons.

errors. The niobium reaction is better in this respect and its 10-day half-life enables samples to be sent world-wide. The zirconium reaction was added to give a measure of the mean energy [10] and this system was the basis of a recent (1981) international intercomparison involving nine countries. The results demonstrate the feasibility of intercomparing such fluences with a precision of ±1% without the necessity of sending around personnel and instrumentation, and with the minimum of effort on the part of the participants.

All d + T fields have a low-energy component owing to (n,2n) reactions with the target, and the indium reaction gives the most sensitive indication of its magnitude. This is shown by the variation of the ratio of indium to niobium activity with low-energy fluence fraction (Fig.3) as calculated for targets of different construction used at NPL and elsewhere [11]. The linearity of the fit to the data is surprisingly good, considering the approximate nature of the calculations. The zinc reaction is less sensitive but its greater half-life means that it can be used at more remote establishments. The addition of the fission reactions gives a useful system that can be used at other establishments relatively easily. At NPL the $^{27}Al(n,p)$ and $^{115}In(n,\gamma)$ reactions are also included

(Table I). In neutron dosimetry work, where the low-energy and scattered neutron components are greater, the activation techniques are used primarily for spectroscopy. A least-squares unfolding program is used to derive spectra from the activities and fission reaction count rates, starting from a guessed spectrum. The techniques were first applied to relatively clean fields, then to the collimated beam in air, and then to measure the spectral changes at different depths in the phantom. It is planned to apply the system (with additional reactions) to d + Be neutron spectra.

With activation techniques it is possible to measure fluence and spectra of neutron fields, even in outside establishments, and derive the kerma. For a relatively monoenergetic d + T field the accuracy can be higher than those of other techniques. However, the aim must be to achieve a reasonable accuracy in fields more relevant to therapy. At AvL reasonable agreement was obtained [12] between activation and ion chamber/GM techniques and the Glasgow technique [13] of using two fission chambers. The latter is very interesting but need not be restricted to only two fission reactions. In principle one could obtain a better fit to the fluence to kerma-in-tissue function with combinations of the excitation functions of any number of appropriate reactions. One difficulty would be that the data from the reactions would have different weights and the treatment of the uncertainties would be more difficult.

5. W-VALUE MEASUREMENTS

Determination of absorbed dose using ion chambers requires a knowledge of W_n — the mean energy expended in the formation of an ion pair by the secondary charged particles produced by the neutrons. The most significant parameters needed for the calculation of W_n are proton W values. In the most recent calculation of W_n for TE gas used in TE chambers [14], the uncertainty quoted is 3%, and the largest contribution to this comes from the estimated uncertainty of 2% in the proton W values for energies above 20 keV. (It should be noted that the recent ICRU report on W values [15] suggests a larger uncertainty, close to 5%, for this energy range.) Improvements are thus needed in the accuracy of proton W-value measurements if the overall uncertainty in ionization chamber dose determination is to be reduced. The aim of the NPL programme is to measure W for protons in various gases for the range of energies available from the NPL Van de Graaff accelerator (i.e. about 1.0 to 3.5 MeV).

The equipment used is similar to that employed by Larson [16]. It consists basically of two parallel plates, 20 cm wide by 770 cm long and 10 cm apart, enclosed within a cylindrical gas-tight vessel. A proton beam enters the chamber through a thin Mylar window (about 0.15 $mg \cdot cm^{-2}$), and is stopped in the gas region between the plates (see Fig.4). To obtain a proton beam with a

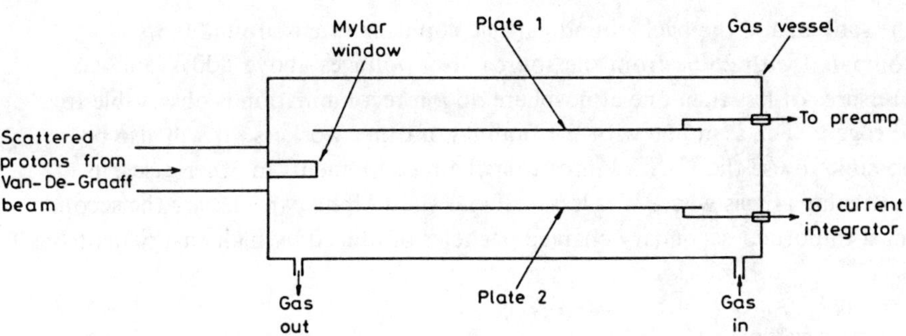

FIG.4. Schematic diagram of the apparatus for W-value measurements.

sufficiently low intensity a thin gold foil has to be used to scatter protons from the main Van de Graaff beam into the chamber. A high voltage applied to one plate (plate 1 in Fig.4) creates a strong electric field, and an ionization current is collected from the opposite plate (plate 2) and measured with an NPL designed current integrator capable of measurements down to a few fA (10^{-15} amp).

A measurement of W also requires a knowledge of the number of particles producing the measured current, and this is obtained from plate 1. A pre-amplifier plus main amplifier combination connected to this plate allows a spectrum to be obtained, from which the number of particles stopped in the chamber can be determined. Gas pressures are arranged so that the protons are stopped between the plates, but with a fairly long path length in order to minimize ion recombination along the path. Pressures between about 0.1 and 1 atmosphere (10 to 100 kPa) can be used, and a flow system is employed to help preserve gas purity in the event of slight leaks or outgassing from the chamber. Specially purified gases may be used if W is sensitive to impurities, and the chamber can be evacuated to about 10^{-3} Pa before admitting the gas. It is intended to make measurements in argon, TE gas, and the three constituents of TE gas (CH_4, CO_2, N_2).

Data acquisition is controlled by an LSI-11 computer. Automatic background and dead-time corrections are made, and a statistical analysis of the data is performed on-line. Measurements are made at a series of plate voltages and gas pressures in order to measure any ion recombination.

It is possible to test much of the equipment by employing a radioactive alpha particle source; and an ^{241}Am source has been acquired for this purpose. A comparison of results obtained in nitrogen using this source with the accurately known value for alpha particles of this energy [15] provides a stringent test of the equipment. Initial results agree with the accepted value provided that allowance is made for ionization due to the americium recoils. The background ion current is around 8 fA compared with that of about 6 pA due to the source, and the background particle counting rate is around 0.5/s compared with 280/s from the source. For voltages above 800 V and gas pressures of less than one atmosphere no ion recombination is observable for nitrogen. This is in line with the findings of other workers. It will also be possible to use the ^{241}Am source to make measurements in other gases, in particular TE gas where W is less well known. (Alpha particles are the second most important secondary charged particles produced by high-energy neutrons.)

6. DISCUSSION

The main purpose of the NPL programme is to provide an accurate and useful standard of neutron absorbed dose against which the instrumentation and techniques of establishments in the UK can be compared and checked. In doing this it is necessary to pursue an active research programme to improve the accuracy of the standards and to make a relevant contribution to this field. It is important to intercompare with other standards laboratories or similar institutions engaged in this work, as is the practice in other fields. This was recognized by CCEMRI – section III (Bureau International des Poids et Mesures) and NPL are to organize an intercomparison involving their facilities in 1983. It will be open to establishments, such as hospitals, as well as to standards laboratories. There will, however, be a charge for machine time. Details will be circulated. In the following year it is expected that BIPM will be circulating a transfer instrument.

Eventually there should be a network of national standards laboratories operating in this field in a manner similar to those in the older established fields of photon dosimetry and neutron fluence measurements. Apart from the inherent practical difficulties of fast neutron dosimetry there is the problem of relating a calibration in a standard field at one neutron energy with that needed for a field with a different spectrum. With the technology of dT therapy machines not enjoying the anticipated improvements, and the future appearing to lie more with higher energy cyclotrons, it is necessary to consider how well the national standards will relate to future needs, and plan accordingly.

REFERENCES

[1] Dosimetry Practice in Neutron Radiotherapy Centres in Great Britain (BCRU 1973), NPL Rep. RS1, National Physical Laboratory, Teddington (1974).

[2] THOMAS, D.J., LEWIS, V.E., Standardisation of neutron fields produced by the ^{3}H(d,n)^{4}He reaction, Nucl. Instrum. Methods **179** (1980) 397–404.

[3] LEWIS, V.E., YOUNG, D.J., Measurement of the fast neutron sensitivities of Geiger-Müller counter gamma dosemeters, Phys. Med. Biol. **22** (1977) 476–80.

[4] MIJNHEER, B.J., GULDBAKKE, S., LEWIS, V.E., BROERSE, J.J., Comparison of the fast neutron sensitivity of a GM counter using different techniques, Phys. Med. Biol. **27** (1982) 91–96.

[5] LEWIS, V.E., HUNT, J.B., Fast neutron sensitivities of Geiger-Müller counter gamma dosemeters, Phys. Med. Biol. **23** (1978) 888–93.

[6] ROSSITER, M.J., LEWIS, V.E., WOOD, J.W., The response of thermoluminescence dosemeters to fast (14.7 MeV) and thermal neutrons, Phys. Med. Biol. **22** (1977) 731–36.

[7] MAKAREWICZ, M., PSZONA, S., Theoretical characteristics of a graphite ionization chamber filled with carbon dioxide, Nucl. Instrum. Methods **153** (1978) 423–28.

[8] WATERMAN, F.M., KUCHNIR, F.T., SKAGGS, L.S., KOUZES, R.T., MOORE, W.H., Energy dependence of the neutron sensitivity of C-CO_2, Mg-Ar and TE-TE ionisation chambers, Phys. Med. Biol. **24** (1979) 721–33.

[9] LEWIS, V.E., "Use of non-hydrogenous ionisation chambers in fast neutron dosimetry", Ion Chambers for Neutron Dosimetry (Proc. Symp. Rijswijk, 1979) (BROERSE, J.J., Ed.), Harwood, Switzerland (1980) 333–36.

[10] LEWIS, V.E., ZIEBA, K.J., A transfer standard for d + T neutron fluence and energy, Nucl. Instrum. Methods **174** (1980) 141–44.

[11] LEWIS, V.E., THOMAS, D.J., Activation and fission detector measurements in d + T neutron fields, NPL Rep., in preparation.

[12] MIJNHEER, B.J., HARINGA, H., NOTHENIUS, H.J., ZIJP, W.L., Neutron spectra and neutron kerma derived from activation & fission detector measurements in a d + T neutron therapy beam, Phys. Med. Biol. **26** (1981) 641–55.

[13] PORTER, D., LAWSON, R.C., HANNAN, W.J., A novel fast neutron dosemeter based on fission chambers Part I, Phys. Med. Biol. **20** (1975) 431–45.

[14] GOODMAN, L.J., COYNE, J.J., W_n and neutron kerma for methane-based tissue-equivalent gas, Radiat. Res. **82** (1980) 13–26.

[15] INTERNATIONAL COMMISSION ON RADIATION UNITS AND MEASUREMENTS, Average energy required to produce an ion pair, ICRU Rep. 31, Washington DC (1979).

[16] LARSON, H.V., Energy loss per ion pair for protons in various gases, Phys. Rev. **112** (1958) 1927–28.

DOSIMETRY OF HEAVY CHARGED PARTICLES IN RADIATION THERAPY AND RADIATION BIOLOGY

BIOLOGICAL DOSIMETRY AND RELATIVE BIOLOGICAL EFFECTIVENESS (RBE)

D.K. BEWLEY
MRC Cyclotron Unit,
Hammersmith Hospital, London,
United Kingdom

Abstract

BIOLOGICAL DOSIMETRY AND RELATIVE BIOLOGICAL EFFECTIVENESS (RBE).

Biological methods of dosimetry are needed when absorbed dose alone is inadequate for correlation with biological effect. This occurs when RBE is unknown or variable and sometimes when irradiation is non-uniform. Biological dosimetry is particularly useful with high-LET radiations such as fast neutrons, heavy ions and Π^- mesons. The RBE of fast neutrons depends on many factors. The effects of neutron energy, dose level and n/γ ratio are discussed. Methods of predicting the effect of mixed radiations $(n + \gamma)$ are analysed in some detail and compared with an experimental result. Mammalian cells have been used as dosimeters to check the validity of physical measurements of dose distributions from beams of fast neutrons and charged particles. This method has its greatest value in the design of filters for velocity modulation of heavy-ion beams. Finally a few comments are made about the use of chromosome aberrations in human lymphocytes, a method which has been found of great value in the assessment of radiation accidents.

The measurement of absorbed dose is a question of physics. The need to bring in biological methods arises when absorbed dose alone is not enough for a prediction of biological effect. An alternative statement is that biological methods of dosimetry are needed when the relative biological effectiveness (RBE) is not known. In fact, biological dosimetry is closely bound up with the measurement of RBE.

RBE depends on the quality and the geometrical and temporal distribution of the radiation, together with the biological end-point and the level of effect. In the context of the present meeting, radiation quality can be described by the neutron spectrum, the n/γ ratio, the spectra and types of secondary (or primary) charged-particles, and of course by microdosimetric parameters. The geometrical distribution can be important on a microscopic scale, for example in the uptake of ^{3}H thymidine in cell nuclei or macroscopically when whole animals are irradiated non-uniformly. The time-scale of the irradiation is also an important factor but not one which I propose to discuss here.

The value of biological dosimeters in neutron therapy lies in the fact that RBE usually varies with neutron energy and with the proportion of the dose given by gamma radiation. With beams of heavy charged particles RBE varies with depth, particularly close to the end of the particles' range.

RBE and neutron energy

Fig.1 is a summary of many investigations of the variations of RBE with neutron energy[1]. With the exception of the P388 leukaemia cells, RBE reaches a maximum at about 0.3 MeV where the effective LET (governed by the energy of the recoil protons) is also a maximum. Most of these experiments were done with monoenergetic neutrons. Neutron beams for therapy usually have broad spectra but RBE is nearly always found to fall as the mean neutron energy rises, at least up to the highest energy in use at present (from 67 MeV p on Be). This is demonstrated in Fig.2 which shows the results of various intercomparisons between neutron therapy centres in USA and Japan[2]. Over the neutron energy range from 15 MeV d/Be to 67 MeV p/Be, RBE changes by around 30%, a very significant factor if one is to compare the doses given to patients at different centres.

RBE and dose level

These examples ignore the factor of level of effect, or dose. When comparing neutrons with X rays, RBE shows a substantial variation with dose, usually according to D_n^{-x} where x is $\leqslant 0.5$[3] (Fig.3). A similar dependence on dose is likely when comparing two neutron energies but the dependence is so weak that it is difficult to demonstrate. Hall and his colleagues made very careful measurements of RBE for V-79 cells as a function of neutron energy with monoenergetic neutrons[4]. After complex statistical adjustments they derived "best values" for parameters of survival curves at each neutron energy. From those I have calculated RBE values as a function of dose at each neutron energy, using 15 MeV neutrons as the standard. The results appear in Fig.4. Only at one neutron energy does the variation go in the opposite way to that expected. Over the range up to 10 Gy the RBE (relative to 15 MeV neutrons) usually varies by only a few percent, an amount easily obscured by experimental uncertainties. The lowest panel compares two beams used for therapy and shows a greater variation, in the expected sense.

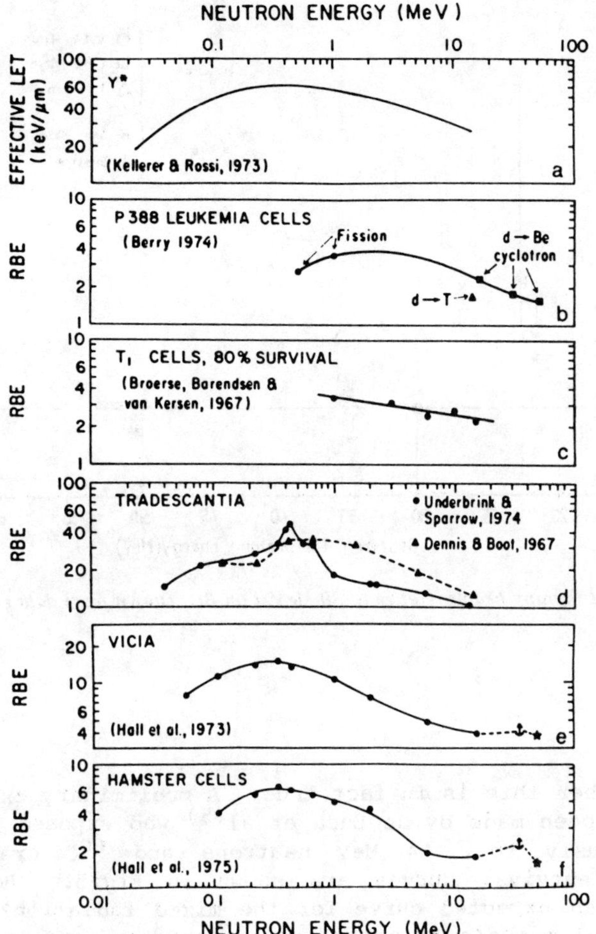

FIG.1. RBE as a function of neutron energy for various end points. Reproduced from Ref.(1).

RBE and the n/γ ratio

For the neutron beam at Hammersmith, the ratio D_γ/D_n varies from 0.04 at the surface with a small field to 0.20 at 20 cm deep with a 20 x 20 cm field(3). If we take an RBE of 3 for the neutron component, the RBE for the mixed radiation would vary between 2.92 and 2.67, a change of 10%.

These figures are based on the assumption that the neutron and gamma components have a constant RBE. There is some doubt

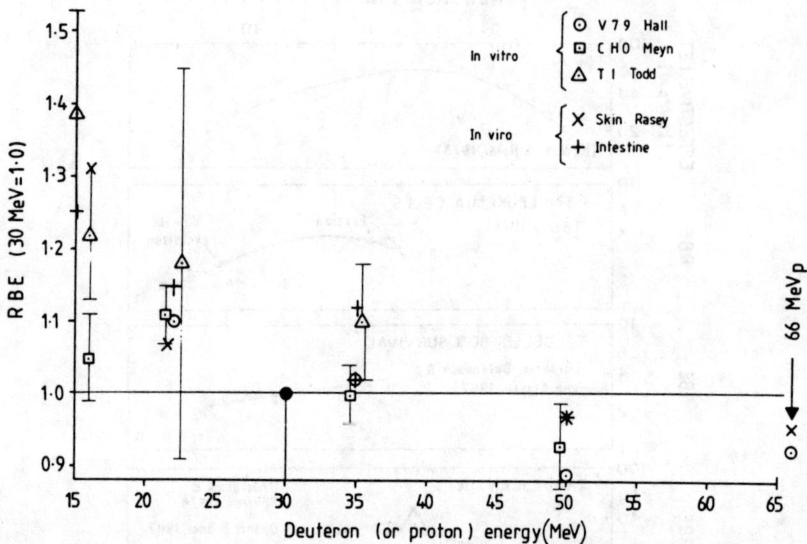

FIG.2. RBE of therapy beams relative to 30 MeV d on Be. Drawn from data given in Ref.(2).

as to whether this is in fact true. A preliminary experimental check has been made by de Luca et al[5] who exposed V-79 cells simultaneously to 14 MeV neutrons and ^{60}Co γ rays. The resultant survival curves are shown in Fig.5. How can we calculate an expected curve for the mixed radiation? I have used several methods to calculate the expected survival after a dose of 1000 rad total (400 n + 600 γ).

De Luca et al fitted their results to equations of the form

$$-\ln S = \alpha D + \beta D^2 \quad (1)$$

If the total dose D is divided into two components D_1 and D_2 with coefficients α_1, α_2, β_1, β_2 and if the actions of the two components are independent of one another

$$-\ln S = \alpha_1 D_1 + \beta_1 D_1^2 + \alpha_2 D_2 + \beta_2 D_2^2 \quad (2)$$

The result of this equation appears in Table I under "independent". This value of survival must be too high because it ignores the effect of the sublethal damage inflicted by one component on survival to the other component.

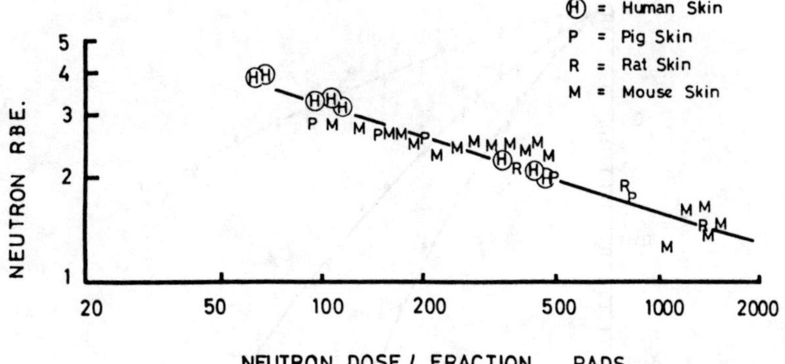

FIG. 3. RBE of 16 MeV d/Be neutrons relative to 250 kV X-rays for acute skin reactions in four species.

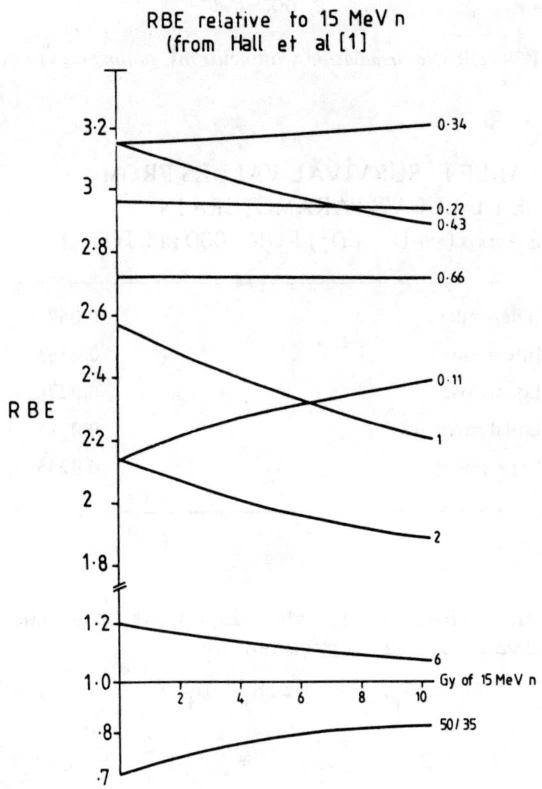

FIG. 4. RBE of monoenergetic neutrons relative to 15 MeV neutrons for V79 cells, calculated from the survival-curve parameters of Ref. (4). The lowest curve shows the RBE of neutrons generated by 50 MeV d on Be relative to 35 MeV d on Be.

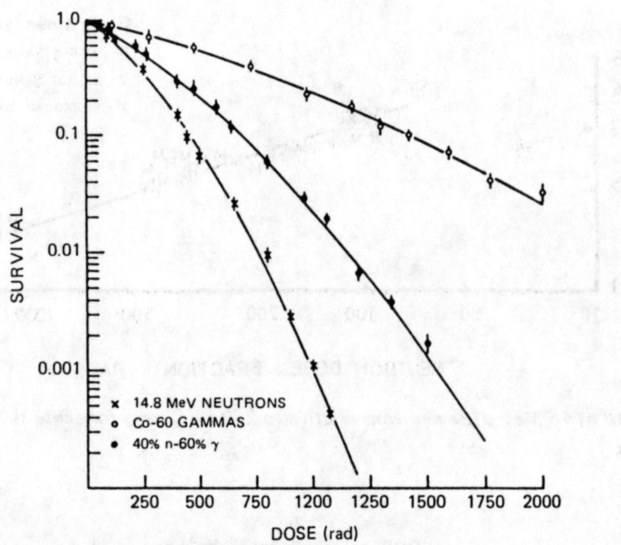

FIG.5. Survival of V79 cells after irradiation with neutrons, gamma rays or both simultaneously.

TABLE I. SURVIVAL VALUES FROM DE DE LUCA'S PARAMETERS IN $S = \exp(-\alpha D - \beta D^2)$ FOR 1000 rad TOTAL

Independent	0.067
Interaction	0.0385
Equal dose	0.0270
Equal survival	0.0385
Experiment	0.0235

If, on the other hand, the doses of the two components interact, survival can be expressed by

$$- \ln S = (\Sigma(\alpha_i D_i) + (\Sigma(\beta_i^{1/2} D_i))^2$$

$$= \alpha_1 D_1 + \alpha_2 D_2 + \beta_1 D_1^2 + \beta_2 D_2^2 + 2(\beta_1 \beta_2)^{1/2} D_1 D_2 \quad (3)$$

If $\alpha_1 = \alpha_2$ and $\beta_1 = \beta_2$, this reduces to equation (1). This method gives the figure shown under "interaction" in Table I.

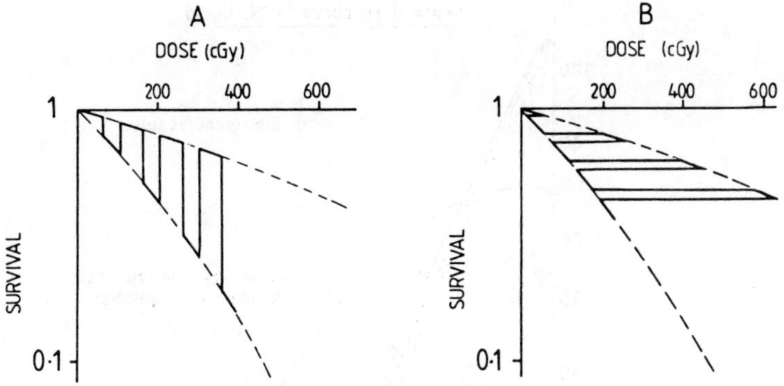

FIG.6. Survival to mixed radiation:
A. Equal sub-lethal damage at equal dose.
B. Equal sub-lethal damage at equal survival.

The next two columns in Table I are based on graphical methods explained in Fig. 6. Fig.6A shows the sum of many small doses of each radiation given sequentially, on the assumption that the sublethal damage at a given dose is the same for each radiation quality. The total of (-lnS) is found by adding solid sections of the two survival curves and is given by

$$-\ln S = 0.4\ (\alpha_1\ D + \beta_1\ D^2) + 0.6\ (\alpha_2 D + \beta_2\ D^2)$$

The result appears in Table I "equal dose". The item under equal survival is based on the idea that sublethal damage depends only on the level of survival, as in Fig.6B. I have not found an analytical method of evaluating this so the result was obtained by successive approximation.

All these methods give survival levels higher than that found experimentally, suggesting that there is some additional interaction beyond what one might expect. It is possible, however, that this is due to a dose-rate effect: the dose-rate of the mixed irradiation was the sum of the dose-rates of the two components when used separately.

Isodose curves and biological effect

As a neutron beam passes into tissue there are changes in neutron spectrum and in the n/γ ratio as a function of depth and lateral position. There is therefore some doubt about how well isodose distributions represent the distribution of biological effectiveness. The problem can be investigated by using mammalian cells as biological dosimeters.

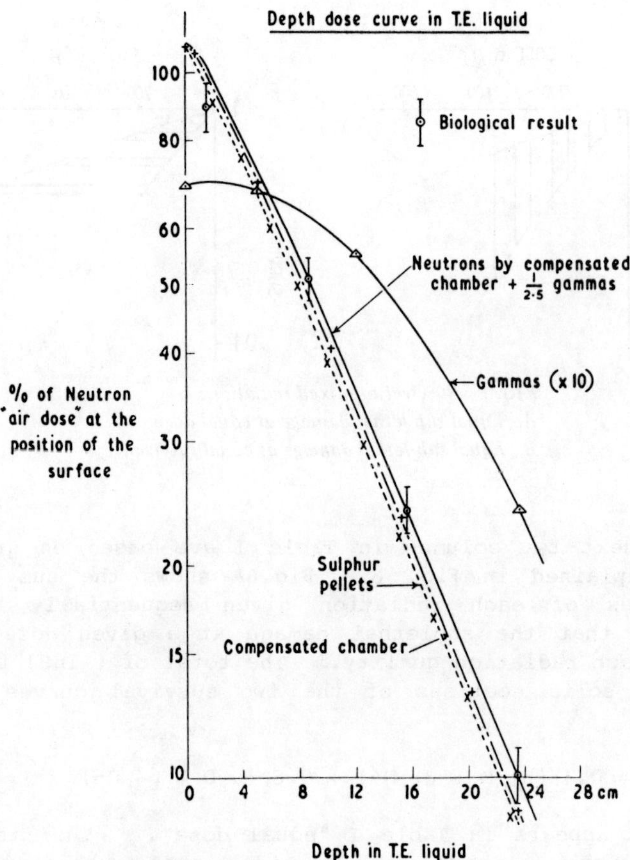

FIG. 7. A comparison of central axis depth doses in tissue equivalent (TE) liquid, measured by physical (sulphur pellets and ion chambers) and biological methods.

The central axis depth-dose curve of neutrons from the Hammersmith cyclotron was investigated by McNally and Bewley[6]. We exposed Ehrlich mouse ascites cells in suspension, gassed with air or nitrogen. The cells were placed at various depths in the phantom and at the position of the entry surface but in the absence of the phantom. Survival curves were obtained at each position, giving immediately the relative effectiveness of the beam. The result is shown in Fig.7 which is relative to the kerma in air. The biological result agrees closely with the effective dose given by ($D_n + D_\gamma /2.5$). In addition the oxygen enhancement ratio (OER) was found to be identical in the absence of the phantom and at 8.7 cm deep. On the other hand Berry and Bewley [7] found a lower OER when cells were irradiated at a depth in the phantom. This different result may arise from the different response of P-388 cells as a function of LET and neutron energy (see Fig.1, panel 2).

Spectral measurements at depths along the central axis have shown only minor changes in mean neutron energy[8]. Much greater variations have been found at right angles to the axis[9]. Biological dosimetry is more difficult in this situation as the dose varies so sharply across the geometrical edge of the beam. We used a technique developed by Raju from a suggestion of Palcic and Skarsgard[10][11]. Ehrlich ascites cells were suspended in serum mixed with gelatin and drawn into narrow syringes which were placed on ice to solidify the gel[12]. The syringes were then placed in an array at right angles to the beam. After irradiation the gel was slowly extruded and sections 1.7 mm long were cut off and cultured for assay of cell survival. Other cells were exposed to known doses of neutrons in order to derive a standard cell-survival curve. In this way the effective dose of neutrons could be obtained along a profile at right angles to the beam.

The results are shown in Fig.8. Outside the beam, both in air and in phantom, the cells indicate a slightly greater effective dose than was given by physical measurements of (D_n + $D\gamma/3$). This is probably due to the lower mean neutron energy outside the beam found by Bonnett and Parnell both in air and in phantom[8].

Charged-particle beams

The technique in which cells are embedded in gel was developed by Raju to study the biological effect close to the Bragg peak of a beam of charged particles, where both dose and LET are changing very rapidly. To obtain a uniform biological effect over a useful volume it is necessary to use velocity modulation to broaden the peak. The question then arises whether a uniform dose over the target volume is the correct criterion; only biological dosimetry can answer this question. Raju has studied beams of H, He, C, Ne, An and Π^- mesons in this way[13]. Fig.9 shows his results for Ne ions. The deeper part of the modified Bragg peak is at a higher LET than the shallower part as the mean energy of the Ne ions is lower. The dose has been reduced by 30% to allow for this; it would seem that the reduction has been slightly overdone in this instance.

Non-uniform irradiation

I noted at the beginning that RBE depends on the geometrical distribution of radiation. Partial-body exposure

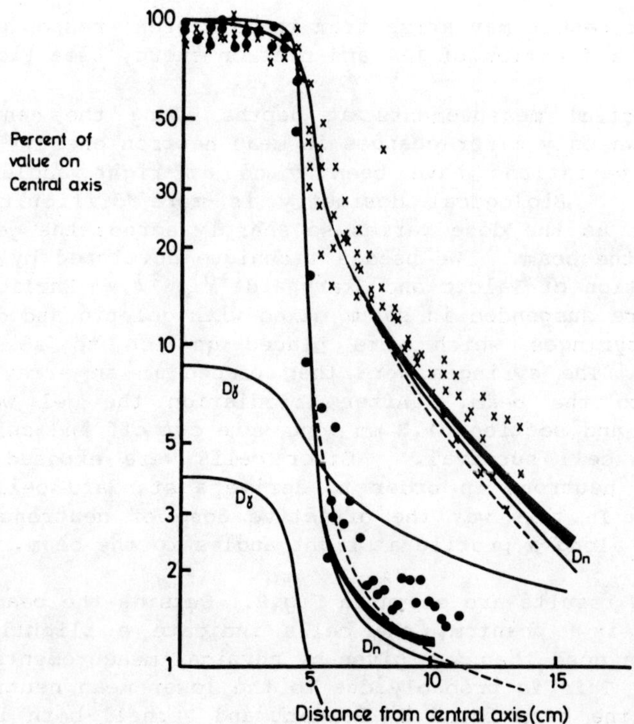

FIG.8. Physical and biological profiles of neutron beam in air and at 10 cm deep in phantom. The full lines represent D_n and D_γ measured with chamber and GM counter. The dotted lines represent activation of $^{115}In(n,n^1)^{115}In^m$. The points relate to measurements with cells. The wide black areas indicate "effective dose" $D_n + D_\gamma/R$ where R varies from 2.8 to 4.5. The upper sets are for measurement in phantom; the lower in air.

produces a smaller biological effect than uniform total-body exposure. This is particularly important in radiation protection, both for accidental exposure and for medical uses of radiation. It is also a significant factor in radiotherapy. Modern developments in radiotherapeutic technique are to a large extent aimed at delivering radiation to the tumour with the greatest specificity, i.e. making the ratio tumour dose/average dose to the rest of the body as large as possible. The use of proton beams is based entirely on this principle.

Analysis of chromosome aberrations in lymphocytes has been used by many workers to indicate the biological effectiveness of non-uniform irradiation. From animal experiments we know that the yield is the same for irradiation in vivo or in vitro.

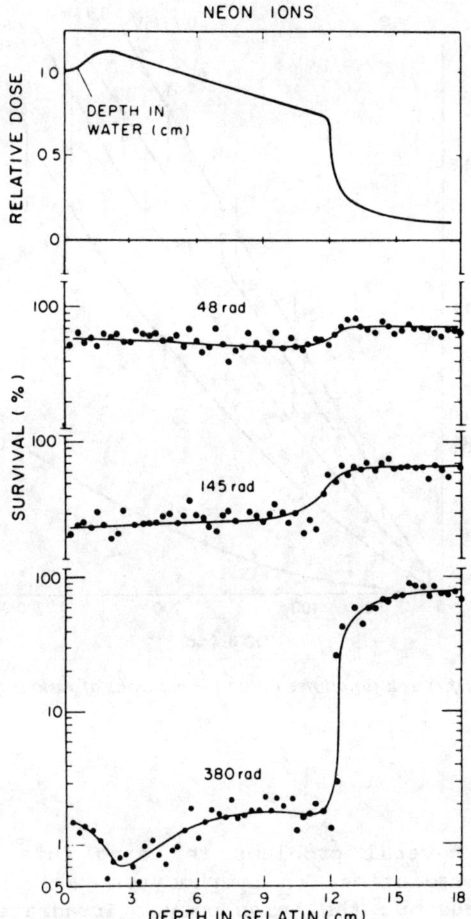

FIG.9. Relative dose and survival of T_1 cells as a function of depth for Ne ions with velocity modulation, reproduced from Ref.(13).

Dose-response curves for X and gamma rays are usually expressed by a relation of the form

$$Y = \alpha D + \beta D^2$$

When $D = \alpha/\beta$ the two terms are equal; this mostly occurs at 1.5 Gy for 200 kV X rays, 0.1 - 0.5 Gy for gamma rays and over 5 Gy for neutrons (14). Some dose-response curves are shown in Fig.10.(15).

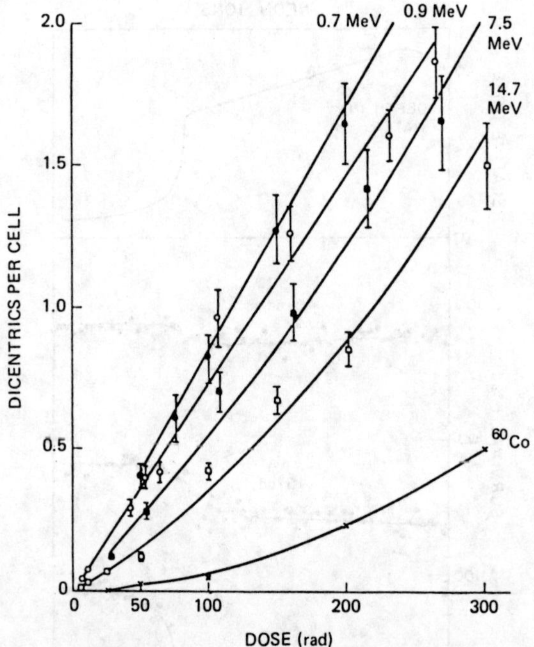

FIG.10. Yield of dicentrics as a function of dose for neutrons of various energies, reproduced from Ref.(15).

There are several problems in using this technique for partial-body irradiation. Lymphocytes are not uniformly distributed throughout the human body. Irradiated lymphocytes suffer intermitotic death before aberrations can be scored at metaphase, leading to a depression of yield when irradiated and unirradiated populations are mixed. The dose-squared term preponderant with high doses of gamma rays means that the result cannot be a true measure of average dose. This last point would not apply with neutrons where the dose-response curves are nearly linear. Useful results have been achieved after accidental irradiations where the doses were fairly low or the degree of non-uniformity not too extreme. The technique could be used for biological dosimetry in a similar way to cell survival as discussed above.

An alternative often suggested for partial-body irradiation with neutrons is measurement of induced radioactivity, in particular ^{24}Na. When fast neutrons are used, as in radiotherapy, this suffers from the disadvantage that production of ^{24}Na depends on the fluence of thermal neutrons whereas the biological effect is overwhelmingly due to fast neutrons. Small changes in the low-energy component of the beam would have a much greater impact on production of ^{24}Na than on biological effects.

Conclusion

Biological dosimetry is of greatest value with high LET radiation where RBE varies between therapy installations and with position in the beam. Under these conditions physical dosimetry alone is not sufficient to predict the biological effect.

REFERENCES

(1) HALL E.J., NOVAK J.K., KELLERER A.M., ROSSI H.H., MARINO S. and GOODMAN L.J. RBE as a function of neutron energy: 1. Experimental observations. Radiation Research, 64 (1975) 245-255.

(2) HALL E.J., WITHERS H.R., GERACI J.P., MEYN R.E., RASEY J., TODD P. and SHELINE G.E. Radiobiological intercomparisons of fast neutron beams used for therapy in Japan and the United States. Int.J.Radiation Oncology Biol.Phys., 5 (1979) 227-233.

(3) CATTERALL,M. and BEWLEY D.J. Fast neutrons in the treatment of cancer. London: Academic Press, (1979).

(4) KELLERER A,M., HALL E.J., ROSSI H.H. and TEEDLA P. RBE as a function of neutron energy: II. Statistical analysis. Radiation Research, 65 (1976)172-186.

(5) de Luca et al. Proc. 4th Symposium on neutron dosimetry, vol.2., p55. (Burger G. and Ebert H.G.,Eds†, Munich, 1-5 June 1981. Pub: EEC,EUR 7448 EN.

(6) McNALLY N.J. and BEWLEY D.K. A biological dosimeter using mammalian cells in tissue culture and its use in obtaining neutron depth dose curves. Br.J.Radiol., 42 (1969) 289-294.

(7) BERRY R.J.and BEWLEY D.K., Br.J.Radio.49,(1976)458.

(8) BONNETT D.E. and PARNELL C.J. Effect of variation in the energy spectrum of a cyclotron-produced fast neutron beam in a phantom relevant to its application in radiotherapy. Br.J.Radiol., 55, 48-55, 1982.

(9) BONNETT D.E. and PARNELL C.J. Proc. 4th Symposium neutron on dosimetry, vol.2., pl95. (Burger G. and Ebert H.G.,Eds), Munich, 1-5 June 1981. Pub: EEC,EUR 7448 EN.

(10) RAJU et al., Radiology, 116,(1975) 191.

(11) SKARSGARD L.D. Pretherapeutic research programmes at meson facilities. In: Proc.XIII International Congress of Radiology, Madrid, Spain, October 15 - 20, 1973.

(12) BEWLEY D.K., CULLEN B. and PAGE B.C. Proc.4th Symposium on neutron dosimetry, vol.2., p227. (Burger G. and Ebert HG.,Eds), Munich, 1-5 June 1981. Pub:EEC,EUR 7448 EN.

(13) RAJU M.R. et al. A heavy particle comparative study. Br.J.Radiol., 51 (1978) 699-727.

(14) EVANS H.J., and LLOYD D.C. Mutagen-induced chromosome damage in man. Edinburgh University Press (1978).

(15) LLOYD et al, Int.J.Rad.Biol., 29 (1976) 169.

CONTRIBUTION TO A PRE-STUDY OF HEAVY IONS USED FOR RADIOBIOLOGICAL EXPERIMENTS

NGUYEN VAN DAT
Institut de Protection et de Sûreté Nucléaire.
Laboratoire de Dosimétrie Sanitaire,
Fontenay-aux-Roses, France

Abstract

CONTRIBUTION TO A PRE-STUDY OF HEAVY IONS USED FOR RADIOBIOLOGICAL EXPERIMENTS.

Because of their well-defined macroscopic range for each initial energy, the heavy ions are ideally suited for therapy applications. In the perspective of clinical uses, radiobiological experiments are made with high-energy charged particles at the SATURNE synchrotron in SACLAY. This report briefly reviews a study of heavy ions and illustrates their potential in biology investigations. The experimental set-up and its on-line data treatment system have been already presented by the author elsewhere.

DOSIMETRY SYSTEM

When the thickness of the water phantom varies, the dose delivered by heavy ions varies continuously but it can have also a very large gradient. For a correct dose measurement, the wall of the ion chamber should not alter the particle spectrum and particles must have the same path before arriving in the cavity, and the thickness of this cavity must be as small as possible.

The wall of the ion chamber is a plane and made of 1 mm SHONKA plastic. The distance between electrodes is 2 mm. The sensitive volume is limited by a ring guard also made of SHONKA plastic. Under these conditions, the tissue equivalent thickness of the cavity is equal to 2.26 μm [1].

The sensitive volume must be exactly defined within the measurement conditions. For this purpose, it has been determined by two different methods:

- geometrically calculated; and
- deduced from exposure with a calibrated ^{60}Co source.

The volume values given by these two methods are 47.52 ± 2.38 mm^3 and 47.87 ± 0.47 mm^3, respectively.

For a dose-rate exposure of 39 R/min ^{60}Co, the current value on the plateau is equal to 10^{-11} A.

EXPERIMENTAL RESULTS

The ion chamber was used to determine the Bragg curves of a 649 MeV helion beam. The experimental set-up and its on-line data treatment system were described in an earlier paper [2].

In the case of a collimated beam, the dose gradient between 167 and 177 mm is equal to 2 cm^{-1}, it varies up rapidly to 12 cm^{-1} between 177 and 180 mm. The relative dose variations in the cavity are, respectively, 4.5×10^{-4} and 2.7×10^{-3}.

These results show that this ion chamber is suited for measurements under large dose gradient conditions.

PROVISIONAL CALCULATION FOR N^{7+}

Radiobiological results obtained with He^{2+} indicated a higher RBE value for all biological systems. So, investigations are planned with heavier ions, for example N^{7+}, with a view to study potential therapeutic advantages with high ionization density particles.

In the energy range from 2.8 to 6 GeV, the stopping powers of N^{7+} vary from 147 to 226 $MeV \cdot cm^2 \cdot g^{-1}$.

The influence is reduced by about a factor of 5 at the Bragg peak so an initial fluence of about 2×10^8 cm^{-2} is necessary to produce a dose of 1 Gy.

CONCLUSION

Radiobiological experiments with He^{2+} are the first step in the study and investigation of the possibility of heavy ions for clinical use. Basic experiments on N^{7+} and Ne^{10+} are planned for the next two years.

REFERENCES

[1] INTERNATIONAL COMMISSION ON RADIATION UNITS AND MEASUREMENTS, ICRU Rep. 10 (1964).

[2] NGUYEN, V.D., et al., in Biomedical Dosimetry: Physical Aspects, Instrumentation, Calibration (Proc. Symp. Paris, 1980), IAEA, Vienna (1981) 93–109.

DOSIMETRY FOR RADIATION SURGERY USING NARROW 185 MeV PROTON BEAMS*

B. LARSSON, B. SARBY
Department of Physical Biology,
The Gustaf Werner Institute,
Uppsala, Sweden

Abstract

DOSIMETRY FOR RADIATION SURGERY USING NARROW 185 MeV PROTON BEAMS.

The purpose of the work described in this paper was to optimize and standardize the irradiation conditions and dosimetry for narrow proton beams, so as to permit their safe, routine use in cerebral radiation sutgery and radiation therapy of small tumours. The beams were thus collimated and led to the place where the irradiations were performed, 25 m from the synchrocyclotron, with a system of focusing quadrupole magnets and bending magnets. The final collimation of the beam was arranged with a system of accurately aligned cylindrical and plane-parallel metal apertures. The energy of the protons in the beam was 185 ± 0.2 MeV and the maximum total fluence was 5×10^{10} protons$\cdot$s^{-1}. A method of ^{11}C activation dosimetry was developed to determine the fluence and, indirectly, the dose, by using small polystyrene cylinders on the beam axis at the isocentre. The activity induced was measured in a well-type crystal detector, the efficiency of which had been determined with standard preparations of ^{22}Na and ^{60}Co. Dose monitoring during irradiation was done with a thin-window chamber with its electrodes parallel to the beam. The chamber could not be used for absolute measurements, however, since there was a contribution to the ionization in the chamber from scattered protons and secondary radiation from the walls of the collimator defining the cross-section of the ionization volume. Correction factors for the ionization chamber readings were determined for various collimator geometries in the beam by means of ^{11}C activation measurements. The properties of the narrow proton beam, as it passed through the brain, were studied by measuring the density distributions of photographic films exposed in phantom material. It was found that contributions to the dose outside the ideal geometrical beam path, caused by geometrical penumbra and scattering in the collimator walls, were comparable to the contributions from diffusing Coulomb electrons and protons undergoing multiple scattering in the brain tissue. The dose distributions were considered favourable, from a clinical aspect, in view of the steep gradient in the transverse dose distribution, the small secondary radiation dose to the brain tissue outside the geometrical beam, and the excellent central axis depth dose distribution.

* The full text of this paper is to be published in Acta Radiobiologica.

HEAVY CHARGED-PARTICLE BEAM DOSIMETRY*

J.T. LYMAN
Lawrence Berkeley Laboratory,
University of California,
Berkeley, California,
United States of America

Abstract

HEAVY CHARGED-PARTICLE BEAM DOSIMETRY.

A computational description of the physical properties and the beam composition of a heavy charged-particle beam is presented. The results with this beam model have been compared with numerous sets of experimental data and it appears to provide an adequate representation of the major features of a heavy charged-particle beam. Knowledge of the beam composition aids in the identification of regions of the beam where special dosimetry problems may be encountered.

1. Introduction

For heavy charged particles, the absorbed dose can be determined from a knowledge of the charged particle fluence spectrum ϕ and the stopping power S of the absorber material at the point of interest [1]. If the energy of the particles is denoted by E, and if delta ray equilibrium is established, the dose in a small mass inside a homogeneous medium is given by

$$D = \frac{1}{\rho}\int_0^{E_m} \phi\,(E)\;S\,(E)\;dE \qquad \text{Eq. 1.}$$

where E_m is the maximum kinetic energy of the particles and ρ is the density of the medium.

The knowledge of the charged particle fluence spectrum is a valuable aid to identifying the possible errors associated with the dosimetry. Both experimental and theoretical approaches are used to determine the beam composition and thereby identify the major contributors to the absorbed dose. Experimentally, beam properties are determined and particles are identified through

* This work was supported by the United States Department of Energy under Contract No. DE-AC03-76SF00098.

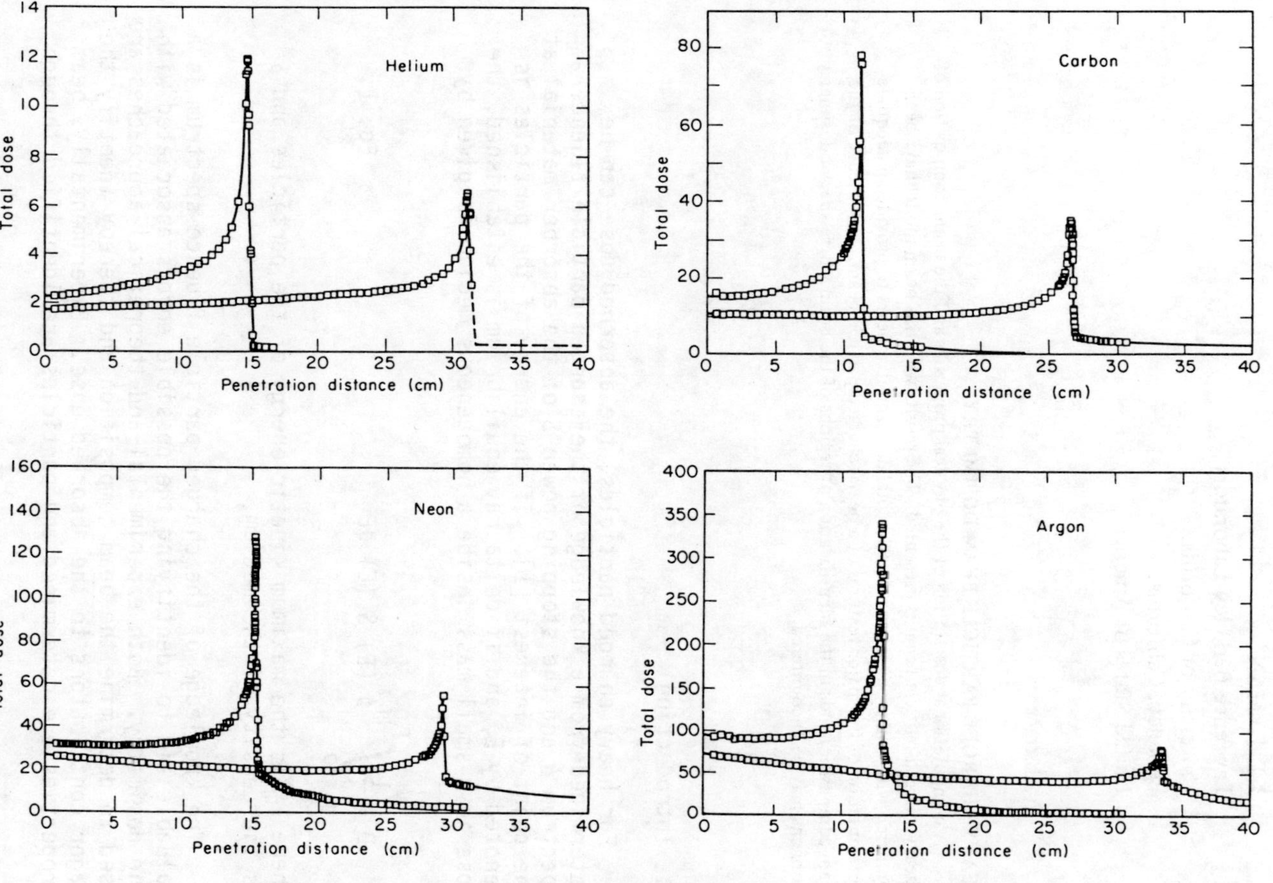

FIG.1. Bragg curves of helium, carbon, neon and argon ion beams in water [5].

the use of devices such as ionization chambers, Faraday cups, secondary emission monitors, plastic scintillators, silicon or germanium semiconductor detectors, calorimeters, thermoluminescent materials, nuclear emulsions and particle track detectors, [2-8]. Theoretical and empirical studies of the range-energy relationships, multiple scattering and fragmentation aid in the development of the computational beam model [9,10] which can be used to assist in interpretation and to supplement the available beam composition data.

2. Methods and Materials

The heavy charged-particle beams that have received the most interest in the biomedical program at Berkeley are the beams of helium, carbon, neon, silicon and argon ions. The energies and ranges in water of these beams varies from initial energies between 225 and 900 MeV/u[1] and residual ranges between 3 and 30 cm.

Bragg curves of four different types of heavy particles beams (helium, carbon, neon, and argon) are shown in Figure 1. Two initial energies were used for each ion and the energy spread of beams is typically less than 0.5 percent. The Bragg curve is the average specific ionization as a function of penetration distance into an absorber. The differences in the shapes of the curves are due mainly to the probability of the primary ion being fragmented by a nuclear collision before it has completely given up its kinetic energy, and to the number and type of secondary particles produced.

The average energy of the primary ions at the Bragg peak depends upon the initial energy spread of the beam and the range straggling [10]. For protons, the average energy at the Bragg peak is typically about 10 percent of the initial energy [2] while for the heavier ions it is less because of the decrease in the range straggling of the heavier charged particles.

If the nuclear interactions are of a negligible importance, the Bragg curve can be obtained from the specific ionization of a single particle and the particle straggling distribution [11]. When nuclear interactions are important, the Bragg cruve is composed of the contributions of the primary ions and of the fragments which are produced by the nuclear interactions. The fragment contribution is composed of the ionization of the secondary particles and any of their fragments which may be

---

[1] u = unified atomic mass unit.

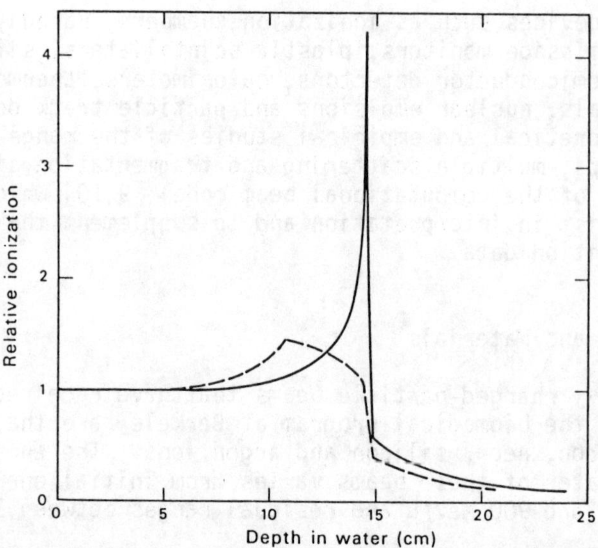

FIG.2. Bragg curve of a 400 MeV/u neon ion beam and a range modulated dose distribution of the same beam in water [5].

produced by additional nuclear interactions. The probability for the loss of a primary particle depends on the size of the primary nucleus and the nuclei of the absorbing media [12]. The probability of producing a particular fragment is given by the empirical formulation of Silberberg and Tsao [13]. Based upon these principles, a beam model has been developed with which it is possible to predict many of the characteristics of a heavy charged-particle beam.

The Bragg peak is generally too narrow for most radiotherapy applications. To effectively utilize these beams for cancer therapy the stopping region of the particles is modulated to produce a high-dose region broader than the Bragg peak, but still a region in which the biological effectiveness of the particles is greater than on either side of this region [10] (Fig. 2). This is accomplished by the superposition of Bragg peaks of beams with different ranges of penetration [5,14,15]. The positions of the least penetrating and most penetrating Bragg peaks are designated as the proximal and distal peaks and define the high dose region. If one understands the beams composition of a single Bragg curve, the more complicated dose distributions can also be understood if the distributions are separated into the individual Bragg curves.

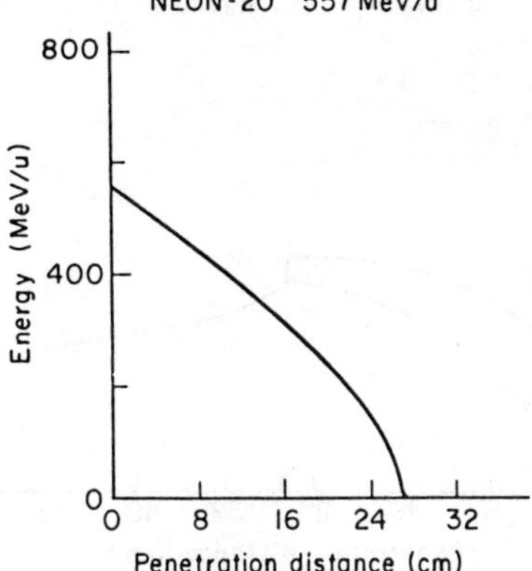

FIG. 3. Energy of primary ions relative to the depth of penetration in a water medium.

3. Results

The beam model has been used to calculate many of the properties of a 557-MeV/u neon ion beam and some of the results are present in figures 3-11. As the primary ion loses energy to the absorbing medium, the kinetic energy decreases until the particle comes to rest (Fig. 3). The calculations of the range-energy and energy loss relations over this energy range are probably good to better than few percent [16] and curves of this type are easily verified by measurement.

The primary ions are either exponentially removed from the beam or lose all their kinetic energy and therefore come to rest (Fig. 4). The rate at which the ions are lost from the beam depends upon the ion mass. Lighter ions survive better than the heavier ions. The build-up of the secondary ions depends upon their rate of production and their rate of removal. The most numerous ions will be protons because they have the highest production probability, the lowest rate of loss.

The beam charge is the sum of the products of the number of each type of charged particle and its charge. This is a quantity that can be measured with a Faraday cup. If the charge carried in the beam is partitioned into two components (the primary and the fragment charge), it can be seen (Fig. 5) that the primary current is exponentially decreased as was the primary fluence.

If the number of particles in the beam is multiplied by the stopping power of particles (Eq. 1) then one obtains the dose as

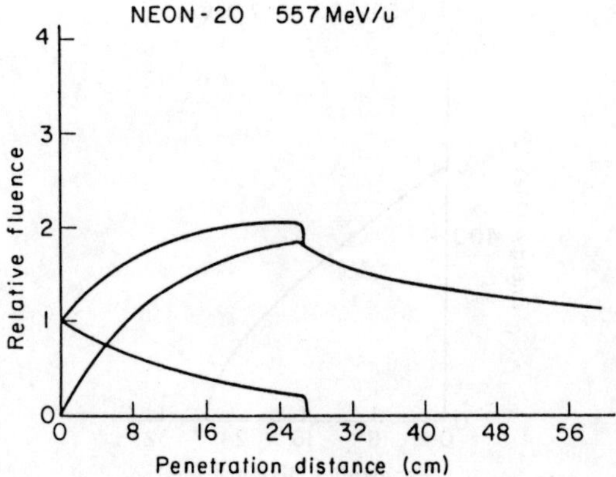

FIG.4. Number of primary, fragment and total particles relative to the depth of penetration.

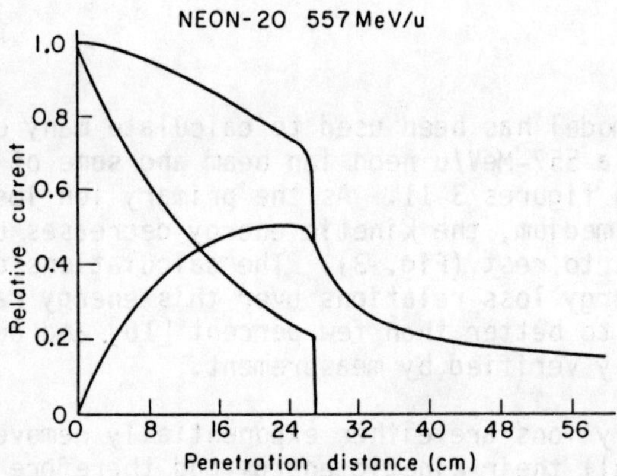

FIG.5. Beam current relative to the depth of penetration.

a funtion of penetration distance (Fig. 6). The experimental Bragg curve (points) includes the contribution of both the primary ions and the fragment particles which are produced in the nuclear collisions between the incident projectiles and the nuclei of the absorbers. The primary dose contribution is obtained from the number of surviving primary ions and their energy loss in the absorber. The major uncertainty in the result is from the number of primary ions in the beam. This can be measured with a particle telescope or any other particle detector which can distinguish the primary ions from the secondary ions.

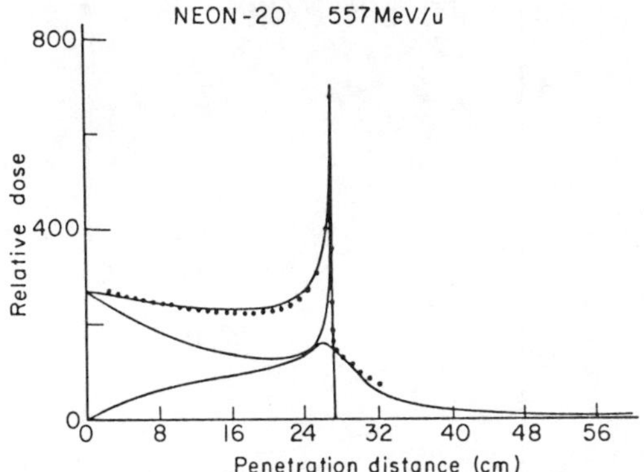

FIG. 6. Bragg curve of a neon ion beam. Also shown are the contributions of the primary ions and of the fragment particles.

The secondary dose contribution, which starts at zero and monotonically increases to a broad maximum near the Bragg peak and then decreases, is calculated in the same manner as the primary dose but it is the sum of the Bragg curve of the most probable charged fragments.

The secondary heavy charged particles in the beam are produced in nuclear interactions of the primary ions and the absorbing nuclei. Both nuclei may fragment in the collision. If the primary particle fragments, the fragments formed generally continue in the same direction and with the same velocity as the incoming ion [17]. The types of heavy charged particles that are produced can be any of the neotron deficient isotopes of the primary ion or isotopes of any lighter element with a mass less than that of the incoming projectile. That is for a neon-20 projectile, the expected fragments would be isotopes of elements with $Z = 1$ to 9 with $A \leqslant 20$ or isotopes with $Z=10$ and $A < 20$. Generally the secondaries will have longer ranges than the primaries, however there will be some secondaries which will have a shorter ranges, (i.e., neutron deficient isotopes of the primary ion). In the calculations the energy per nucleon of each fragment is taken to be equal to that of the primary particle at the point of the collision. The contribution of each fragment to the dose is determined by the production probability. At a point near the Bragg peak the secondary dose approaches the primary dose.

Figure 7 shows permissible energy and energy loss values that can be expected for the most probably charged particles in the beam at a point a few centimeters before the Bragg peak. The

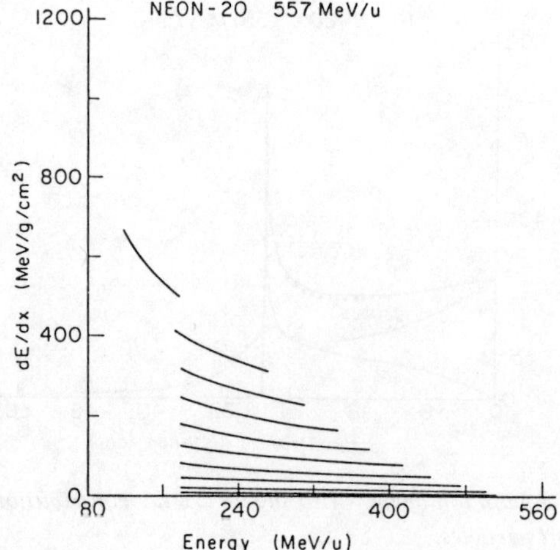

FIG.7. Range of allowable rate energy loss and energy per nucleon for charged particles expected 4.1 cm upstream of the Bragg peak. Each curve represents particles of a single element with an atomic number between and including 1 and 10.

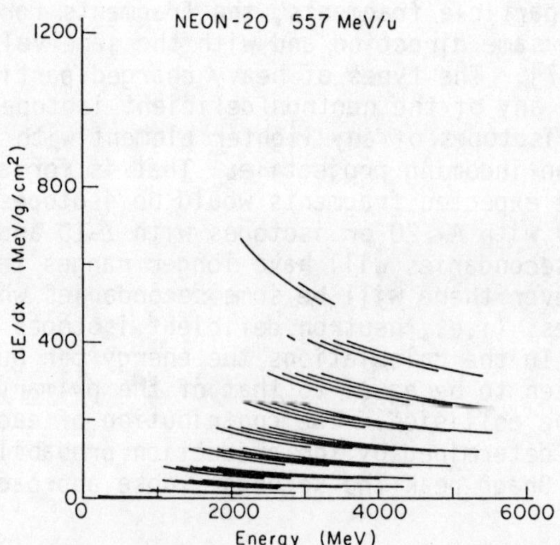

FIG.8. Range of allowable rate of energy loss and energy for charged particles expected 4.1 cm upstream of the Bragg peak.

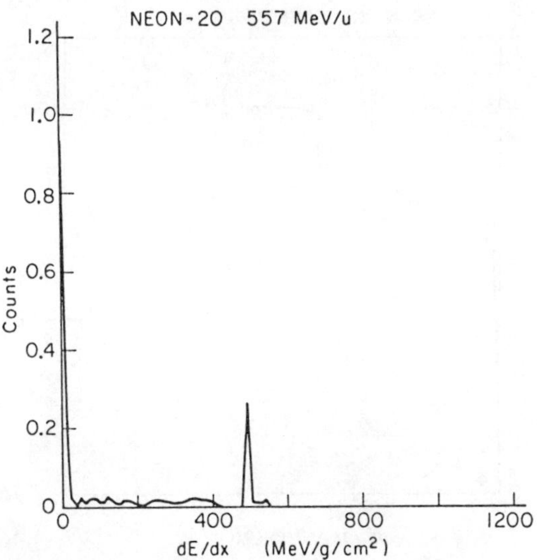

FIG.9. Fluence relative to the rate of energy loss 4.1 cm upstream of the Bragg peak.

energy per nucleon scale is useful because all particles with the same energy per nucleon will have the same velocity. Particles with nuclear charges of 1 to 10 are shown. The lighter fragments are distributed from some high energy/nucleon down to the energy/nucleon of the primary ions. For each element, this energy is a clue to the origin of the fragment. When produced the fragments, in general, will have the same energy per nucleon as the primary ions. Those with the lowest energy are usually closest to their point of origin; higher energies are associated with a greater distance to their origin. This relationship can be inverted for some neutron deficient isotopes becue of the range and energy loss relationships. Neutron deficient neon isotopes appear on the plot as the lower energy and higher rate of energy loss ions. If the data is presented as the rate of energy loss vs. total kinetic energy (Fig. 8) the plot takes on a different appearance and the isotopes of the different elements can be appreicated.

While the previous two figures do not give the relative probability of having a given particle with a specified energy, figure 9 shows the probability of having a particle with a specified average rate at energy loss. Of more interest may be the number of particles with rates of energy loss below a given value (Fig. 10) or how the dose contribution varies with the average rate of energy loss (Fig. 11). Table 1 summarizes this data.

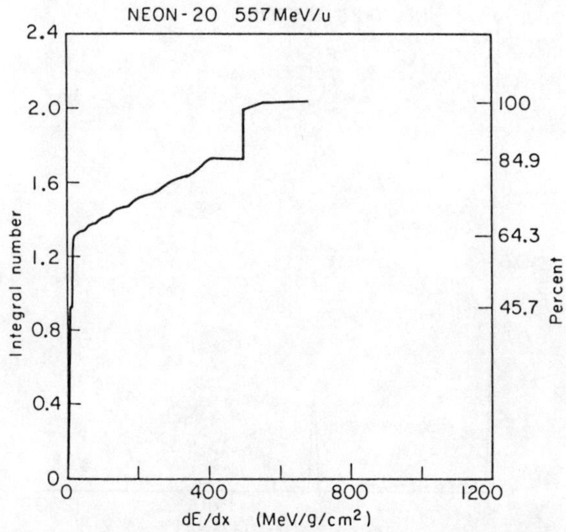

FIG.10. Cumulative fluence relative to the rate of energy loss 4.1 cm upstream of the Bragg peak.

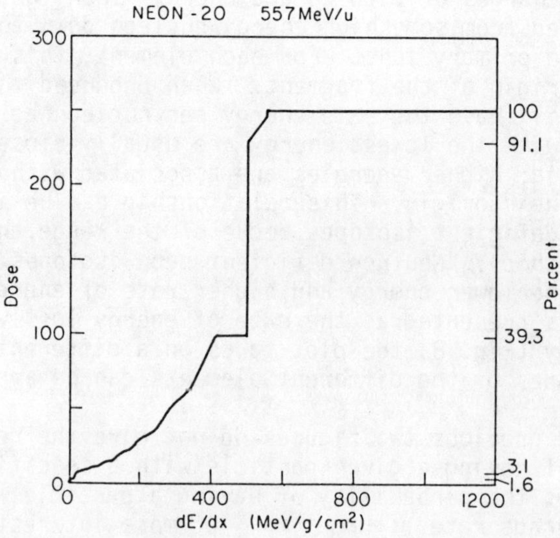

FIG.11. Cumulation dose relative to the rate of energy loss 4.1 cm upstream of the Bragg peak.

TABLE 1. RELATIVE PARTICLE ABUNDANCE AND CONTRIBUTION TO ABSORBED DOSE 4.1 cm UPSTREAM OF THE BRAGG PEAK OF A 557 MeV/u NEON ION BEAM

PARTICLE	PERCENT OF TOTAL PARTICLES	PERCENT CONTRIBUTION TO DOSE
protons	45	1.6
proton plus helium	65	3.1
1 < = Z = > 9	85	40
Z = 10	15	60

4. Discussion

The dose to a small mass of tissue can be calculated if all the particles passing through the small mass are known (Eq. 1). Therefore it is desirable to identify all the particles in the beams. This is a formidable job considering the number of different ion beams and energies involved.

In practice, the dose is usually obtained from a measurement with an ionization chamber located within a phantom. Ideally if the phantom ionization chamber and ionization chamber gas are tissue equivalent (TE) no corrections would be required to convert from the ionization chamber dose to the tissue dose. Water, polystyrene or perspex are commonly substituted for a TE phantom. The relative dose and particle range in the phantom materials can be scaled by the relative electron densities [18]. For these ion beams, the carbon to oxygen ratio in the TE material is not as critical to the dosimetry as it is with neutron and pion beams. If a TE ionization chamber is used not with TE gas but with air as the filling gas, the required correction is determined from:

1) the change in the mass of the filling gas (ratio of densities is sufficient because the volume is the same),
2) the relative mass stopping powers of the two gases (these two items determine the energy lost in the gases), and
3) the ratio of the average energy expended to make an ion pair (w).

The mass (or volume) of the gas is determined by a calibration of the ionization chamber in a cobalt-60 beam.

The relative mass stopping power of gas to tissue can be determined from tabulated range-energy data. The stopping of the various materials for particles above a few MeV is assumed to be accurate to better than a few percent and the relative stopping power should have even higher accuracy.

The energy to make an ion pair (w) in the gas is taken to be the same as for a higher energy electron to make an ion pair [19]. Since the change in w occurs primarily at low velocities (< 2 MeV/u), it is relatively unimportant for these heavy-charged particle beams at most depths of penetration. At the Bragg peak, a correction to w might be appropriate

The large accelerators that produce these heavy charged-particle beams are all pulsed machines. Pulse to pulse variations in shape and intensity are common. Corrections need to be considered for recombination in the ionization chambers [20]. Ideally any correction necessary should be applied on a per pulse basis because of the pulse shape variations normally encountered.

The determination of a heavy charged-particle absorbed dose, based upon an ionization measurement, may have an uncertainty as great as 10 percent. The large uncertainty is mainly from lack of information on:

1) the beam composition at various depths in the absorbing material,
2) the energy to make an ion pair for each particle passing through the ionization chamber,
3) the stopping power ratios of the different materials for the various particles,
4) and corrections for recombination of the ions produced in densely ionized tracks by the pulsed beams.

This beam model is used to supplement the experimental determination of the beam composition. Because of the uncertainties in the conversion factors the charged particle beam dosimetry task group of the American Association of Physicist in Medicine [1] recommends that the ionization chamber be calibrated with a TE calorimeter [21].

---

[1] Draft protocol for charged-particle beam dosimetry, being prepared by Task Group No.20 of the American Association of Physicists in Medicine.

REFERENCES

[1] ROESCH, W. C. and ATIX, F. H., "Basic concepts of dosimetry." In: Radiation Dosimetry Vol I 2nd ed., (Attix, F.H. and Roesch, W.C, Eds.), Academic Press, New York (1968).

[2] RAJU, M. R., LYMAN, J. T., BRUSTAD, T. and TOBIAS, C. A., "Heavy-charged particle beams." In: Radiation Dosimetry Vol III 2nd Ed., (Attix, F. H., Roesch, W. C., and Tochilin, E., Eds.), Academic Press, New York (1969).

[3] TODD, P.W., LYMAN, J. T., ARMER, R., SKARSGARD, L. D. and DEERING, R. A., Dosimetry and apparatus for heavy ion irradiation of mammalian cells in vitro. Radiat. Res. 34:1-23 (1970).

[4] SCHIMMERLING, W., VOSBURGH, K. G., TODD, P. W. and APPLEBY, A., Apparatus and dosimetry for high-energy heavy-ion beam irradiations. Radiat. Res. 65:389-413 (1976).

[5] LYMAN, J. T. and HOWARD, J., Dosimetry and Instrumentation for Helium and Heavy Ions. Int. J. Radiat. Oncol. Biol. Phys. 3:81-85 (1977).

[6] ACETO, H., JOLLY, R. K. and BUCKLE, D., Biophysical aspects of the Space Radiation Effects Laboratory (SREL) 710-MeV helium ion beam: Physical measurements and dosimetry. Radiat. Res. 77:5-20 (1979).

[7] LARSSON, B., "Dosimetry and radiobiology of protons as applied to cancer therapy and neurosurgery." Gustaf Werner Institute Report, GWI R 1/79 (1979).

[8] VERHEY, L. J., KOEHLER, A. M., MCDONALD, J. C., GOITEIN, M., MA, I., SCHNEIDER, R. J., and WAGNER, M., The determination of absorbed dose in a proton beam for purposes of charged-particle radiation theapy. Radiat. Res. 79:34-54 (1979).

[9] LITTON, G., LYMAN, J. AND TOBIAS, C., "Penetration of high-energy heavy ions with the inclusion of coulomb, nuclear and other stochastic processes," Lawrence Berkeley Laboratory Report UCRL-17392 rev., (1968).

[10] LYMAN, J. T. "Computer modeling of heavy charged-particle beams." Proceedings, International Workshop on Pion and Heavy Ion Radiotherapy: Pre-clinical and Clinical Studies, Vancouver,July, 1981 (to be published).

[11] WILSON, R. R., Radiological use of fast protons. Radiol. 47:487-491 (1946).

[12] CHATTERJEE, A., TOBIAS, C. A. and LYMAN, J. T., "Nuclear fragmentation in therapeutic and diagnostic studies with heavy ions." In: Spallation Nuclear Reactions and their Applications (Shen and Merker, Eds.), D. Reidel, Boston 161-191 (1976).

[13] SILBERBERG, R. and TSAO, C. H, Partial cross-sections in high-energy nuclear reactions, and astrophysical applications. I. Targets with $Z \leqslant 28$. Astrophys. J. Suppl. 25:315-333 (1973).

[14] KARLSSON, B. G., Methoden zur Berechnung und Erzielung einiger fur die Tiefentherapie mit hochenergitischen Protonen gunstiger Dosisverteilunger. Strahlentherapie 124:481-492 (1964).

[15] KOEHLER, A. M., SCHNEIDER, R. J. and SISTERSON, J. M. Range modulators for protons and heavy ions. Nucl. Inst. Meth. 131:437-440 (1975).

[16] AHLEN, S. P., Theoretical and experimental aspects of the energy loss of relativistic heavily ionizing particles. Rev. Mod. Phys. 52:121-173 (1980).

[17] Greiner, D. E., Lindstrom, P. J., Heckman, H. H., Cork, B., and Bieser, F. S., Momentum distributions of isotopes produced by fragmentation of relativistic ^{12}C and ^{16}O projectiles. Phys. Rev. Lett. 35:152-155 (1975).

[18] GOODMAN, L. J. and COLVETT, R. D., Biophysical studies with high-energy argon ions. 1. Depth dose measurements in tissue-equivalent liquid and in water. Radiat. Res. 70:455-468 (1977).

[19] ICRU. "Average energy required to produce an ion pair." ICRU Report 31, Washington (1979).

[20] BOAG, J. W., "Ionization chambers." In: Radiation Dosimetry Vol II 2nd ed., (Attix, F. H. and Roesch, W. C., Eds.), Academic Press, New York (1966).

[21] MCDONALD, J. C., LAUGHLIN, J. S. and FREEMAN, R. E., Portable tissue equivalent calorimeter. Med. Phys. 3:80-85 (1976).

IAEA-AG-371/6

DOSIMETRY FOR PION THERAPY

Myriam SALZMANN
Swiss Institute for Nuclear Research (SIN),
Villigen, Switzerland

Abstract

DOSIMETRY FOR PION THERAPY.

The dosimetric system for the pion therapy, as it is in use at SIN (Swiss Institute for Nuclear Research), is presented. As the radiation quality of pions is mixed and varies in space, dosimeters which respond differently to different LET mixtures are necessary. Therefore, several methods for dose measurements are used: ^{7}LiF TLDs, films, activation of aluminium foils and $CaF_2:Tm$ TLDs. The experience gained with these dosimeters for three-dimensional dose mapping, for the verification of treatment plans in solid phantoms, and for the in-vivo dosimetry is described.

1. INTRODUCTION

The advantages to be expected in cancer therapy with pi-mesons are due to their specific dose distribution and to the high LET component in the stopping region of the pions. It is known that this high LET component causes an increased cell-killing effectiveness and a reduced radioresistance of anoxic tumour cells.

Only three centres, LAMPF (United States of America), TRIUMF (Canada) and SIN (Switzerland) have accelerators which produce pion beams with a dose rate sufficiently high for radiotherapy.

SIN is the only centre in the world providing simultaneous multi-port irradiation with high LET particle beams. A ring cyclotron accelerates protons to an energy of 600 MeV. From the 140 μA proton beam 20 μA are separated by an electrostatic beam splitter, allowing the biomedical facility to be almost independent of the physics research programme.

The π-mesons are produced through the collision of the protons with the material of a pencil-shaped target (Fig. 1). The pions are collected, bent and focused on to the axis of the treatment chamber by two toroidal magnetic fields. Each torus consists of 60 superconducting coils, and 60 convergent pion beams of identical momentum, and hence of identical range, are obtained. The very complex and new technology of superconductivity turned out to be necessary in order to collect the pions in a very large solid angle (1 sterad), having at the same time a short flight path of the π^- [1].

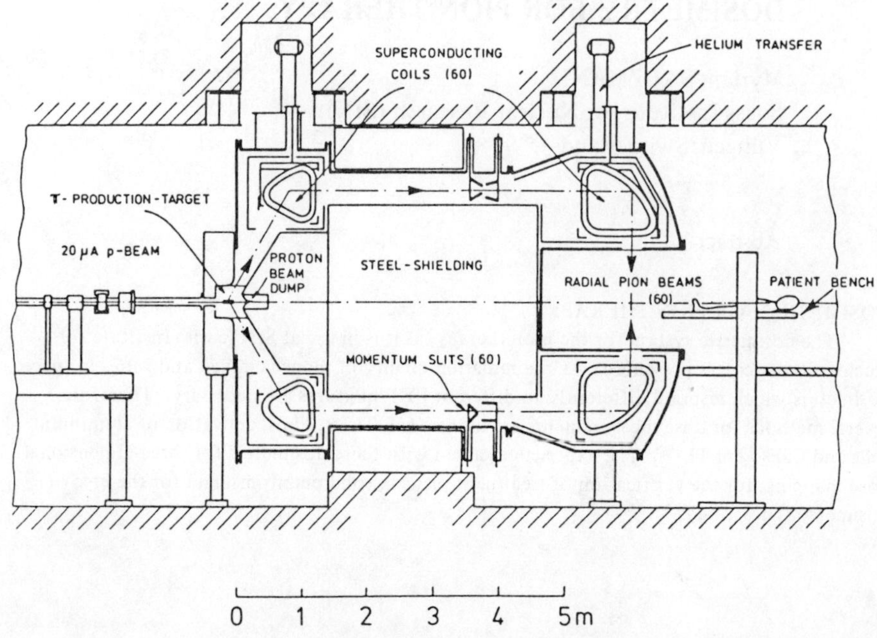

FIG.1. Cross-section view of the Piotron.

The facility at SIN, called Piotron, has been operating since summer 1980, and in summer 1981 the clinical programme for cancer therapy was started.

To make it clear why a rather sophisticated and complex dosimetry system is required, the properties of a pion single beam, the treatment modalities, and the treatment planning, are discussed first.

1.1. Properties of a single π^- beam

Figure 2 shows the main properties of a π^- beam. The integral dose of a single beam is plotted as a function of depth in water. The two main components which contribute to the total physical dose are the dose due to the slowing down of the pions (mainly low LET), and the most important high LET contribution which is due to neutrons and heavier secondary particles, produced at the pion stop position.

These heavy secondary products deposit their energy more or less locally, and this is called star dose.

In addition there is a low LET contribution from electrons and muons, which are always present in a pion beam as contamination. This dose is clearly visible in the data points behind the peak.

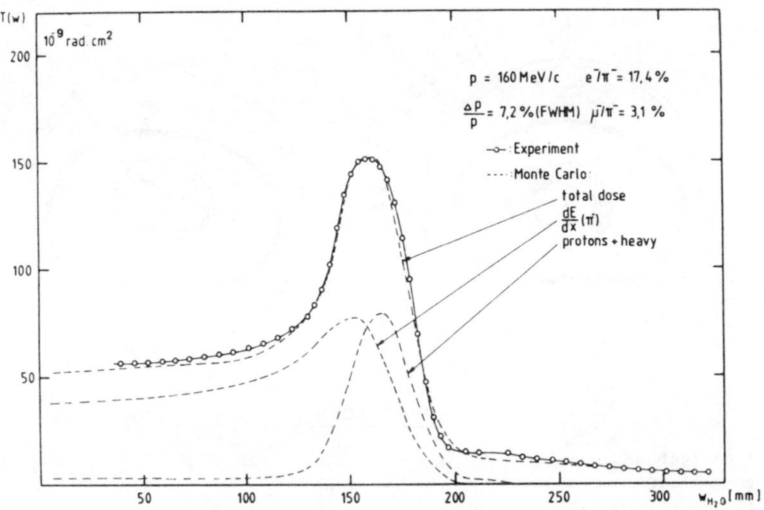

FIG.2. Pion single beam: The dose as a function of depth in water.

1.2. Treatment modalities

As all the 60 beams of the Piotron have the same range and fixed beam optics, it is necessary to shape the dose field by using dynamic treatment. With the Piotron two different modes are possible – "ring scanning" and "spot scanning". For both methods material must be added around the patient so as to 'cylindrize' him and to compensate for inhomogeneities. Depending on the momentum, the pions will stop either in a spot at the central axis or in a concentric ring.

In the case of ring scanning (Fig. 3) the dose is distributed over the target volume by changing the momentum of the beams, sweeping the dose peak region radially. This leads to a superposition of rings or ring segments. By switching beams on and off individually, and by scanning radially at different positions along the axis of the patient, the dose is shaped in all three dimensions.

For spot-scanning (Fig. 4) the patient is, in addition, surrounded by a rubber bag filled with water. The outer cylinder of this water bolus is kept fixed with respect to the Piotron axis. The patient can be moved in all directions inside the water bolus. The pion momentum is chosen in such a way that all peaks overlap on the geometrical axis of the Piotron, producing a very intense dose spot. Assuming that the patient has about the same density as water, any displacement of the patient will leave the shape of the spot and its position with respect to the Piotron unchanged. In a good approximation the spot is only translated with respect to the patient when he is moved. In this way a three-dimensional shaping of the dose according to the given target volume is achieved by appropriate patient

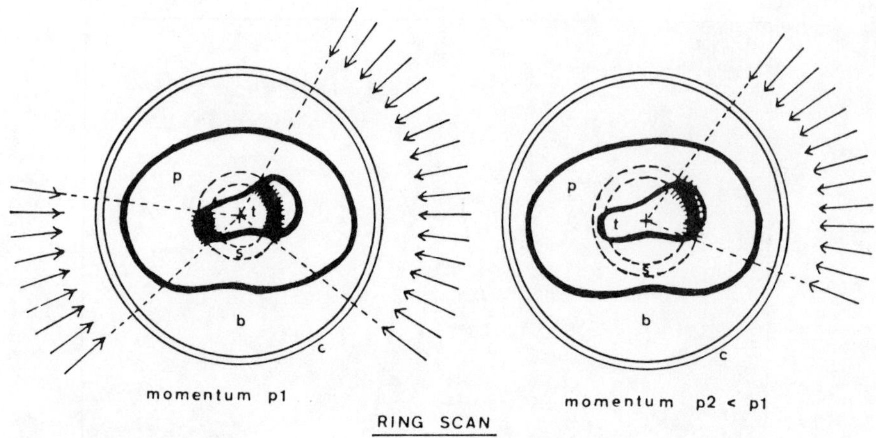

FIG.3. Treatment method of ring scan.

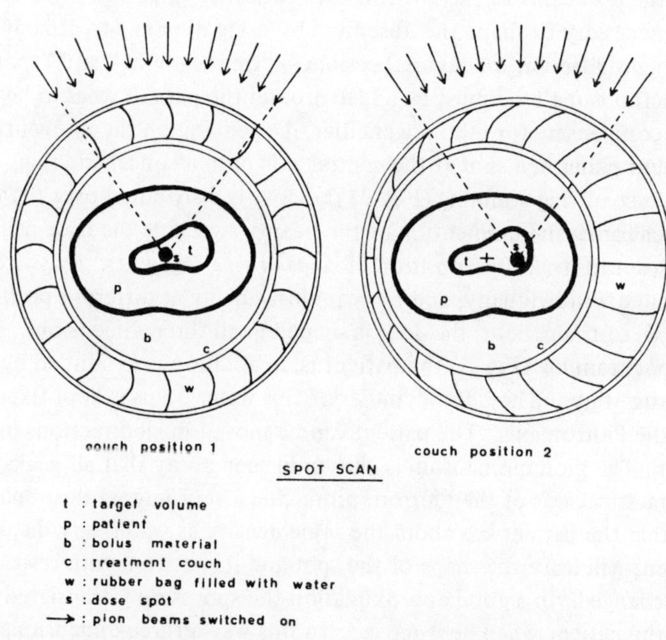

FIG.4. Treatment method of spot scan.

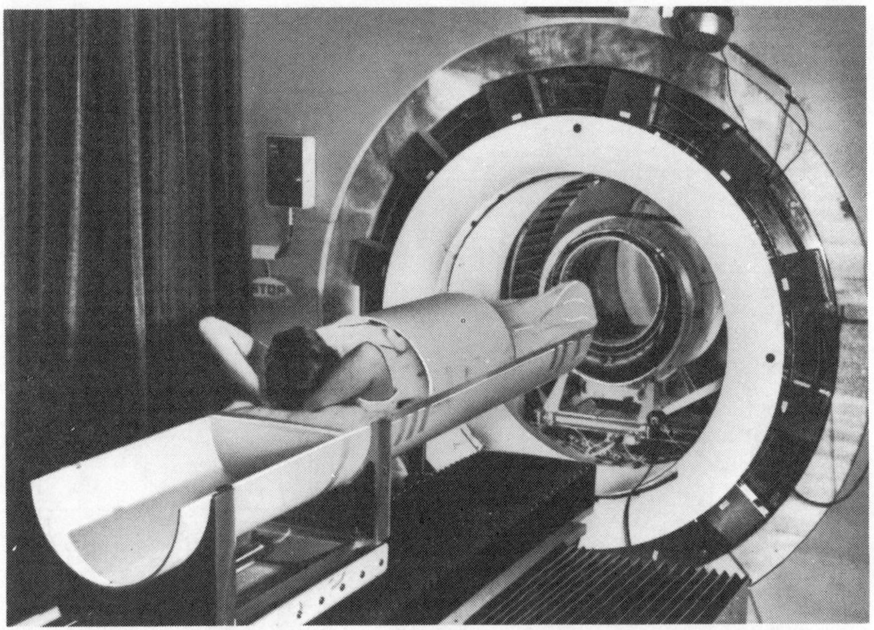

FIG.5. Patient on the couch in front of the Piotron.

movement in all three directions. Figure 5 shows the patient on the couch in front of the Piotron.

1.3. Treatment planning

The basis for treatment planning is a series of equidistant CT slices of the treatment region. The CT scans are taken with the patient placed in its plexiglass cast, surrounded with the material necessary for cylindrization.

These CT scans define the patient's position and give the three-dimensional density distribution in the region of interest. A three-dimensionally shaped target volume is defined by implementing a contour in each CT slice.

The goal of the treatment planning is to deliver a homogeneous dose inside the target volume and to minimize the dose to the tissue outside of this region. This is achieved by choosing the appropriate treatment modality, by selecting the beams and by using compensation if necessary. The computer program calculates and optimizes automatically the dose distribution [2]. The basic information for these dose calculations are dosimetric measurements. They consist of three-dimensional dose distributions of single pion beams measured in a water phantom with an ionization chamber.

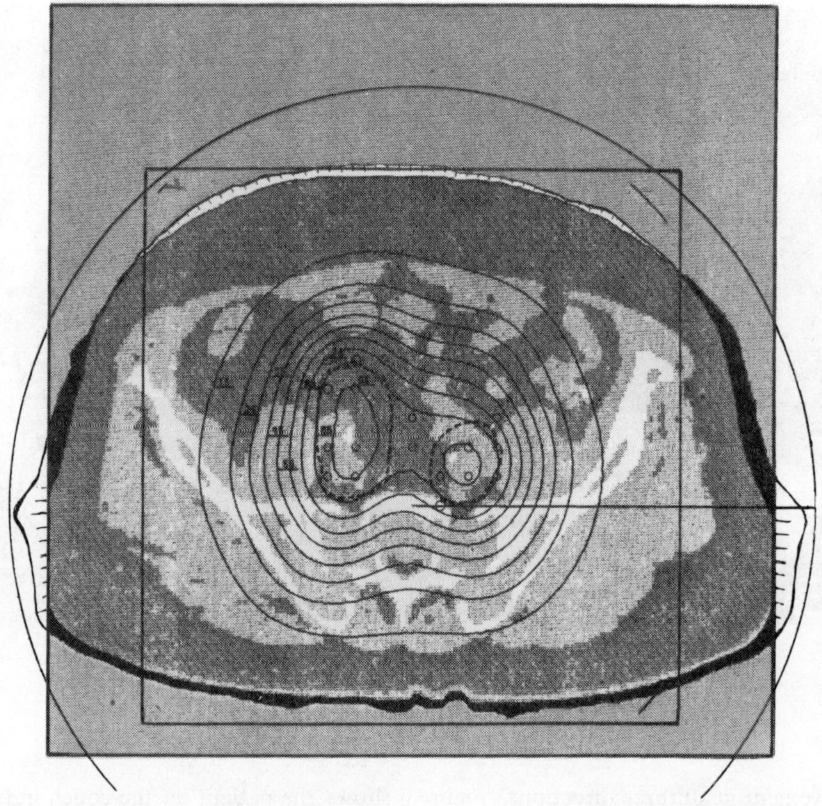

FIG. 6. Treatment planning for a bladder case. Dashed lines show 85% contours as implemented in this CT slice by the physician. Solid lines show isodose levels calculated with the optimization procedure of the therapy planning programme.

Figure 6 gives an example of therapy planning for a bladder case. The calculated isodose lines in one of the CT slices are shown together with the 85% contour of the target volume (dashed lines).

2. DOSIMETRY

As we have seen, the most important complications for the dosimetry at the Piotron are the specific dose distribution of a pion beam, the variation of LET mixture, the fact that we have 60 beams and, finally, the problems which occur with the dynamic treatment. Therefore, our dosimetry system has to fulfil the following requirements:

The detector system has to measure the low LET and the high LET dose component separately;
Special equipment is necessary to map the dose distribution in three dimensions for a configuration of 1 to 60 beams;
To simulate dynamic treatments for deep-seated tumours in a realistic way, integrating dosimeters and solid phantoms have to be used;
The monitoring system should check all 60 beams individually; and
The in-vivo dosimeters have to be small, integrating, and their sensitivity should be in the range of the dose delivered in one fraction.

2.1. Monitoring and calibration

For monitoring two independent systems are used, which are based on completely different measuring principles.

2.1.1. The pion clock

The pion clock consists of scintillation counter telescopes at four positions just upstream of the slits. The coincidence rate of the two plastic scintillators in each telescope gives the time base for the irradiations.

2.1.2. The monitor chambers

Near the window through which the pions enter the treatment tank 60 parallel plate ionization chambers are used. They monitor the intensity of the 60 beams individually. Figure 7 shows one of these monitor chambers. The frame is made of lucite, the walls are 0.25-mm-thick Mylar foils coated with copper and gold. The size of the chamber is 30×5 cm^2, the gap is 8 mm on each side. It is an open chamber, flushed with dry air. These chambers have been calibrated individually in an electron beam. The ionization current, which is proportional to the individual pion rate in each beam, is converted into a digital signal and is stored in the PDP 11/45, or in the microprocessor of the control system. This monitoring system is of extreme importance during the treatments because it checks the proper functioning of the slit system by measuring the 60 individual beam intensities right at the positions where the beams enter the treatment region.

2.1.3. Absolute calibration

For the absolute dose calibration a tissue-equivalent ionization chamber (Spokas) inside a solid phantom is used. The chamber is calibrated in a Co source. For a specific setting the deposited dose with respect to the pion-clock rate is measured.

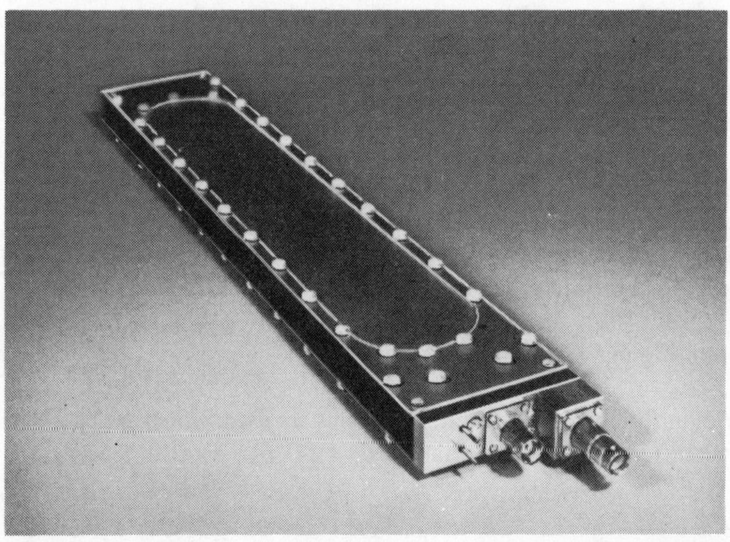

FIG.7. One of the 60 parallel plate ionization chambers used for monitoring.

2.2. Detector systems

As has already been mentioned, it is necessary to measure more than just the total physical pion dose. We hope to get a good correlation with the biological effects if we know two parameters at each point of the dose distribution. These two parameters would ideally be the low LET and the high LET dose. But from the experimental aspect it is much easier to measure the total physical dose and the star dose. Therefore, the following dosimeters are used:

– ionization chambers;
– ^{7}LiF TLDs;
– films;
– activation of aluminium foils; and
– CaF_2:Tm TLDs

2.2.1 Ionization chambers

Ionization chambers are used to map the dose distributions of a single beam and of overlapping multiple beams.

To verify the total dose of dynamic treatments ionization chambers are used in a solid phantom at selected positions only. For dose distribution measurements small and integrating dosimeters are preferable because we have to deal with dynamic treatment.

2.2.2. ^{7}LiF TLDs

^{7}LiF powder encapsulated in glass is used routinely for distribution measurements in solid phantoms and for the in-vivo dosimetry. These TLDs show normally the expected level of reproducibility of ± 5% or better. But, despite the fact that all TLDs are calibrated individually, errors of the order of 15% are sometimes observed. This problem is not yet completely understood but it is being overcome at the moment by using more than one dosimeter at once.

2.2.3. *Films*

Films have an efficiency which drops rapidly for increasing LET. Therefore, we use a combination of a film in close contact with a rare-earth intensifying screen. In the presence of this scintillator the film records mainly the light from the scintillating material and this corresponds quite closely to the applied physical dose. The main advantage of this system is the possibility of measuring a two-dimensional distribution of the physical dose in one irradiation. The present accuracy of our film dosimetry is of the order of ± 10%. But it is hoped to improve the reproducibility of the film development and to find a better control of the non-linearity of the film response as a function of the applied dose. The result of a film irradiation is shown in Fig. 8. In this case the pion momentum was chosen in such a way that the pions stop on a ring.

2.2.4. *Activation of aluminium foils*

The main purpose of the aluminium activation is to determine the spatial distribution of the radiation quality or, in other words, the high LET component in the pion radiation field. The specific complication in pion treatment is the fact that the high LET dose component can change very rapidly in space.

The activation of aluminium has proved to be a good method for measuring the pion stop distribution. This in turn reflects the distribution of the high LET component because the main contribution to the high LET is produced by the short-ranged fragments emitted after the stopped pion capture.

We use 0.1-mm-thick aluminium foils for activation. With a very high probability the stopped pions produce β-active ^{24}Na nuclei in aluminium – approximately 7.5% of the pions which stop in aluminium produce a ^{24}Na nucleus [3].

Results of the aluminium activation in a single pion beam and the correlation to the stop distribution have been published [4].

A special problem is the aluminium activation of the emitted secondary neutrons which do not deposit their energy locally at the pion stop position. But their dose is rather flat in space and can be calculated as a separate contribution. However, we plan to measure the n-dose separately at selected points in a phantom.

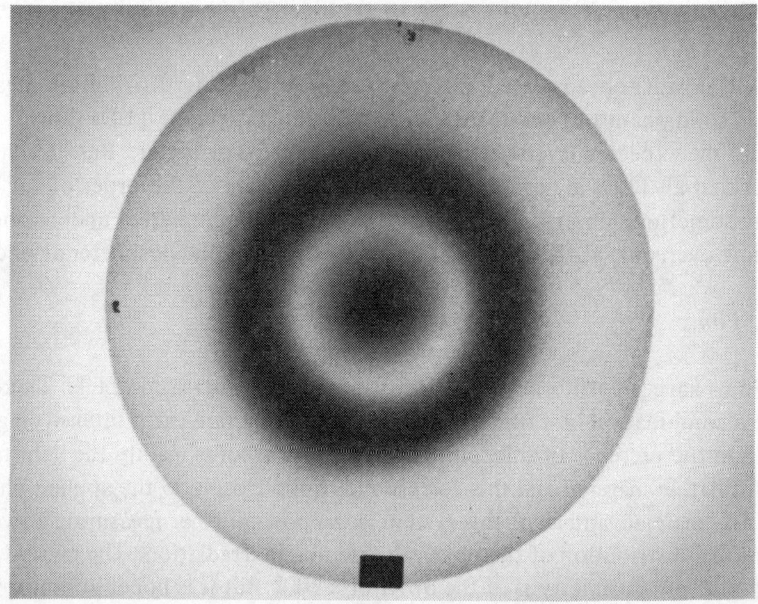

FIG.8. Result of a film irradiation. The ring indicates the pion stops and the spot in the centre is due to the dose of the electron contamination. (Phantom diameter: 46.7 cm; target: 30 mm Be; momentum: 170 MeV/c; slits: 100% open, all 60.)

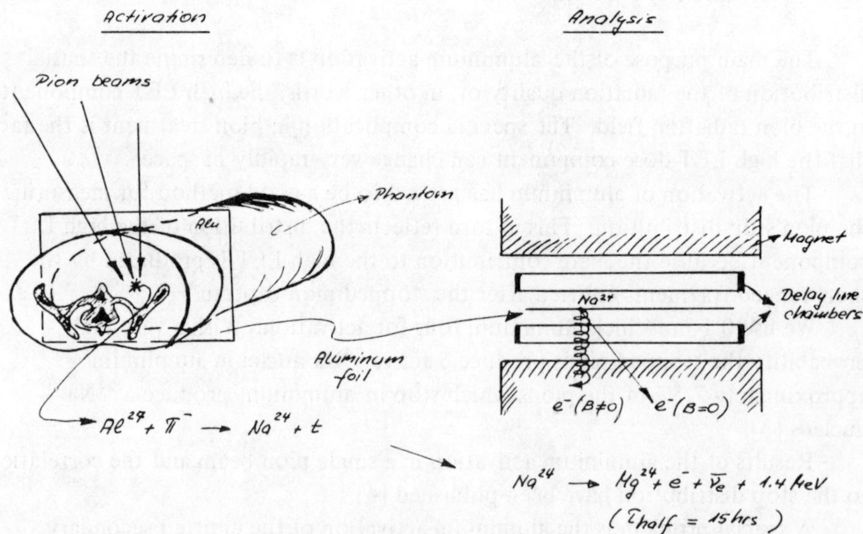

FIG.9. Set-up for the activation of aluminium foils and for analysis of two-dimensional activity distributions.

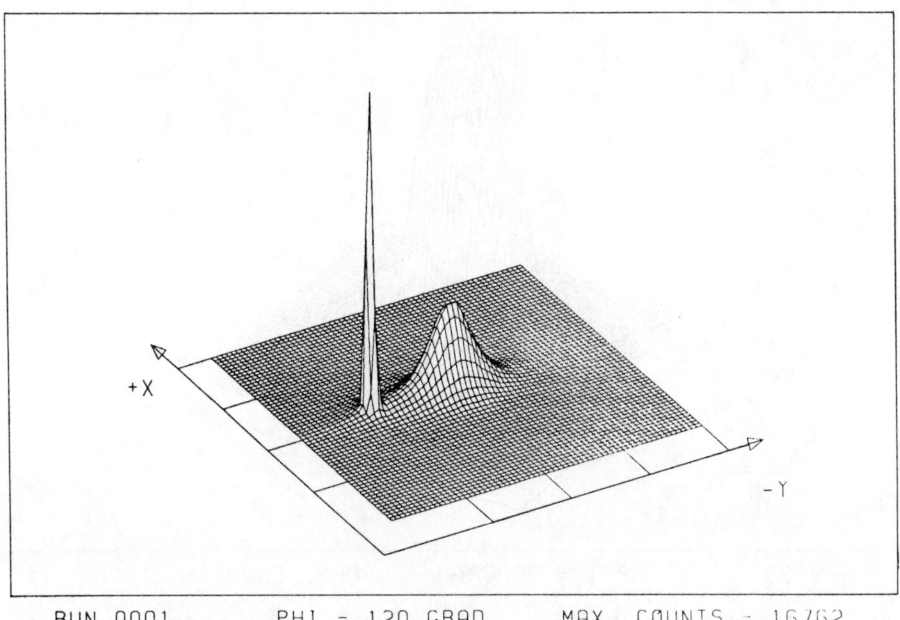

FIG.10. Effect of the magnetic field on the spatial resolution of the activity. The image of a point-like source is shown with and without the magnetic field (distance of the two positions of the source: 10 cm).

To get three-dimensional activity distributions a number of large aluminium foils placed between the slices of a phantom are irradiated (Fig. 9). The activity within one foil has to be analysed in two dimensions. We have developed a new technique [5] which allows a rapid and simple measurement of such two-dimensional distributions; about 10 h after the irradiation the short-lived isotopes are negligible. The foil is placed between two stacks of multiwire chambers and the electrons emitted by the ^{24}Na nuclei are detected. As the electrons are emitted isotropically and only one pair of co-ordinates is measured at some distance from the foil, it was necessary to find a method to improve the spatial resolution. This problem was solved by applying a large magnetic field perpendicular to the foil and the wire chamber planes. This leads to a resolution of the order of 1 mm.

Figure 10 shows the effect of the magnetic field on the spatial resolution. The image of a point-like radioactive source is shown with and without the magnetic field. The distance between the two distributions corresponds to 10 cm. The activation in one plane of the realistic phantom, resulting from a dynamical treatment of a large volume, is shown in Fig. 11.

We are able to analyse foils of 35 × 35 cm^2. The system is now working very reliably and it is used routinely for in-vivo dosimetry. Furthermore,

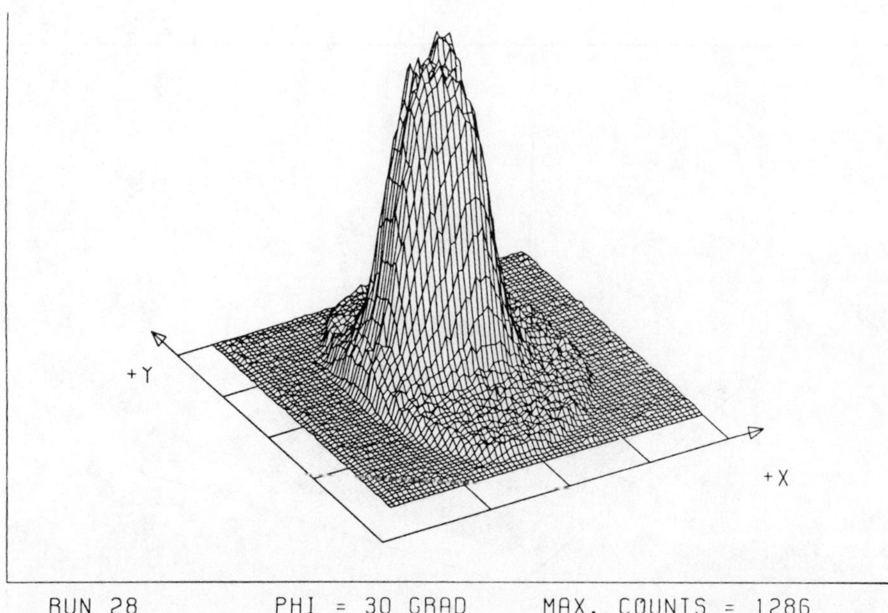

FIG.11. Activation distribution in one plane of the realistic phantom after the simulation of a large-volume dynamic treatment.

distributions of induced activity in aluminium are calculated in treatment planning, providing an additional possibility in planning confirmation.

The aluminium activation is now being correlated to cell survival experiments and microdosimetry. However, more experiments are necessary before this correlation can be used routinely in clinical applications. Our final goal is to optimize an effective equivalent biological dose rather than a physical dose.

Another use of the aluminium activation is the correction of TLD and film readings. These dosimeters have a reduced efficiency for high LET radiation. In the pion peak region the correction is of the order of 10% [4].

2.2.5. CaF_2: Tm TLDs

The CaF_2: Tm TLD is another very interesting system, perfectly suited for dosimetry in radiation fields with mixed LET. The glow curve of CaF_2 : Tm has two distinct peaks. They have a strong and differing dependence on LET. It has been shown by Hoffmann et al. [6] that the ratio of the two maximum values is proportional to $\overline{y}_D$ (dose average lineal energy transfer).

The sensitivity of this material is so high that 1 mg is enough to measure therapeutic relevant doses to ± 3%. It is planned to use CaF_2-Tm routinely as soon as enough experience has been gained.

FIG.12. Water phantom with scanner and ionization chamber.

2.3. Measurements and results

Input into the therapy planning programme: To measure the three-dimensional dose distributions, which is necessary for therapy planning, lucite cylinders of different diameters are used. They can be mounted on a water tank and filled with water (Fig. 12). The tank contains a scanner which can be driven in all three directions. The scanner holds an ionization chamber on a long arm. With this apparatus the very time-consuming dose mapping in three dimensions is controlled automatically with a computer. Figure 13 shows the result of such a mapping in one plane.

Verification of treatment plans in solid phantoms: To verify treatment plans we have two different solid phantoms. In a homogeneous modular phantom made from polyethylene we use the ionization chamber at selected points and TLDs, aluminium foils and films to measure distributions.

All treatments have been checked at specific points with the ionization chamber. Figure 14 summarizes this verification for all patients of one treatment period. The maximum deviation between measurements and calculations lies within a ± 5% limit of the maximum dose. The reproducibility of the ionization chamber data is of the order of 1%.

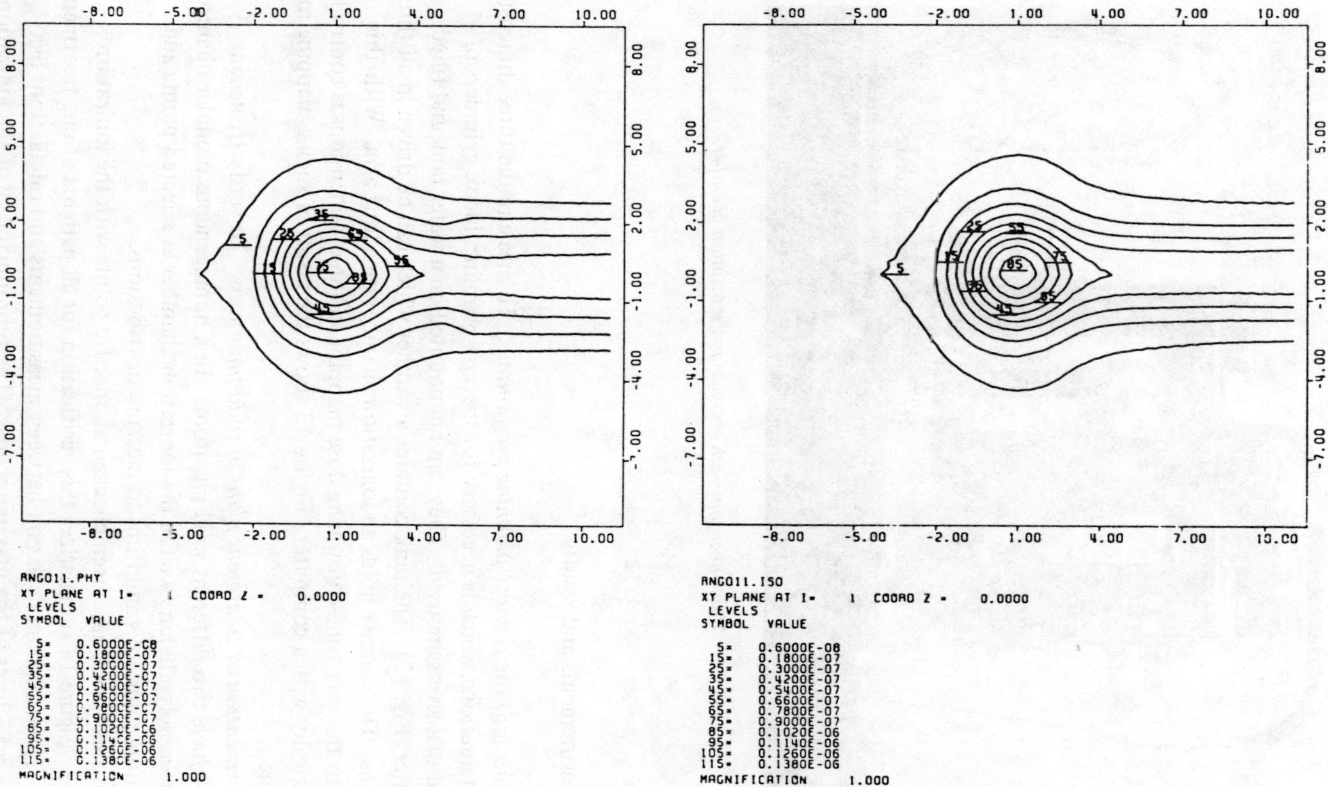

FIG.13. Isodose lines of a single pion beam in a plane along the beam axis. The measurements (left) are taken in the water phantom. For comparison the result of the calculation is shown on the right.

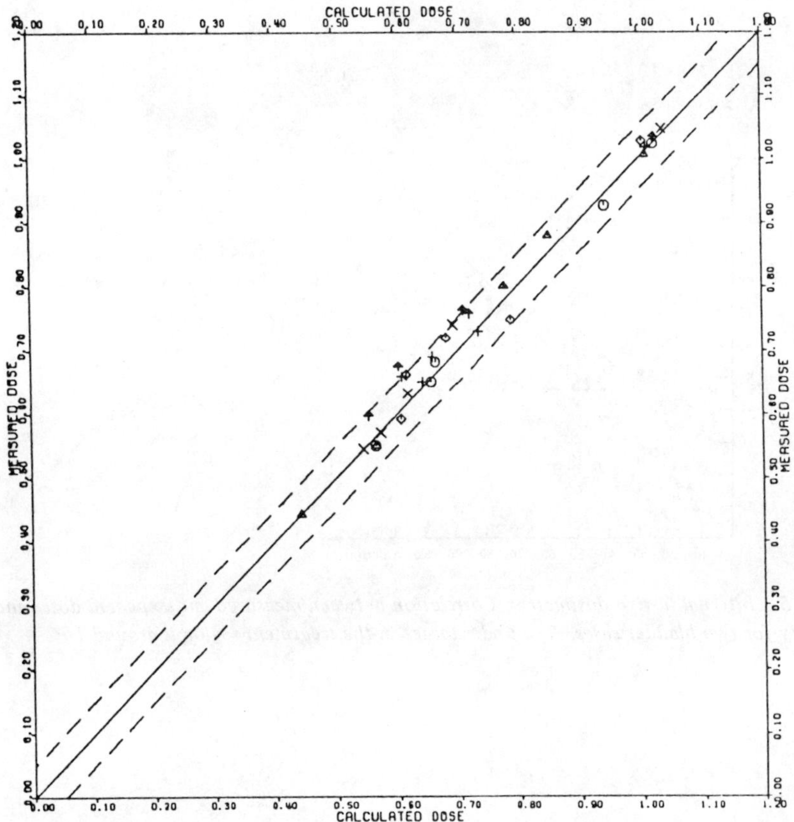

FIG.14. Verification of the treatments with the ionization chamber. The measured and the calculated doses are shown for all treatments of one period. The maximum deviation between calculations and measurements are within ±5% of the maximum dose.

To simulate the treatments in a realistic way, a phantom was built from solid tissue substitutes described by Constantinou [7]. The materials are based on epoxy resins. The phantom is built up in 1.5-cm-thick slices. The shape of the boundaries between the different tissues is taken from CT scans of a patient.

The aluminium foils and the films can be placed between the slices, and the TLDs are put into holes within the slices. For the case of spot scan with 60 beams no significant difference could be found between irradiation in the homogeneous phantom and irradiation in the realistic phantom. However, the present level of accuracy for these tests is around ± 10%.

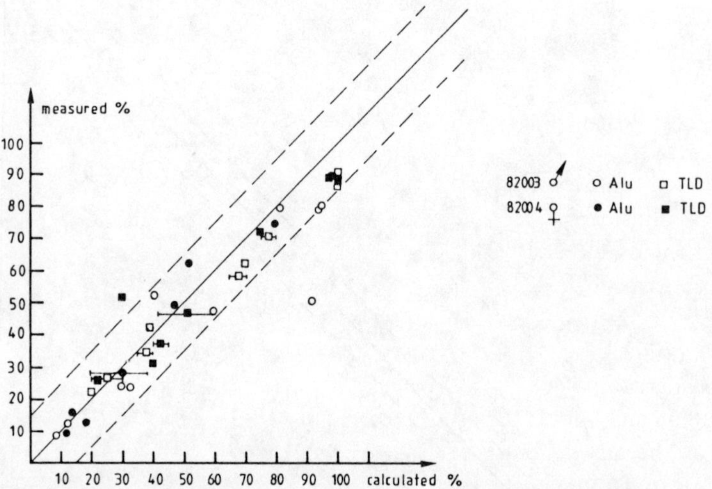

FIG.15. Internal in-vivo dosimetry: Correlation between measured and expected doses and activity for two bladder cases. The underdosage in the treatment region is around 10%.

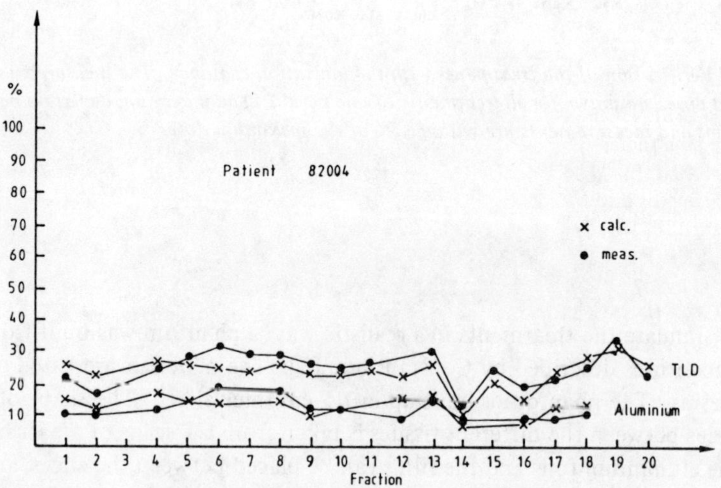

FIG.16. Skin in-vivo dosimetry: Results for one of the bladder cases with 20 fractions.

2.4. In-vivo dosimetry

To have the information about total dose and star dose, TLDs and aluminium activation are used simultaneously for in-vivo dosimetry.

For skin dosimetry small boxes of polyethylene, which contain two or more TLDs and one folded aluminium foil, are used. The most interesting part of our in-vivo dosimetry, however, is the internal dosimetry for the dynamic treatment of deep-seated tumours. TLDs and aluminium foils, arranged in a tube ($\phi \cong 5$ mm), can be placed inside the patient's body, inside cavities or surgically. For treatment of bladder tumours the dose distribution in the bladder and in the rectum was measured in vivo.

In Fig. 15 one can see the correlation between measured and calculated dose and activity, including all measurements for two patients. Both kinds of dosimeters show an underdosage in the treatment region of the order of 10%. Imperfect cylindrization and air gaps between the patient and the cast could explain this discrepancy, because there are cases where the deviation is less than 5%.

Figure 16 demonstrates the agreement of measured and expected dose and activity at the skin for one of the patients and 20 fractions.

Apart from the need for more tests on the reproducibility and some work on calibration, this in-vivo dosimetry works very satisfactorily and it is a very important means for gaining confidence in the dynamical treatment of deep-seated tumours with pi-mesons.

REFERENCES

[1] VON ESSEN, C.F., et al., The Piotron: Initial performance, preparation and experience with pion therapy, Int. J. Radiat. Oncol., Biol. Phys. **8** (1982) in press.

[2] PEDRONI, E., "Therapy planning system for the SIN-pion therapy facility", Treatment Planning for External Beam Therapy with Neutrons (BURGER, G., BREIT, A., BROERSE, J.J., Eds), Urban und Schwarzenberg (1981) 60.

[3] HOGSTROM, K.R., AMOLS, H.I., Pion in vivo dosimetry using aluminium activation, Med. Phys. **7** (1980) 55.

[4] SALZMANN, M., Measurements at the π E3 single beam correlated to dosimetry and therapy planning for the pion applicator at SIN, Radiat. Environ. Biophys. **16** (1979) 219.

[5] SEILER, P.G., DIETLICHER, R., WEMMERS, G., SALZMANN, M., MOLINE, A., Two-dimensional measurement of pion-induced beta activity in extended foils, Phys. Med. Biol. **27** (1982) 709.

[6] HOFFMANN, W., MOLLER, G., BLATTMANN, H., SALZMANN, M., Pion dosimetry with thermoluminescent materials, Phys. Med. Biol. **25** (1980) 913.

[7] CONSTANTINOU, C., Tissue substitutes for particulate radiation dosimetry and radiotherapy, Thesis, University of London, 1978.

LIST OF PARTICIPANTS AND DESIGNATING MEMBER STATES AND ORGANIZATIONS

BELGIUM

Wambersie, A.	Tour Claude Bernard, UCL – Cliniques Universitaires Saint Luc, Avenue Hippocrate 54, B-1200 Brussels

FRANCE

Huynh, V.D.	Bureau International des Poids et Mesures, Pavillon de Breteuil, F-92310 Sèvres
Nguyen Van Dat	Radiation Dosimetry, Commissariat à l'énergie atomique, F-92260 Fontenay-aux-Roses
Caumes, J. (Observer)	Primary Laboratory for Ionizing Radiation Metrology, Boîte Postale 21, F-91190 Gif-sur-Yvette
Fukuda, A. (Observer)	4, rue Alphonse Pécard, F-91190 Gif-sur-Yvette

GERMANY, FEDERAL REPUBLIC OF

Burger, G.	Gesellschaft für Strahlen- und Umweltforschung mbH, Institut für Strahlenschutz, Ingolstädter Landstrasse 1, D-8042 Neuherberg
Dietze, G.	Physikalisch-Technische Bundesanstalt, Postfach 3345, D-33 Braunschweig
Menzel, H.G.	Fachrichtung 3.6 – Biophysik und Physikalische Grundlagen der Medizin, Universität des Saarlandes, D-665 Homburg
Rassow, J.	Abteilung für Med. Strahlenphysik, Hufelandstrasse 55, D-43 Essen 1

JAPAN

Ito, A. — Cyclotron Unit,
The Institute of Medical Science,
4–6–1 Shirokanedai Minato-ku, Tokyo 108

NETHERLANDS

Broerse, J.J. — Radiobiological Institute,
Organisation for Health Research TNO,
151 Lange Kleiweg, P.O. Box 5815,
2280 HV Rijswijk

Mijnheer, B.J. — Antoni van Leeuwenhoekhuis,
Plesmanlaan 121, 1066 CX Amsterdam

POLAND

Makarewicz, M. — Radiation Protection Department,
Institute of Nuclear Research, Świerk

SWEDEN

Larsson, B. — Department of Physical Biology,
University of Uppsala,
Box 531, S-75121 Uppsala

SWITZERLAND

Salzmann, M. — Swiss Institute for Nuclear Research,
CH-5234 Villigen

UNITED KINGDOM

Bewley, D.K. — MRC Cyclotron Unit, Hammersmith Hospital,
Ducane Road, London, W12 0HS

Lewis, V.E. — Division of Radiation Science and Technology,
National Physical Laboratory,
Teddington, Middlesex TW11 0LW

UNITED STATES OF AMERICA

Goodman, L.J. — United States Department of Commerce, National Bureau of Standards, Washington, D.C. 20234

Lyman, J.T. — University of California, Lawrence Berkeley Laboratory, Berkeley, California 94720

Smathers, J.B., — Department of Radiation Oncology, B3109, Center for Health Sciences, University of California, Los Angeles, California 90024

INTERNATIONAL ATOMIC ENERGY AGENCY

Boyd, A.W. *(Scientific Secretary)* — Dosimetry Section, Division of Life Sciences, P.O. Box 100, A-1400 Vienna, Austria

Eisenlohr, H.H. — Dosimetry Section, Division of Life Sciences, P.O. Box 100, A-1400 Vienna, Austria

Nam, J.W. — Dosimetry Section, Division of Life Sciences, P.O. Box 100, A-1400 Vienna, Austria

HOW TO ORDER IAEA PUBLICATIONS

An exclusive sales agent for IAEA publications, to whom all orders and inquiries should be addressed, has been appointed in the following country:

UNITED STATES OF AMERICA	UNIPUB, P.O. Box 433, Murray Hill Station, New York, NY 10157

In the following countries IAEA publications may be purchased from the sales agents or booksellers listed or through your major local booksellers. Payment can be made in local currency or with UNESCO coupons.

ARGENTINA	Comisión Nacional de Energía Atomica, Avenida del Libertador 8250, RA-1429 Buenos Aires
AUSTRALIA	Hunter Publications, 58 A Gipps Street, Collingwood, Victoria 3066
BELGIUM	Service Courrier UNESCO, 202, Avenue du Roi, B-1060 Brussels
CZECHOSLOVAKIA	S.N.T.L., Spálená 51, CS-113 02 Prague 1 Alfa, Publishers, Hurbanovo námestie 6, CS-893 31 Bratislava
FRANCE	Office International de Documentation et Librairie, 48, rue Gay-Lussac, F-75240 Paris Cedex 05
HUNGARY	Kultura, Hungarian Foreign Trading Company P.O. Box 149, H-1389 Budapest 62
INDIA	Oxford Book and Stationery Co., 17, Park Street, Calcutta-700 016 Oxford Book and Stationery Co., Scindia House, New Delhi-110 001
ISRAEL	Heiliger and Co., Ltd., Scientific and Medical Books, 3, Nathan Strauss Street, Jerusalem 94227
ITALY	Libreria Scientifica, Dott. Lucio de Biasio "aeiou", Via Meravigli 16, I-20123 Milan
JAPAN	Maruzen Company, Ltd., P.O. Box 5050, 100-31 Tokyo International
NETHERLANDS	Martinus Nijhoff B.V., Booksellers, Lange Voorhout 9-11, P.O. Box 269, NL-2501 The Hague
PAKISTAN	Mirza Book Agency, 65, Shahrah Quaid-e-Azam, P.O. Box 729, Lahore 3
POLAND	Ars Polona-Ruch, Centrala Handlu Zagranicznego, Krakowskie Przedmiescie 7, PL-00-068 Warsaw
ROMANIA	Ilexim, P.O. Box 136-137, Bucarest
SOUTH AFRICA	Van Schaik's Bookstore (Pty) Ltd., Libri Building, Church Street, P.O. Box 724, Pretoria 0001
SPAIN	Diaz de Santos, Lagasca 95, Madrid-6 Diaz de Santos, Balmes 417, Barcelona-6
SWEDEN	AB C.E. Fritzes Kungl. Hovbokhandel, Fredsgatan 2, P.O. Box 16356, S-103 27 Stockholm
UNITED KINGDOM	Her Majesty's Stationery Office, Publications Centre P.O. Box 276, London SW8 5DR
U.S.S.R.	Mezhdunarodnaya Kniga, Smolenskaya-Sennaya 32-34, Moscow G-200
YUGOSLAVIA	Jugoslovenska Knjiga, Terazije 27, P.O. Box 36, YU-11001 Belgrade

Orders from countries where sales agents have not yet been appointed and requests for information should be addressed directly to:

Division of Publications
International Atomic Energy Agency
Wagramerstrasse 5, P.O. Box 100, A-1400 Vienna, Austria

84-00551